BIOTECHNOLOGY BY OPEN LEARNING

Functional Physiology

PUBLISHED ON BEHALF OF :

Open universiteit and **University of Greenwich (formerly Thames Polytechnic)**

Valkenburgerweg 167
6401 DL Heerlen
Nederland

Avery Hill Road
Eltham, London SE9 2HB
United Kingdom

Butterworth-Heinemann Ltd
Linacre House, Jordan Hill, Oxford OX2 8DP

 PART OF REED INTERNATIONAL BOOKS

OXFORD LONDON BOSTON
MUNICH NEW DELHI SINGAPORE SYDNEY
TOKYO TORONTO WELLINGTON

First published 1993

British Library Cataloguing in Publication Data
A catalogue record for this book is
available from the British Library

Library of Congress Cataloguing in Publication Data
A catalogue record for this book is
available from the Library of Congress

ISBN 0 7506 0563 4

Composition by University of Greenwich
(formerly Thames Polytechnic)
Printed and Bound in Great Britain by
Thomson Litho, East Kilbride, Scotland

Functional Physiology

BOOKS IN THE BIOTOL SERIES

The Molecular Fabric of Cells
Infrastructure and Activities of Cells

Techniques used in Bioproduct Analysis
Analysis of Amino Acids, Proteins and Nucleic Acids
Analysis of Carbohydrates and Lipids

Principles of Cell Energetics
Energy Sources for Cells
Biosynthesis and the Integration of Cell Metabolism

Genome Management in Prokaryotes
Genome Management in Eukaryotes

Crop Physiology
Crop Productivity

Functional Physiology
Cellular Interactions and Immunobiology
Defence Mechanisms

Bioprocess Technology: Modelling and Transport Phenomena
Operational Modes of Bioreactors

In vitro Cultivation of Micro-organisms
In vitro Cultivation of Plant Cells
In vitro Cultivation of Animal Cells

Bioreactor Design and Product Yield
Product Recovery in Bioprocess Technology

Techniques for Engineering Genes
Strategies for Engineering Organisms

Principles of Enzymology for Technological Applications
Technological Applications of Biocatalysts
Technological Applications of Immunochemicals

Biotechnological Innovations in Health Care

Biotechnological Innovations in Crop Improvement
Biotechnological Innovations in Animal Productivity

Biotechnological Innovations in Energy and Environmental Management

Biotechnological Innovations in Chemical Synthesis

Biotechnological Innovations in Food Processing

Biotechnology Source Book: Safety, Good Practice and Regulatory Affairs

The Biotol Project

The BIOTOL team

OPEN UNIVERSITEIT, THE NETHERLANDS
Prof M. C. E. van Dam-Mieras
Prof W. H. de Jeu
Prof J. de Vries

UNIVERSITY OF GREENWICH (FORMERLY THAMES POLYTECHNIC), UK
Prof B. R. Currell
Dr J. W. James
Dr C. K. Leach
Mr R. A. Patmore

This series of books has been developed through a collaboration between the Open universiteit of the Netherlands and University of Greenwich (formerly Thames Polytechnic) to provide a whole library of advanced level flexible learning materials including books, computer and video programmes. The series will be of particular value to those working in the chemical, pharmaceutical, health care, food and drinks, agriculture, and environmental, manufacturing and service industries. These industries will be increasingly faced with training problems as the use of biologically based techniques replaces or enhances chemical ones or indeed allows the development of products previously impossible.

The BIOTOL books may be studied privately, but specifically they provide a cost-effective major resource for in-house company training and are the basis for a wider range of courses (open, distance or traditional) from universities which, with practical and tutorial support, lead to recognised qualifications. There is a developing network of institutions throughout Europe to offer tutorial and practical support and courses based on BIOTOL both for those newly entering the field of biotechnology and for graduates looking for more advanced training. BIOTOL is for any one wishing to know about and use the principles and techniques of modern biotechnology whether they are technicians needing further education, new graduates wishing to extend their knowledge, mature staff faced with changing work or a new career, managers unfamiliar with the new technology or those returning to work after a career break.

Our learning texts, written in an informal and friendly style, embody the best characteristics of both open and distance learning to provide a flexible resource for individuals, training organisations, polytechnics and universities, and professional bodies. The content of each book has been carefully worked out between teachers and industry to lead students through a programme of work so that they may achieve clearly stated learning objectives. There are activities and exercises throughout the books, and self assessment questions that allow students to check their own progress and receive any necessary remedial help.

The books, within the series, are modular allowing students to select their own entry point depending on their knowledge and previous experience. These texts therefore remove the necessity for students to attend institution based lectures at specific times and places, bringing a new freedom to study their chosen subject at the time they need and a pace and place to suit them. This same freedom is highly beneficial to industry since staff can receive training without spending significant periods away from the workplace attending lectures and courses, and without altering work patterns.

Contributors

AUTHORS

Dr K. T. Bohme, Manchester Polytechnic, Manchester, UK

Dr P. C. Foster, University of Central Lancashire, Preston, UK

Dr V. E. Hick, Manchester Polytechnic, Manchester, UK

Dr I. Kay, Manchester Polytechnic, Manchester, UK

EDITOR

Dr A. R. Waller, De Monfort University, Leicester, UK

SCIENTIFIC AND COURSE ADVISORS

Prof M. C. E. van Dam-Mieras, Open universiteit, Heerlen, The Netherlands

Dr C. K. Leach, De Montfort University, Leicester, UK

ACKNOWLEDGEMENTS

Grateful thanks are extended, not only to the authors, editors and course advisors, but to all those who have contributed to the development and production of this book. They include Mrs A. Allwright, Miss K. Brown, Miss J. Skelton and Professor R. Spier.

The development of this BIOTOL text has been funded by **COMETT, The European Community Action Programme for Education and Training for Technology**. Additional support was received from the Open universiteit of The Netherlands and by University of Greenwich (formerly Thames Polytechnic).

Contents

How to use an open learning text

An open learning text presents to you a very carefully thought out programme of study to achieve stated learning objectives, just as a lecturer does. Rather than just listening to a lecture once, and trying to make notes at the same time, you can with a BIOTOL text study it at your own pace, go back over bits you are unsure about and study wherever you choose. Of great importance are the self assessment questions (SAQs) which challenge your understanding and progress and the responses which provide some help if you have had difficulty. These SAQs are carefully thought out to check that you are indeed achieving the set objectives and therefore are a very important part of your study. Every so often in the text you will find the symbol Π, our open door to learning, which indicates an activity for you to do. You will probably find that this participation is a great help to learning so it is important not to skip it.

Whilst you can, as an open learner, study where and when you want, do try to find a place where you can work without disturbance. Most students aim to study a certain number of hours each day or each weekend. If you decide to study for several hours at once, take short breaks of five to ten minutes regularly as it helps to maintain a higher level of overall concentration.

Before you begin a detailed reading of the text, familiarise yourself with the general layout of the material. Have a look at the contents of the various chapters and flip through the pages to get a general impression of the way the subject is dealt with. Forget the old taboo of not writing in books. There is room for your comments, notes and answers; use it and make the book your own personal study record for future revision and reference.

At intervals you will find a summary and list of objectives. The summary will emphasise the important points covered by the material that you have read and the objectives will give you a check list of the things you should then be able to achieve. There are notes in the left hand margin, to help orientate you and emphasise new and important messages.

BIOTOL will be used by universities, polytechnics and colleges as well as industrial training organisations and professional bodies. The texts will form a basis for flexible courses of all types leading to certificates, diplomas and degrees often through credit accumulation and transfer arrangements. In future there will be additional resources available including videos and computer based training programmes.

Preface

Multicellular organisms are highly organised structures and maintaining their functional structure requires a continuous input of energy. This energy must be obtained from the environment. Organisms capable of carrying out photosynthesis (plants and some micro-organisms) can make direct use of radiation energy but all other organisms must import fuels. Putting it in more physiological terms; they must obtain food from their environment. The characteristic features of living organisms can be related to this boundary condition. Organisms must be able to take up food from their environment and convert it to a form usable by their cells. In nature all organisms are competing for energy. This has been the major driving force for the development of highly ordered structures in which groups of specialised cells co-operate in a strictly controlled way. Humans are an example of such an ordered structure and their physiology is the subject of this text.

Human physiology is a very large, and rapidly expanding topic. Here the emphasis is placed on how groups of cells, tissues and organs interact to fulfil important functions. The authors have approached each topic from a functional point of view, by which we mean that they have first posed the question 'What is the system supposed to be doing?' and then explaining how the different components of the system work together to realise these objectives. They have embedded their explanation of how each functional system operates into a description of the essential anatomical features of each of the systems.

The first chapter describes how cells are organised into tissues. Chapters two and three deal with the nervous and endocrine systems. These two chapters thus focus on the coordination of vital functions. Subsequently the following functional systems are described; the digestive tract, the circulatory system, the respiratory tract and the kidneys. In the final part of the book, the nutritional demands, the different types of musculature and reproduction are described. These chapters integrate, to some extent, subjects dealt with in earlier chapters.

Inevitably, with such a large subject, careful selection of material has had to be made. We particular draw attention to the omission of discussion of the defence mechanisms of living systems, an essential feature in a highly competitive environment. This aspect of physiology is the topic of the companion BIOTOL texts 'Mechanisms of Defence' and 'Cellular Interactions and Immunobiology'. These, together with the lists of texts suggested for further reading provided at the end of this text, will enable readers to elaborate and extend their understanding of this important area.

The author and editor team have combined to produce a stimulating and informative text whose content, organisation, intext activities and readability provides an excellent opportunity to learn the essential elements of physiology.

Scientific and Course Advisors: Professor M. C. E. van Dam-Mieras
Dr C. K. Leach

Cell organisation

Cell organisation

1.1 Introduction

cell specialisation

In this book you will be studying mammalian physiology in terms of the various functions of the body. There are several ways in which the material could be arranged, but we believe it is more understandable if dealt with on a function by function, rather than on an organ by organ basis. The most important thing to remember is that all the functions are integrated - none of them is independent of the others. This chapter describes the organisations of cells into tissues, organs, systems and the whole organism, and relates specialised structural features of cells to their function. The specialisation of cells has the advantage that they can be better adjusted to carry out their function, making them more efficient at it, while at the same time relying on the other cells for support. An analogy would be the specialistion of people into particualr jobs, eg in an assembly plant, leading to greater productivity than would appear if all workers did all jobs. (Fortunately, specialised cells do not seem to suffer from boredom!). Of course, the specialisation means that individual cells then become very dependent on other cells carrying out their particular function correctly. Problems in one part of the system can lead to disaster in another. For this reason, there needs to be close co-ordination of all the different systems, and this necessitates a sophisticated system of communication between them.

1.2 Levels of organisation

organismic level

Mammals consist of several levels of structural organisation. The highest level, the organismic level, is the whole living individual. This could be considered to be made up of a number of systems, each with its own defined function, for example, the respiratory system. Each system is composed of a number of organs. For example the respiratory system comprises the nose, pharynx, larynx, trachea, bronchi and lungs. Each of these organs has a recognisable shape and function and is made up from two or more different tissues.

Π List the names of the organs of the digestive system?

digestive system

The most obvious ones are probably the stomach, small intestine, and large intestine but you could also include the mouth, pharynx, oesophagus, rectum and anus as parts of the gastrointestinal tract and also associated structures including the teeth, tongue, salivary glands, liver, gall-bladder and pancreas.

Tissues are groups of similar cells and their intercellular material. The trachea, for example consists of epithelial tissue which covers the inner surface, connective tissue which binds and supports other tissues, and muscular tissue, which changes the shape of the trachea during swallowing. Each of these tissues is itself made up from one or more types of cells. In the trachea for instance the epithelial tissue contains columnar epithelial cells which bear cilia which are used to move mucus up the trachea towards the mouth; the goblet cells which secrete mucus to trap foreign particles that are inhaled and basal cells which grow to replace the others (Figure 1.1).

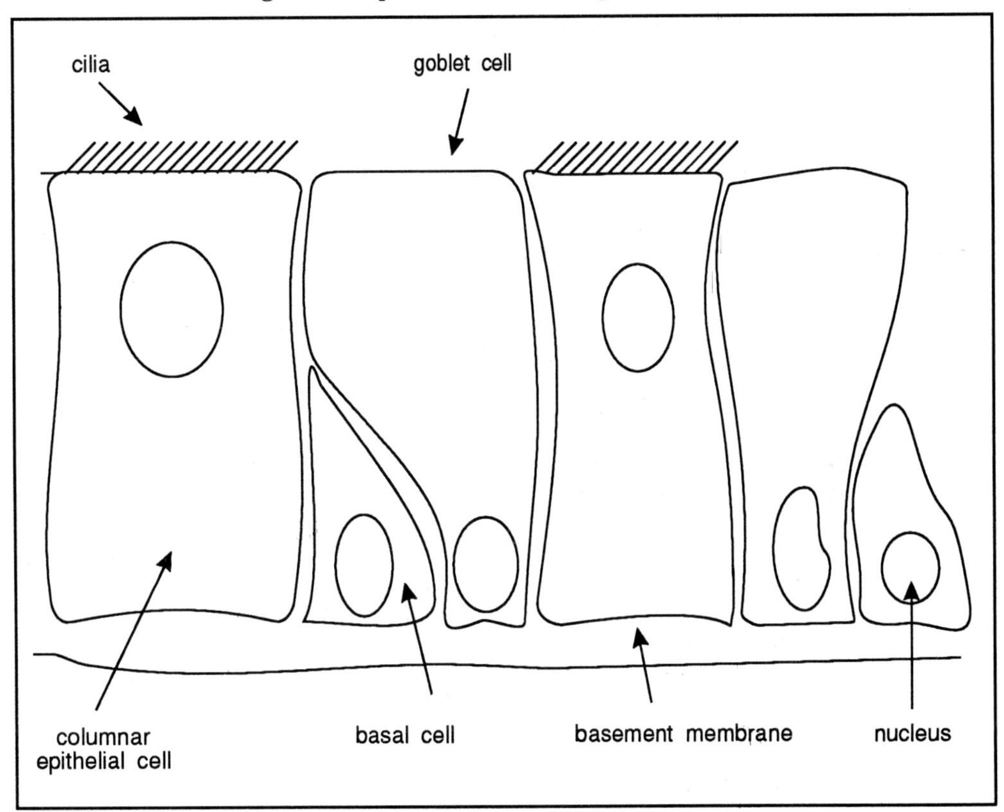

Figure 1.1 Epithelial tissue of the trachea.

System	Representative organs	Functions
Integumentary	Skin, hair, nails	Protects body, helps to regulate body temperature, receives environmental stimuli.
Skeletal		
Nervous		
Muscular		
Endocrine		
Cardiovascular		
Lymphatic		
Respiratory		
Digestive		
Urinary		
Reproductive		

Table 1.1 See SAQ 1.1

SAQ 1.1

Table 1.1 shows the major systems of the body, representative organs, and their functions, but it is incomplete. From the following list choose the most appropriate response and write the number in the table. (Alternatively, draw out Table 1.1 onto a fresh piece of paper and include the items below in their appropriate place. This will provide you with an excellent summary sheet).

1) Bones, cartilages, joints.
2) Testes, ovaries and organs that transport and store reproductive cells.
3) Regulates the chemical composition of blood, eliminates wastes and toxins, regulates fluid and electrolyte balance and volume, helps to maintain acid-base balance.
4) Gastrointestinal tract, salivary glands, liver, gall-bladder, pancreas.
5) Lungs, trachea, bronchi.
6) Lymph, lymph vessels, spleen, thymus gland, lymph nodes, tonsils.
7) Blood, heart, blood vessels.
8) Glands that produce hormones.
9) Brain, spinal cord, nerves and sense organs.
10) All muscle tissue.
11) Organs that produce, collect and eliminate urine, (kidney, bladder).
12) Distributes oxygen and nutrients to cells, carries CO_2 and wastes from cells, maintains acid-base balance, protects against disease, helps to regulate body temperature.
13) Regulates body activities through nerve impulses.
14) Regulates body activities through hormones transported in the blood.
15) Breaks down food, physically and chemically, and absorbs nutrients.
16) Produces white blood cells, protects against disease, helps to maintian blood composition.
17) Enables the organism to reproduce.
18) Regulates the composition of blood, eliminates wastes.
19) Brings about movement, maintains posture.
20) Supports and protects body, provides leverage for movement, produces blood cells.

1.3 Types of tissues

Earlier, you read that the trachea consists of three types of tissue, epithelial, connective and muscular tissue.

Π Can you suggest one type of tissue that will be present throughout the gastrointestinal tract?

By analogy with the respiratory system, you could suggest epithelial tissue would cover the inner surface of the gastrointestinal tract. Connective tissue is usually present in organs to hold cells together and, of course, muscular tissue must be present in most of the gastrointestinal tract to aid passage of food. A fourth kind of tissue, nervous tissue, is also present and its function is to regulate the overall activity of the system.

All the tissues of the body can be classified into these four principal types according to their structure and functions, and these types can be further subdivided.

- **Epithelial tissue** covers body surfaces or tissues, lines body cavities and forms glands;
- **Connective tissue** protects, supports and holds organs together;
- **Muscular tissue** is responsible for movement;
- **Nervous tissue** sends and conducts electrical impulses to co-ordinate body activities.

1.3.1 Epithelium

Epithelium is divided into two sub-types:

- covering and lining epithelium;
- glandular epithelium.

Their names show their functions. Covering and lining epithelium forms the outer covering of body surfaces and some internal organs, lines the body cavities and all tubes within the body. It consists mainly of closely packed cells in single or multilayers forming continuous sheets, with no blood supply (unless the epithelium forms the inside of a blood vessel). The cells are attached to connective tissue by a basement membrane. The cells can be replaced if damaged by division of stem cells.

Glandular epithelium is involved in the active secretion of substances into ducts, onto a surface, or into the blood. Glandular cells usually lie in clusters below the covering and lining epithelium. A gland may consist of only one cell or a group of epithelial cells (Figure 1.2).

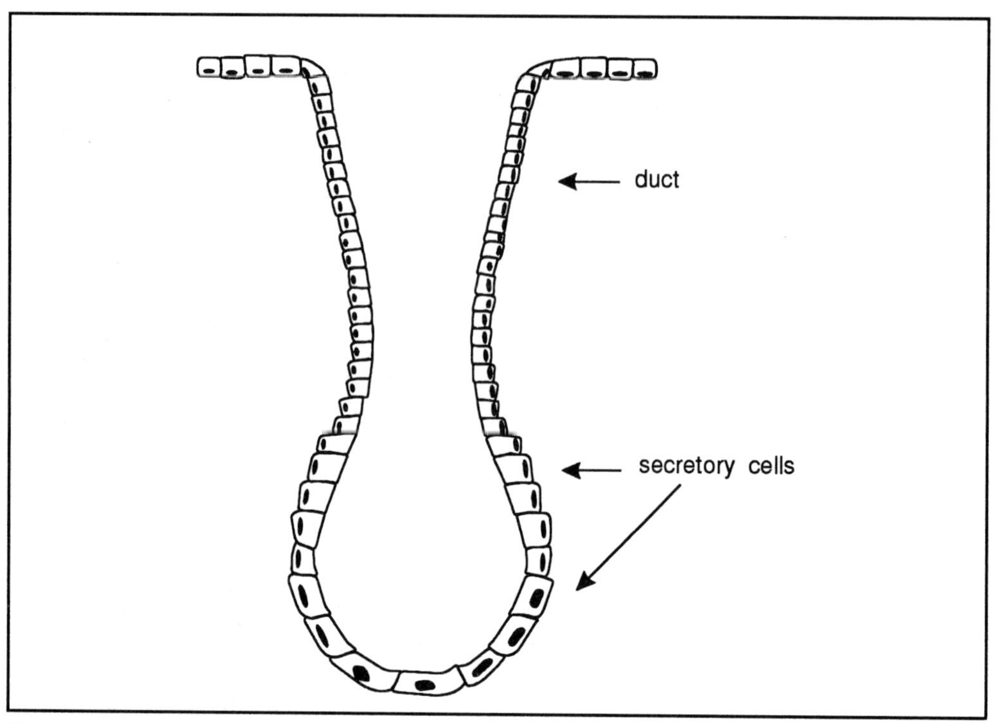

Figure 1.2 Glandular epithelium, a simple acinar gland (highly stylised).

exocrine glands

Exocrine glands secrete their products into ducts that empty at the surface of the covering or lining epithelium. For example, the goblet cells (single celled glands), secrete mucus into the repiratory and gastrointestinal tracts, and the sweat glands secrete sweat through ducts onto the surface of the skin.

endocrine glands

Endocrine glands secrete their products which are always hormones (chemicals that regulate the activities of various organs) directly into the blood. For example, the thyroid gland (an organ) secretes the hormone thyroxine from numerous glandular cells into the blood.

You should note that the word gland can be used for organs, glandular cells, or groups of cells.

pancreas

The pancreas is an interesting organ, because it is both an endocrine and an exocrine gland. Clusters of cells called pancreatic islets (or islets of Langerhans) contain different cells that secrete either glucagon, insulin or somatostatin directly into the blood. The islets themselves are surrounded by cells (acini) that secrete a mixture of digestive enzymes into small ducts which unite and empty into the small intestine. Thus the islets are endocrine and the acini are exocrine.

1.3.2 Connective tissue

Connective tissue is the most abundant tissue in the body. It is characterised by usually having a rich blood supply, and the cells are widely scattered in a matrix of intercellular material produced by the cells (Figure 1.3).

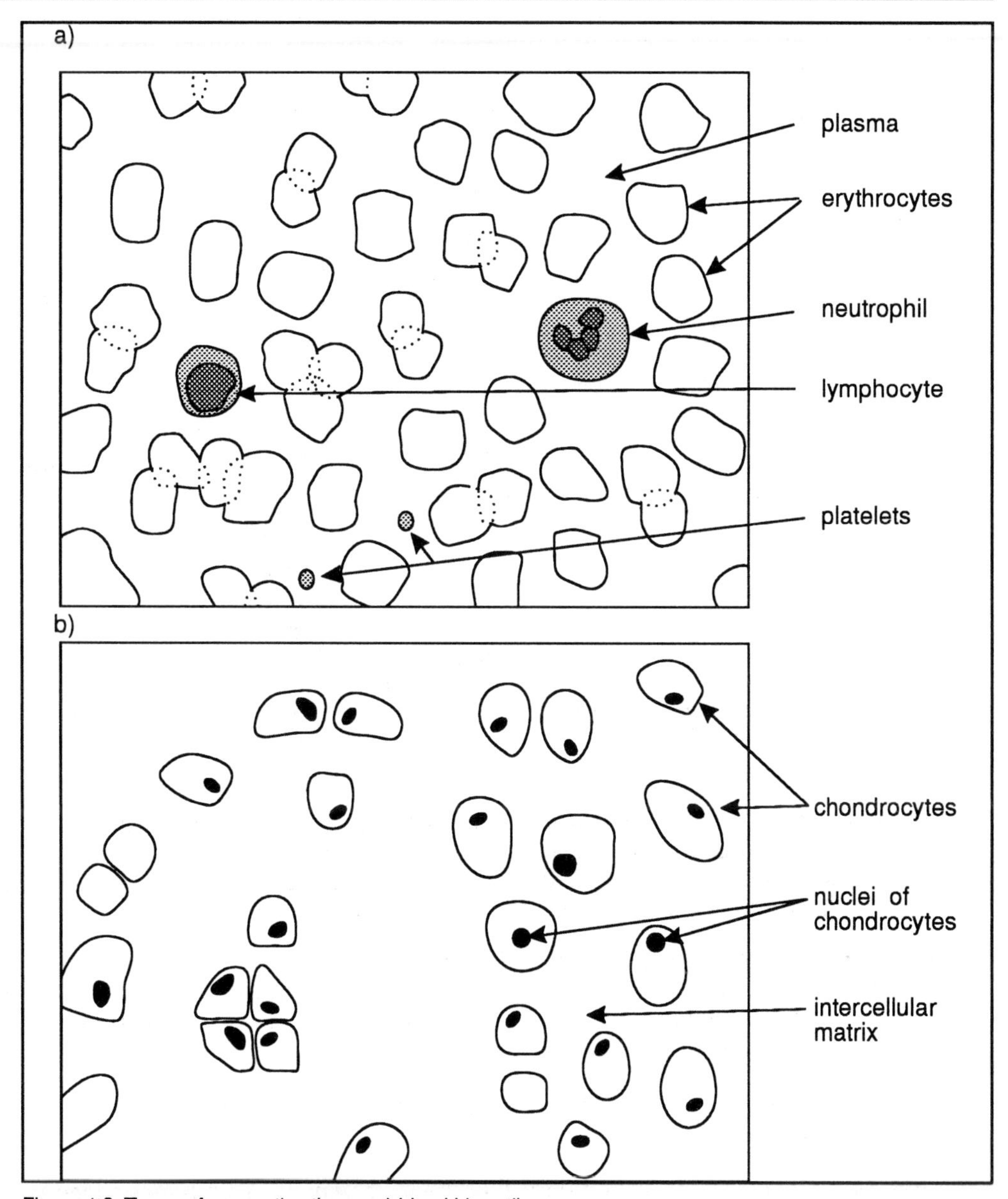

Figure 1.3 Types of connective tissue. a) blood b) cartilage.

cells of connective tissues

The intercellular matrix material can range in composition from the hard material of bone to the liquid material (plasma) in the blood. Connective tissue may also contain fibres and/or other cells. Connective tissues are involved in protecting, supporting and binding together other tissues and organs. The cells found in connective tissue may include fibroblasts, which produce the fibres of the intercellular material, and also cells involved in the provision of defence against infection: macrophages, plasma cells and mast cells. Adipose tissue contains cells called adipocytes which are specialised cells for fat storage. The adipocytes consist mainly of a large droplet of fat with the nucleus and cytoplasm pushed to the edge of the cell. Tendons and ligaments are mainly connective tissue containing a great deal of fibrous material. Cartilage is another type with both fibres and a jelly-like substance, chondroitin sulphate, the combination of the two providing both strength and resilience.

SAQ 1.2

Classify the tissue named or described below into one of the four main types of tissue.

1) Adipose tissue.

2) Tissue that lines body cavities.

3) Blood.

4) Tissue whose cells actively secrete substances into the blood.

5) Tissue with a rich blood supply, widely scattered cells and much intercellular material.

6) Tissue containing cells that produce fibres in the intercellular material.

1.3.3 Muscular tissue

three types of muscle

The role of muscular tissue is to enable motion, maintain posture, and sometimes to produce heat to help maintain body temperature. In order to carry out this role, muscle tissue cells are highly specialised for contraction. There are three types: skeletal muscle, cardiac muscle and smooth muscle (Figure 1.4).

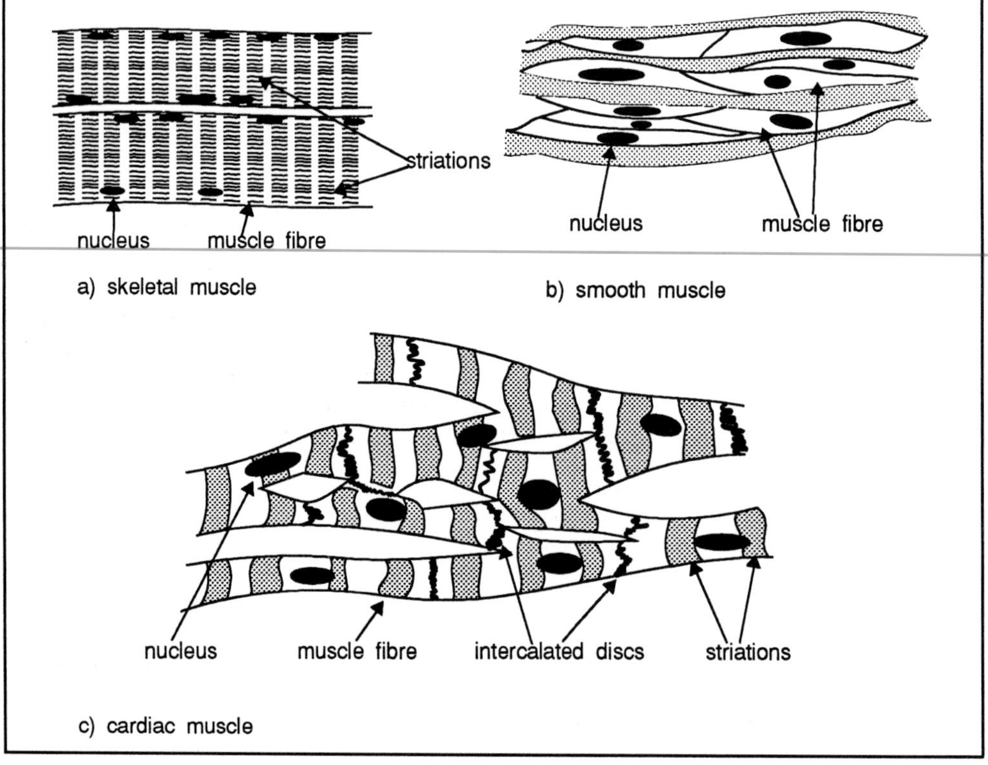

Figure 1.4 Types of muscle tissue. a) skeletal b) smooth c) cardiac.

striations, voluntary muscle

Skeletal muscle is the muscle attached to bones. Under a light microscope each muscle cell (fibre) appears cylindrical and striated, ie there are alternating light and dark bands at right angles to the long axis of the fibre. The different cells lie parallel to each other and are arranged in bundles surrounded by connective tissue. Each cell has a plasma membrane called the sarcolemma, and is usually close to small blood vessels (capillaries) and the end of a nerve cell (synaptic end bulb). Each muscle cell contains many myofibrils (the structure of which gives the muscle its striated appearance) and several nuclei lying close to the sarcolemma. A skeletal muscle fibre has several nuclei because the fibre is a collection of cells, each with its own nucleus, which have become fused together. Skeletal muscle tissue is said to be voluntary, because it can be made to contract by conscious control. (We will discuss muscular activity in detail in Chapter 9).

involuntary muscle

Cardiac muscle is the muscle forming the walls of the heart. It is also striated but the fibres are usually quadrangular and branched to form networks. In contrast to skeletal muscle tissue, each cell usually contains only one nucleus, centrally located. Transverse thickenings of the sarcolemma, called intercalated discs, separate the fibres. Cardiac muscle is said to be involuntary because it is not under conscious control.

Smooth muscle tissue is found in the walls of all the tubes and "bags" of the body, eg intestine, stomach and blood vessels. The cells are usually small, involuntary, non-striated, often spindle-shaped and each contains a single central nucleus.

All three types of muscle tissue share the following characteristics of:

- **contractability** - ability to shorten and thicken;
- **extensibility** - ability to be stretched;
- **elasticity** - ability to return to their original shape;
- **exitability** - ability to respond to electrical stimulation.

The different structural features of the three types are related to their function. Skeletal muscle needs to contract over the whole length of the muscle, and the muscles are effectively acting as self-shortening ropes joining pieces of bone together. Thus all the cells need to be parallel and in order to be able to vary the strength and extent of contraction, each cell is under separate nervous control. Cardiac muscle acts to reduce the size of two types of bags, the atria and the ventricles of the heart. Since contraction is in various directions, the fibres are branched into networks, and those for each bag work together, they are not separately controlled. Smooth muscle fibres may be controlled either singly or in groups, but in both cases contractions tend to be slower and more prolonged than in skeletal or cardiac muscle. Because of the different arrangement of myofibrils within the cells, smooth muscles are able to stretch without developing tension. For example, as the stomach fills with food, the pressure within the stomach remains the same.

1.3.4 Nervous tissue

Nervous tissue consists mainly of two kinds of cells, neurons (Figure 1.5) and neuroglia.

The neurons consists of:

- **a cell body** containing the nucleus and other organelles;

- **dendrites** - highly branched fingers of cell membrane and cytoplasm which conduct nerve impulses towards the cell body;
- **an axon** which is a long finger of cell membrane and cytoplasm that conducts nerve impulses away form the cell body and may or may not be branched.

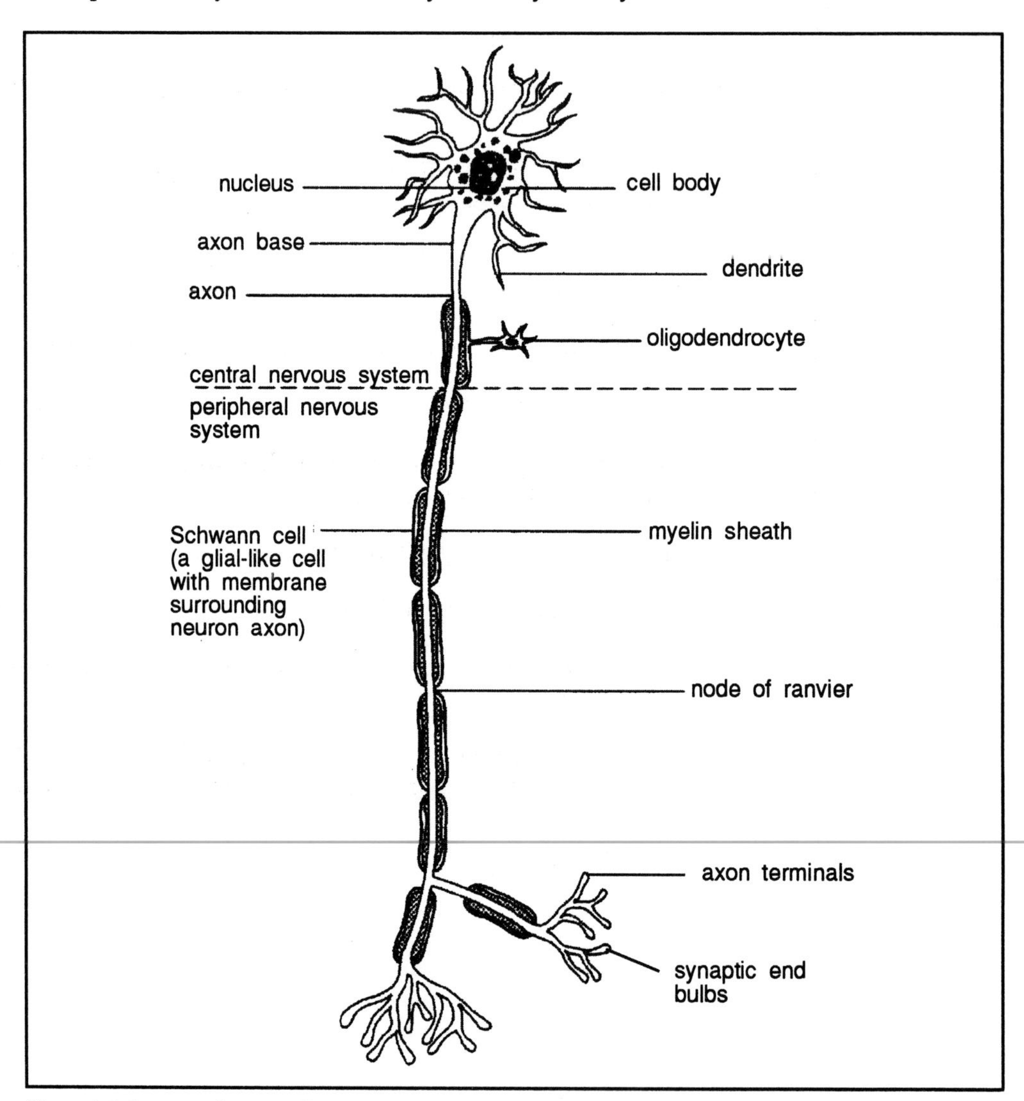

Figure 1.5 A neuron (nerve cell).

neuroglia

Neuroglia are cells that support the neurons, for example by attaching them to blood vessels, or by producing a fatty sheath (myelin) around axons.

The cell bodies of the neurons carry out the normal metabolic functions of the cells, whereas the dendrites allow the cell to receive stimuli from a number of cells. The axon conducts impulses from the cell body to one or more neurons, or to a muscle or gland. At the ends of the axon are fine branched filaments (axon terminals)which themselves end in synaptic end bulbs containing membrane-enclosed sacs (synaptic vesicles) which store the neurotransmitters that convey the signals to other cells. We will examine this in more detail in the next chapter.

Try the following SAQ.

SAQ 1.3

Answer the following by circling the true statement:

1) Glandular epithelium secrete substances actively, whereas lining epithelium may be involved in active absorption.

2) Exocrine glands are collections of glandular epithelial cells that secrete their products into the blood.

3) Smooth muscle tissue is not striated, made up a spindle-shaped cells, and is usually voluntary.

4) Cardiac muscle forms most of the heart wall, is striated, and involuntary.

5) Skeletal muscle is striated, attached to bones, and is involuntary.

6) The nerves supplying skeletal muscle stimulate all the cells of a particular muscle simultaneoulsy.

7) Cardiac muscle cells are arranged in parallel.

8) Dendrites convey electrical impulses away from neuron cell bodies.

9) The nature of the interceullular substances determines the properties of intracellular connective tissue.

10) The ending of axons contain substances that when released are responsible for the transmission of impulses from one nerve to another.

Summary and objectives

In this chapter we have examined the levels of organisation of cells, the different systems of the body and some ways in which cells are specialised.

Now that you have completed this chapter you should be able to:

- explain the organisation of cells into tissues, organs and systems using several examples;
- classify tissues according to structure;
- relate structural features of various tissues to their functions.

The nervous system

The nervous system

2.1 Introduction

2.1.1 Role of the nervous system

nervous system

The nervous system (together with the endocrine system) is the major control and co-ordinating system of the mammalian body. Thus, it conveys information about our immediate internal and external environment , in the form of electrical signals along nerves to the brain. The brain, together with the spinal cord, constitutes the central nervous system (CNS), and is the region where this information is processed. In turn, the CNS generates electrical signals which are passed back to the rest of the body, such that appropriate responses are made to the initial information.

This chapter will examine the physiology of the nervous system beginning at the level of single nerve cells (neurons) and finishing at the level of the brain (which is composed of millions of nerve cells). Let us begin by looking at the overall structure of the mammalian nervous system.

2.1.2 The basic plan of the nervous system

The basic plan of the nervous system is outlined in Figure 2.1. Examine it carefully so that you have this basic plan in mind while you read the rest of the chapter.

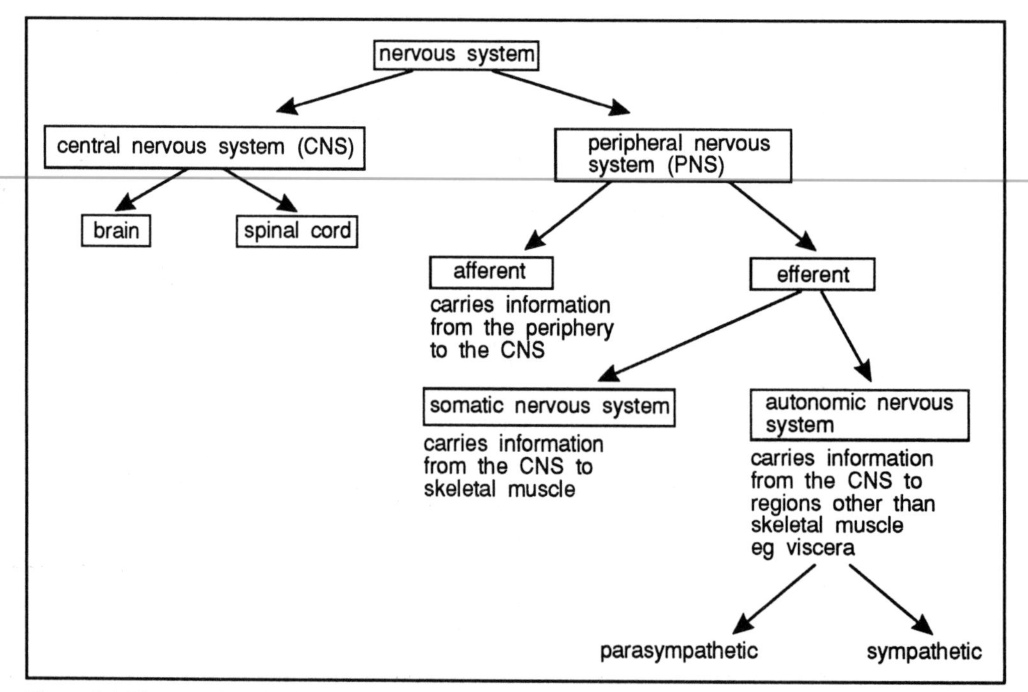

Figure 2.1 Diagram showing the basic organisation of the mammalian nervous system.

2.2 Functional anatomy of the nervous system

We remind you that the nervous system comprises of two major cell types:

- neurons;
- glial cells.

Let us take a look at the role of each in turn.

2.2.1 Neurons

neuron

The neuron is the basic building block of the nervous system. In the body, many individual neurons come together to form nerves. The structure of a typical mammalian neuron is shown in Figure 2.2.

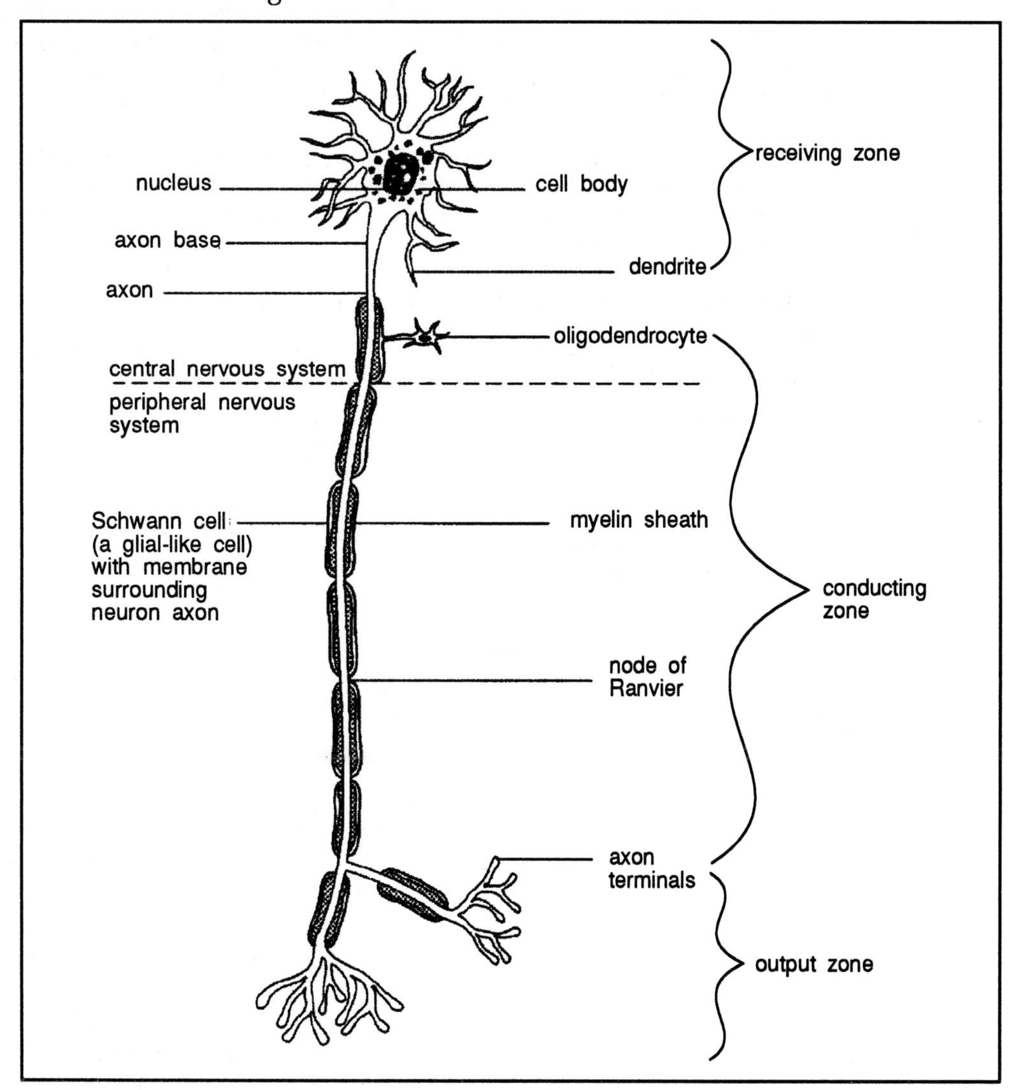

Figure 2.2 Diagram showing the principal features of a typical mammalian neuron.

Neurons function by the generation and transmission of electrical signals called action potentials. This will be discussed further in Section 2.3. It is convenient to consider that the neurons consist of three zones:

- a receiving zone;
- a conducting zone;
- an output zone.

The receiving zone consists of the cell body (soma) and the dendrites. Dendrites are simply a series of branched outgrowths of the cell body. The function of the receiving zone is to act as an input region where signals from other neurons are received. The conducting zone consists of the axon, which is a single long outgrowth of the cell body, which, except for its terminals, is unbranched. It is the axon which conveys the action potentials to another cell, which can be either another neuron or a non-neuronal cell. The initial segment of the axon, ie the portion of the axon nearest to the cell body together with the region of the cell body from which it originates, is the site where action potentials are generated. The output zone of the neuron are the axon terminals which are simple branchings of the axon. It is from this zone that an action potential may influence the activity of other neurons or other cell types, eg muscle, endocrine glands etc. Neurons do not actually make physical contact with each other or with other cells. They are separated by a very small gap called a synapse. The way in which signals are transferred from one cell to another across a synapse is via the release of chemical substances called neurotransmitters from the axon terminals. These substances diffuse across the synapse and thereby influence the activity of other cells. We will look at this in more detail later.

synapse

neuro-transmitters

myelin sheath

The myelin sheath is formed by the wrapping of other cells around the axon. It can be regarded as a fatty sheath enveloping the axon of the neuron. Not all neurons have myelinated axons. We shall see the significance of myelination in Section 2.3.3. The myelination of axons is not complete, there are breaks in the myeline sheath at points called nodes of Ranvier. Although, all neurons conform to this basic pattern, there is considerable variation in their precise structure. Such variation permits a simple classification system. This is the division into unipolar, bipolar and multipolar cells. This classification system is simply based upon the number of projections on the cell body. Some examples are shown in Figure 2.3.

nodes of Ranvier

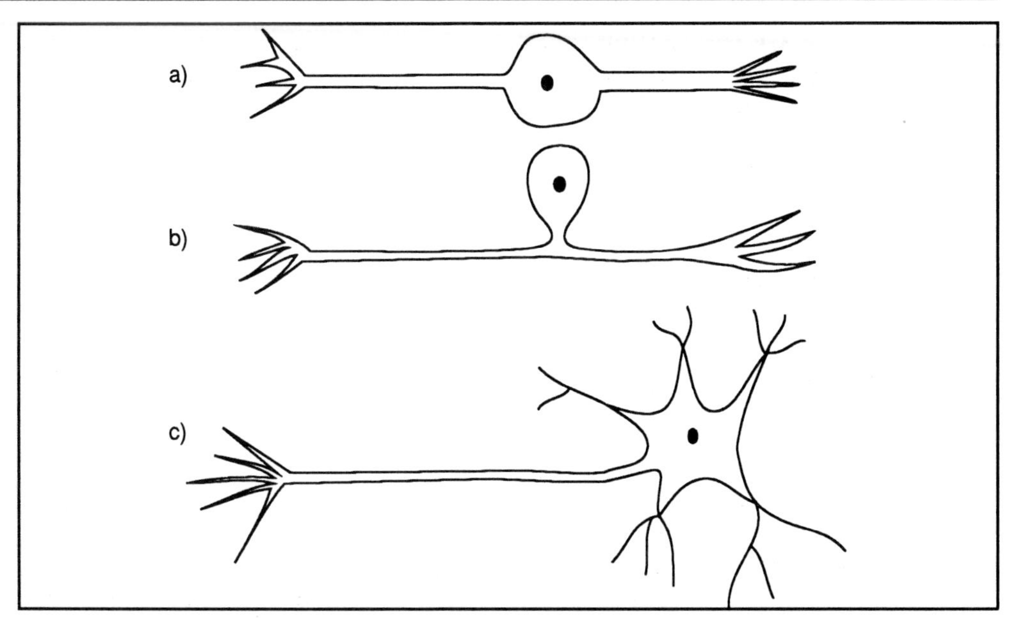

Figure 2.3 Diagram showing examples of three types of neurons from the top down. a) bipolar neuron, b) pseudo-unipolar neuron, c) multipolar cell eg motor neuron. Axons are to the left, dendrites to the right.

2.2.2 Glial cells

glial cells

In the central nervous system, glial cells outnumber neurons by 10:1. In a similar manner to neurons, it is possible to classify them into one of three groups:

- astroglial cells;
- oligodendroglial cells;
- microglial cells.

astroglial, oligodendroglial, Schwann cells, microglia

Astroglial cells consists of a central cell body from which a number of projections radiate in a star-like fashion. These cells have a number of possible functions such as the regulation of the extracellular fluid surrounding neurons by the removal of waste products. Oligodendroglial cells are smaller and have fewer radiating projections than do astroglial cells. A major function of these cells is to provide the myelin sheaths that envelope axons. In the peripheral nervous system, specific glial cells known as Schwann cells provide axons with their insulating myelin sheath. The final group of glial cells are the microglia. Again, these cells have a number of possible functions. These include maintenance of the extracellular ionic environment of neurons and disposal of the metabolites of neuronal activity, this includes the ability to phagocytose (engulf and remove) necrotic tissue.

2.3 Neuronal function

action potentials

Nerve cells function by the generation and transmission of action potentials. It is a common property of all animal cells, that there exists across their cell membrane a small potential difference (voltage). An action potential is simply a transient reversal of the polarity of this potential difference. This section will consider the ionic basis of this potential difference and, more importantly, how it alters to generate an action potential.

2.3.1 The resting membrane potential

resting membrane potential

All animal cells have a potential difference across their membranes. This is called the resting membrane potential, with the inside of the membrane being negative with respect to the outside. Its value varies from -5 mV to -100 mV depending upon the cell concerned, but in neurons it is usually of the order of -70 mV to -80 mV. Figure 2.4 shows how it is possible to measure the resting membrane potential in a neuron.

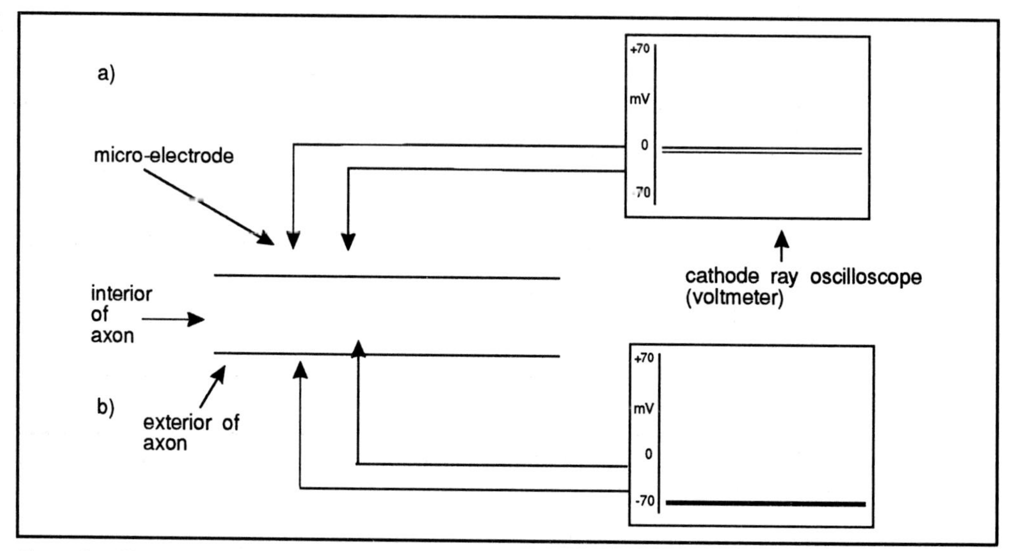

Figure 2.4 Diagram showing the way in which resting membrane potential is measured with a) both electrodes on the exterior of the cell no voltage is recorded. However, b) when one electrode penetrates the cell, a voltage of -70 mV is recorded. Such studies were aided by the use of giant axons found in squid.

The potential that is generated is determined principally by two factors:

- the unequal distribution of ions in the intracellular and extracellular fluid;
- the relative permeability of the neuronal cell membrane to different ions.

The principal ions and their concentration in intracellular and extracellular fluid are shown in Table 2.1.

Ion	Concentration (mM)	
	Intracellular	Extracellular
Na^+	15	150
K^+	150	5
Cl^-	10	110
Pr^-	155	-

Table 2.1 Principal ions of intra and extra-cellular fluids. Pr^- = Negatively charged protein molecules. Note a range of values are reported in the literature reflecting differences in different cell systems. The concentrations of intracellular and extracellular ions reported here give a membrane potential of about -70mV. The membrane potential can be modified by changing the relative intra and inter cellular concentrations of these ions.

In order for us to see how this unequal distribution of ions can give rise to a potential difference across the nerve cell membrane, let us consider the hypothetical situation depicted in Figure 2.5.

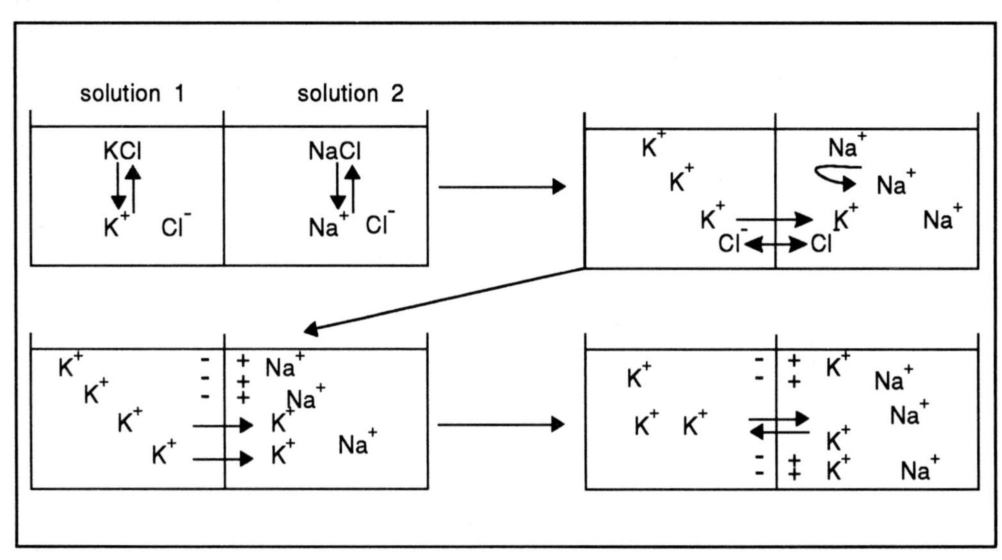

Figure 2.5 Diagram showing the generation of a potential difference caused by the unequal distribution of K^+ ions across an artificial membrane (see text for description).

At the top of Figure 2.5 the illustration depicts a solution of KCl separated from a solution of NaCl by a selectively permeable membrane, permeable to K^+ ions only. The solutions are of equal concentration. Initially, K^+ ions will diffuse down their concentration gradient from solution 1 to solution 2. Na^+ is unable to move in the opposite direction since the membrane is impermeable to sodium. As diffusion of the K^+ ions continues, there is a build up of positive charge in solution 2, and because of the loss of K^+, there is a build up of negative charges in solution 1. Eventually, the build up of positive charge in solution 2, will prevent the net movement of any further K^+ ions from solution 1 to solution 2, ie an equilibrium has been set up.

cell membrane

plasma membrane

ion channels

This is precisely the sort of situation which develops across the cell membrane (plasma membrane) of the neurone. Firstly, there exists a massive K^+ concentration difference between the inside and outside of the cell. Secondly, the membrane is permeable to K^+ ions. Permeability in neurons (or any cells) is usually determined by the presence or absence of specific channels. In this case (neuronal) K^+ channels are present and open. Channels can be considered to be pores which run from one side of the cell membrane to the other, connecting the interior of the cell to the exterior. These pores can be either open or closed.

The potential difference that can be generated across a membrane for a given concentration difference of a particular ion can be calculated using the Nernst equation. The Nernst equation assumes that the membrane is fully permeable to the ion concerned. The Nernst equation is shown below.

Nernst equation

$$E_m = \frac{RT}{zF} \ln \frac{[C_o]}{[C_i]}$$

where:

E_m = potential difference generated for the ion concerned, ie the potential difference at which equilibrium is achieved. This is called the membrane potential or equilibrium potential for the ion.

R = universal gas constant (8.3146 J kg^{-1} K^{-1}); T = absolute temperature (K = °C + 273); z = valence (charge) of the ion; F = Faraday constant (96,486.6 C/mol charge); [C_o] = extracellular concentration of the ion; [C_i] = intracellular concentration of the ion.

K^+ equilibrium potential

Let us use this equation to calculate the K^+ equilibrium potential, at 37°C using data already provided (Table 2.1).

$$E_{K^+} = \frac{310 \times 8.3146}{96{,}486.6} \times \ln\left(\frac{5}{150}\right)$$

= - 0.0908 V

= - 90.8 mV

Immediately we can see that the measured resting membrane potential (-70 mV) is very close to the calculated K^+ equilibrium potential. Thus, we can say that the unequal distribution of K^+ ions across the neuronal membrane is the primary cause of the resting membrane potential.

SAQ 2.1

Use the Nernst equation to calculate what the resting membrane potential would be if the extracellular K^+ ion concentration was increased to 100 mM. Explain your answer.

The fact that we know this intra and extracellular concentration of potassium ions and that the measured resting membrane potential is near to the K^+ equilibrium potential but not equal to it suggests the involvement of other ions. Let us look at the other ions beginning with the involvement of Na^+.

The distribution of Na^+ is exactly the opposite to that of K^+ ie high extracellular concentration and a low intracellular concentration.

Π Use the Nernst equation to calculate the Na^+ equilibrium potential; use the sodium concentration data from Table 1.1.

Your calculation should have shown you that Na^+ equilibrium potential is +61.5 mV, inside positive with respect to the outside. Therefore, it is obvious that the resting neuronal membrane is not completely permeable to Na^+ ions. If it were the measured resting membrane potential would be about -30 mV (ie -90 + (+60)). However, the membrane of the neuron is not completely impermeable to Na^+ ions. Thus some Na^+ ions do enter the neuron from the extracellular fluid through a type of sodium channel. The effect of this is to bring more positive charge into the neuron. This is the reason why the measured resting membrane potential in slightly less negative than the figure obtained for potassium equilibrium potential alone. The positive charge that enters the neuron is sufficient to shift the resting membrane potential from -90 mV as is would be if only potassium were involved, to the value of -70 mV that is actually measured.

SAQ 2.2

Suggest an experiment that could be done to prove that the small influx of Na^+ ions (positive charge) is important in determining the resting membrane potential.

The remaining ion to be dealt with is the Cl^- ion and in neurons this is really of little or no importance. This is because the membrane of the neuron is permeable to the Cl^- ion. Therefore, the resting membrane potential will determine the distribution of Cl^- ions ie a high extracellular concentration and a low intracellular concentration. This generates a concentration gradient for the Cl^- ion which is directed inwards. However, this concentration gradient is exactly opposed by the electrical force across the membrane of the neuron (ie negative inside, positive outside). The outcome of this is that the chloride equilibrium potential (E_{Cl}) is equal to the resting membrane potential. Thus, we can see that the resting membrane potential is in actual fact dependent upon the relative permeability of the neuronal membrane to different ions, primarily potassium and sodium and we can ignore any effect that chloride has. It is possible to predict what membrane potential will be generated by different ions and their relative membrane permeabilities by using a modified form of the Nernst equation. The Goldman equation is shown below.

$$\text{Goldman equation } E_m = \frac{RT}{zF} \ln \frac{[K_o^+] + b\,[Na_o^+]}{[K_i^+] + b\,[Na_i^+]}$$

where:

R, T, z, F - have their usual meanings

$[K_o^+]$, $[K_i^+]$ are the concentrations of potassium ions outside an inside of the membrane.

$[Na_o^+]$, $[Na_i^+]$ are the concentration of sodium ions outside and inside of the membrane.

$b = \frac{PNa^+}{PK^+}$ in which PNa^+ = membrane permeability to Na^+; PK^+ = membrane permeability to K^+.

Since the membrane permeability to Na^+ is only about 1/25th that of K^+ ion permeability then $b = 0.04$.

Π With the data provided, use the modified form of the Nernst equation to calculate the resting membrane potential of a typical neuron. How well does it compare with the measured value?

You should find that the predicted value and measured value of -70 mV are in close agreement with each other. This supports our argument that the resting membrane potential is dependent upon the relative permeability of the neuronal membrane to potassium and sodium ions.

2.3.2 Metabolic activity of the neuron

The measured resting membrane potential of -70 mV is at neither the potassium nor the sodium equilibrium potential. Therefore, there tends to be a small net movement of each ion down its concentration gradient ie K^+ ions are moving out and Na^+ ions are moving in. This movement cannot go unchecked. Therefore, there must be some mechanism to restore the situation back to normal. The mechanism is the Na^+-K^+-ATPase in neurons. This is a pump, driven by the energy provided by the hydrolysis of ATP, which pumps 3 Na^+ ions out of the cell and simultaneously pumps 2 K^+ back into the neuron. This sort of pump is known as an electrogenic pump since it does not pump equal amounts of charge in both directions. However, the contribution that this difference in charge makes directly to the generation of the resting membrane potential is small. Note that ionic pumps are described in the BIOTOL text 'Infrastructure and Activities of Cells'.

SAQ 2.3

Describe the consequences of the inhibition of the Na^+-K^+-ATPase on the resting membrane potential over a long period of time.

2.3.3 The action potential

action potentials

We have stated that the nervous system achieves its function of control and co-ordination by the generation and passage of small electrical signals called action potentials. Action potentials are just short lived changes in the polarity of the resting membrane potential, the inside of neurons becoming momentarily positively charged. The time course of an action potential and its various phases are shown in Figure 2.6.

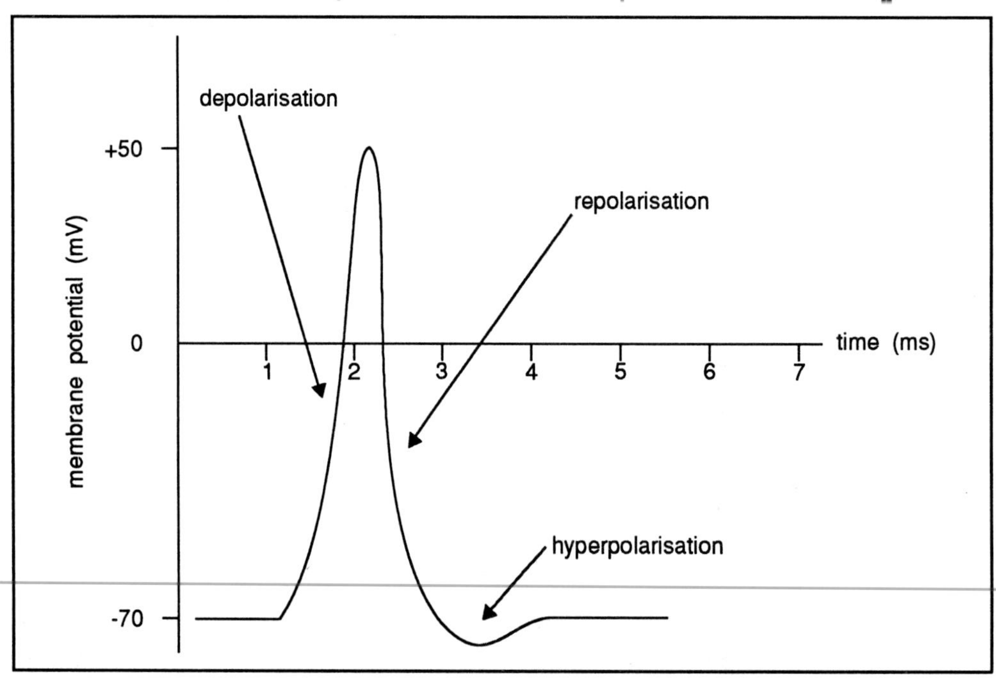

Figure 2.6 Time course of an action potential and its various phases, as recorded by an intracellular electrode in a mammalian motor neuron.

The initial generation of an action potential is usually evoked by electrical stimulation of the neuron in experimental situations. In real life action potentials in a neuron are usually initiated by sensory receptors or by an action potential from another neuron.

The ionic basis of the action potential

In a similar manner to the resting membrane potential, the action potential can be explained by changes in the membrane permeability to Na^+ and K^+ ions.

Π Write down the membrane permeability to Na^+ and K^+ ions of the resting neuron.

During the course of an action potential there are fundamental changes in the permeability of the neuronal membrane. During the depolarising phase, there is a tremendous increase in the permeability of the membrane to Na^+ ions. This is caused by the opening of specific sodium channels, which allows Na^+ ions to enter the neuron from the exterior. This means that the interior of the neuron becomes positively charged whilst the extracellular fluid becomes negatively charged (loss of positive charge in the form of Na^+ ions to the interior). During this phase, because the membrane is more permeable to Na^+ ions than K^+ ions, the membrane potential reaches the sodium equilibrium potential.

The rapid rise of the depolarising phase is aided by a positive feedback mechanism which ensures that as the Na^+ ions enter, the greater the depolarisation that occurs which in turn ensures that more Na^+ ions enter and so on. The positive feedback mechanism is shown in Figure 2.7.

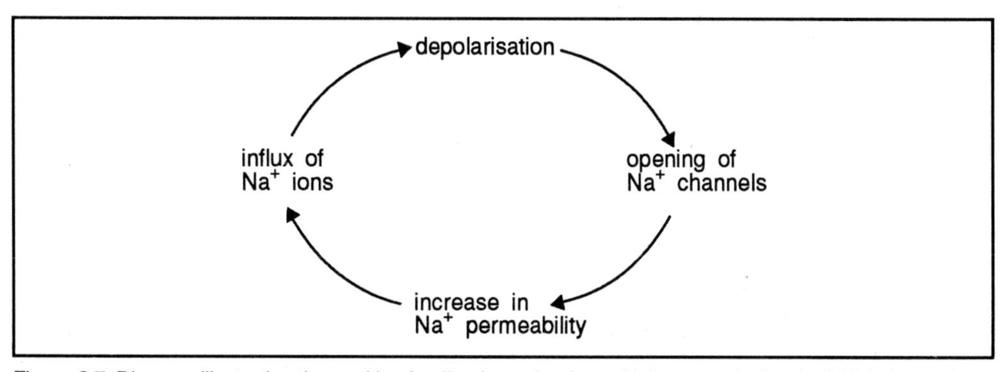

Figure 2.7 Diagram illustrating the positive feedback mechanism which occurs during the initial phase of an action potential, thus evoking a rapid depolarisation.

Virtually as soon as the action potential has reached its peak it returns rapidly back to resting membrane potential. There are two events which cause this to happen. Firstly, the sodium channels which had opened to let Na^+ ions in initially now close. This prevents any further positive charge from entering the neuron. Secondly, a group of potassium channels opens. Thus K^+ ions leave the cell and therefore so does the positive charge. Hence, the neuron returns towards resting membrane potential. In fact, some of the potassium channels stay open even when resting membrane potential has been reached. This is the cause of the hyperpolarisation which occurs at the end of an action potential. Here, the cell membrane potential is moving towards the potassium equilibrium potential since the membrane is more permeable to K^+ ions.

SAQ 2.4

Sketch graphs showing:

1) The normal action potential.

2) The action potential occurring when extracellular sodium concentration is reduced. Explain the shape of this graph.

At the end of the action potential, the neuron has gained some Na^+ ions and lost some K^+ ions. This situation is rectified by the action of the Na^+-K^+-ATPase pump. This will pump Na^+ ions out of the neuron and pump K^+ ions back into the neuron. If this did not occur then ultimately the Na^+ and K^+ ion concentration gradients that exist would disappear and so would the resting membrane potential and action potential.

Threshold and the all or nothing responses of neurons

Not all stimuli, whether they are natural or artificial, will generate an action potential in a neuron. For an action potential to be generated the membrane must be depolarised to such an extent that the positive feedback mechanism shown in Figure 2.7 is engaged. The point at which it is engaged is the point at which there begins the net movement of positive charge (ie Na^+ ions) into the cell. The point at which this occurs is called the threshold potential and the stimuli which produce this change are called threshold stimuli. In most neurons threshold is about 10-15 mV above resting membrane potential ie about -60 to -55 mV. The threshold of all neurons is not identical. Stimuli which do not cause the membrane potential to reach threshold are called subthreshold stimuli. In this case only subthreshold change in membrane potential occur ie no action potential is generated. The subthreshold stimuli and their relationship to the generation of action potentials are shown in Figure 2.8.

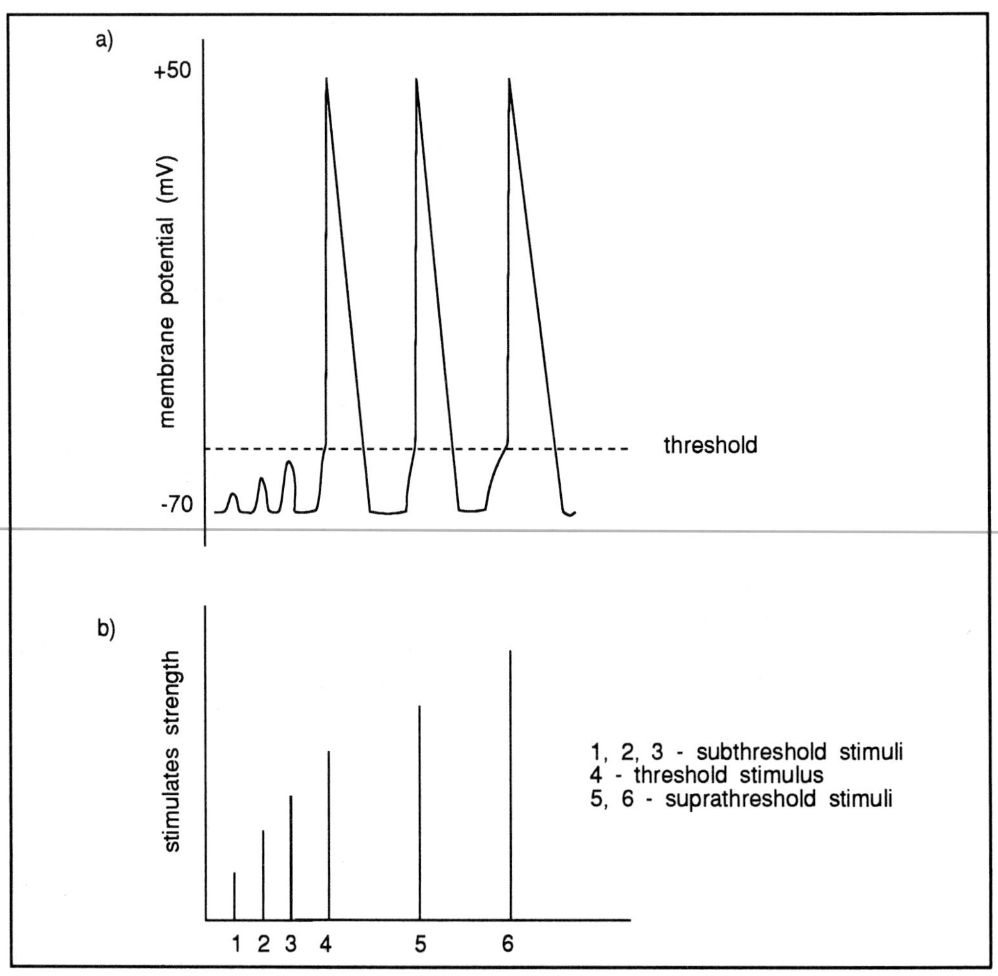

Figure 2.8 Illustration of the effects on subthreshold, threshold and suprathreshold stimuli on changes in membrane potential. a) Membrane potentials for different strengths of stimuli shown in b). Redrawn from Vander, A.J, Sherman, JH, Luciano, DS (1990) Human Physiology: The Mechanisms of Body Function, McGraw Hill, New York.

Π Do you think that the threshold of a single neuron would be the same as that of a nerve (remember a nerve is a collection of individual neurons).

The answer is that since the threshold of different neurons are not identical then there will be a range of thresholds for a nerve, (that is a stimulus of limited strength would cause only those neurons with low thresholds to respond, a stronger stimulus will cause a greater proportion of the neurons to respond). Thus the threshold of a single neuron is not the same as that of a nerve.

all or nothing response

Figure 2.8 also shows the concept of suprathreshold stimuli. Once stimulus threshold has been reached and an action potential generated, then increasing the strength of the stimulus further has no effect on the size of the action potential. Action potentials are generated in an all or nothing fashion ie it is not possible to have half an action potential neither one and a half or a double action potential. The maximum change that can occur in membrane potential is that generated by an action potential.

Refractory period

We have seen that providing a stimulus is of threshold value then an action potential will be generated. How soon after one action potential has been generated can a second be generated? This is important to know since one way in which information is coded in the nervous system is by the frequency with which axons conduct action potentials. Following the generation of an action potential a neuron goes into a refractory period, ie a period of time during which it is either unresponsive or less responsive to a second stimulus similar to that which evoked the first action potential. The refractory period can be divided into two separate sections. The absolute refractory period is the time during the first action potential and for about 1 msec after the finish of this action potential, when a second stimulus, no matter how strong, fails to generate a second action potential. The principal cause of the refractory period during the action potential is due to the presence of open sodium channels. The relative refractory period immediately follows the absolute refractory period and lasts for 10-15 msec. During this time period it is possible to generate a second action potential but to do so the stimulus strength must be increased. Obviously as we move from the beginning to the end of the relative refractory period, the increase in stimulus strength required to evoke a second action potential decreases. The relative refractory period corresponds to the time when potassium channels are open and K^+ ions are leaving the cell. During this period, suprathreshold stimuli are required to generate an action potential. They achieve this by opening increased numbers of the Na^+ channels thus ensuring there is a net inward flow of positive charge to cause a depolarising response.

Propagation of the action potential

Having once generated an action potential, the action potential must be propagated (conducted) along the axon to the axon terminals where it can influence the activity of other cells. Action potentials are propagated along the length of the axon by the creation of local currents between adjacent depolarised and non-depolarised regions of membranes (Figure 2.9).

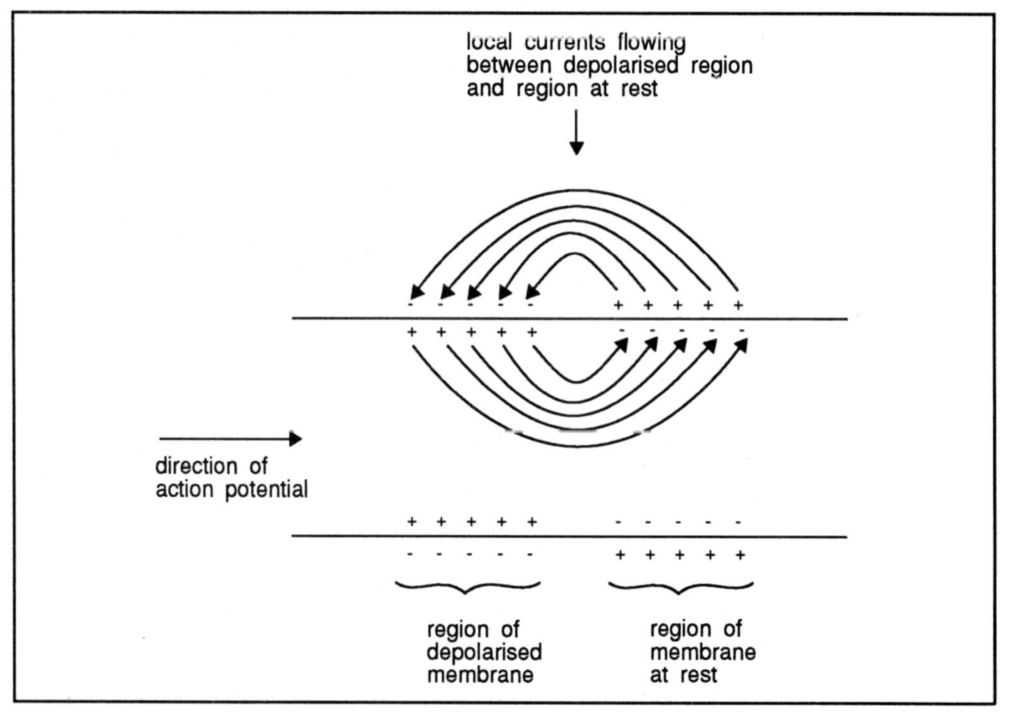

Figure 2.9 Diagram showing the formation of local currents involved in the propagation of the action potential.

As summarised in Figure 2.9 currents flow from positive to negative. Thus on the interior of the neuron positive charge flows from the region of the membrane that is depolarised to regions of the neuronal membrane at resting membrane potential. Charge is moving in the opposite direction on the exterior of the neuron. The effect of positive charge moving towards regions of the membrane at rest in the interior is that the resting membrane potential moves towards threshold. As more positive charge flows and threshold is reached, then an action potential is generated. Thus the action potential appears to have moved from one region to the next. This newly depolarised region then depolarises the resting membrane adjacent to it and so on down the axon.

Π Can you think of an explanation as to why the action potential may travel in one direction only.

The action potential travels in one direction only since for any region of membrane, the region from which local depolarising currents flows, is now in the refractory period. Therefore, at the very minimum suprathreshold stimuli will be required to evoke an action potential to travel in the opposite direction. In addition synapses only permit one way conduction of the action potential (see Section 2.4).

There are two major factors which determines how fast an action potential will travel along an axon. These are:

- the diameter of the axon and;
- the presence or absence of myelin.

different neurons conduct action potentials at different rates

The larger the diameter of an axon, the faster an action potential will travel along it. This is because larger diameter axons offer less resistance to the formation of local currents. Myelin, we have already stated, is the fatty insulating sheath that envelopes the axons of some neurons. It is not continuous but has breaks in it called nodes of Ranvier. It increases the velocity of an action potential since local currents now only flow from node to node. Therefore, action potentials are propagated from node to node along the axon. The fact that individual neurons have different velocities for the conduction of action potentials provides a basis for the classification of neurons eg A fibres are large, myelinated neurons with conduction velocities of up to 100 m s^{-1}. In contrast C fibres are small, non-myelinated axons with a maximum conduction velocity of 2.5 m s^{-1}.

2.4 Synaptic function

Having generated an action potential and propagated it along the axon, the next task that the neuron must undertake is the transfer of the action potential to another cell either neuronal or non-neuronal to influence its activity. We have already stated that there is no physical contact between neurons and other cells, there exists a gap between them called a synapse. A synapse is defined as a region between two neurons, at which the electrical activity of one neuron (the presynaptic neuron) may influence the electrical activity of another neuron (the postsynaptic neuron) via a chemical mediator (a neurotransmitter). Anotomically, synapses are of three types:

types of synapses

- axo-dendritic - between the axon of the presynaptic cell and dendrite of the postsynaptic cell;
- axo-somatic - between the axon of the presynaptic cell and cell body of the postsynaptic cell;
- axo-axonic - between the axon of the presynaptic cell and axon of the postsynaptic cell.

The majority of synapses are axo-dendritic. Let us have a look at synapses in more detail.

2.4.1 Functional anatomy of the synapse

The structure of a typical synapse in shown in Figure 2.10. You might like to add additional information to this figure as you read the following sections.

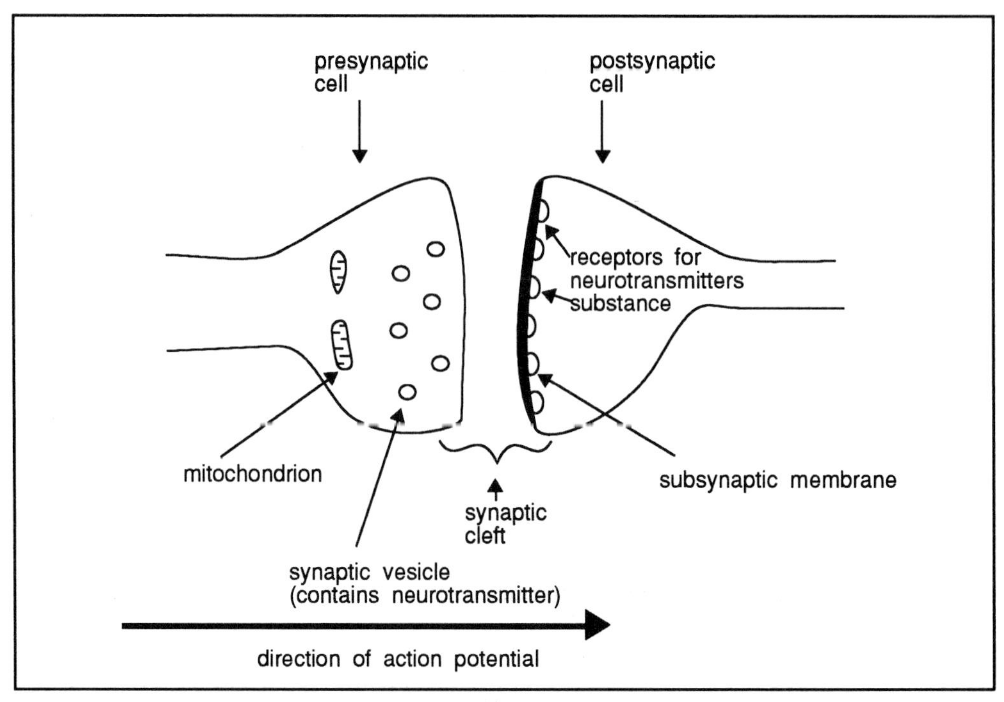

Figure 2.10 Diagram showing the principal features of a mammalian synapse.

The synaptic cleft itself is a space about 20 nm wide. How are messages passed from one neuron to another across the synapse?

calcium channels

role of kinases

synapsin

The first stage in the process is the arrival of the action potential at the axon terminal of the presynaptic neuron. Its arrival and subsequent depolarisation of the axon terminals opens a group of calcium channels which allow Ca^{2+} ions to enter the neuron from the extracellular fluid. The Ca^{2+} ions that have entered the axon terminal activate an enzyme called calcium/calmodulin dependent protein kinase I. As with other kinase enzymes, this enzymes phosphorylates a substrate molecule. In the case of this enzyme, the substrate is a protein called a synapsin which is bound to the synaptic vesicles. The effect of phosphorylation of synapsin is that it detaches from the vesicle. The outcome of this is that the vesicle is now free to fuse with the membrane of the axon terminal. This it does and as a consequence, the contents of the vesicle, the neurotransmitters, are released into the synapse. The neurotransmitters then diffuse across the synapse. Here they interact with the postsynaptic membrane via specific receptors with which they combine. Combination of the neurotransmitters with their receptors influences the postsynaptic membrane's permeability to ions and, as a consequence of this, the electrical activity of the postsynaptic neuron. This whole process takes about 0.5-1.0 msec.

2.4.2 Neurotransmitters

The number of substances that are thought to function as neurotransmitters has increased dramatically over the last few years. Table 2.2 lists some of the substances known to function as neurotransmitters. You may come across some of the substances in other chapters where they also function as hormones.

Esters	Acetylcholine
Amino acids derivatives	Noradrenaline
	Adrenaline
	Serotonin
	Histamine
Amino acids	Glutumate
	Aspartate
	Glycine
Peptides	Substance P
	Bradykinin

Table 2.2 A list of some of the known neurotransmitters in the mammalian central and peripheral nervous system.

However, in spite of the wide diversity of neurotransmitter substances, all have some common features. One such important feature is that once they have initiated electrical activity in the post synaptic cell their action is terminated. This is important, otherwise the neurotransmitter substances would exert a continuous influence on the postsynaptic cell. Acetylcholine, for example, is inactivated by enzymatic degradation. The action of noradrenaline is terminated by its uptake into neurons or non-neuronal cells in the vicinity of the synapse. For a substance to be considered as a neurotransmitter certain criteria must be met. These include:

- the presynaptic neuron must contain the substance and must also be capable of its synthesis;
- appropriate stimulation of the presynaptic neuron must result in the release of the substance;
- application of the substance to the postsynaptic neuron must mimic the effects of the responses evoked by stimulation of the presynaptic neuron;
- the effects of presynaptic stimulation and the application of the substance to the postsynaptic neuron must be affected in the same manner by pharmacological agents.

2.4.3 Initiation of postsynaptic neuronal activity

excitatory response

In this section we will consider how a neurotransmitter influences the electrical activity of a postsynaptic neuron. It is possible to distinguish two types of postsynaptic response, an excitatory and an inhibitory response. Let us consider the excitatory response first.

EPSP

At an excitatory synapse, the response of the postsynaptic neuron is a depolarising one ie a shift in the membrane potential towards threshold. The mechanism behind this response is an opening of Na^+ ion channels allowing ions to enter the postsynaptic neuron. The effect of this is a net flow of positive charge into the neuron which tends to depolarise it. This change in membrane potential is called the excitatory postsynaptic potential (EPSP). The EPSP is an example of a graded or local potential. Unlike the action potential which is propagated without loss, graded potentials radiate with decremental loss away from their point of origin. However, its passage away from its site of origin along dendrites and over the cell body occurs via the formation of local currents in a similar manner to the action potential. Obviously, the larger the EPSP, the further it will travel before fading away. In order for an EPSP to generate an action potential in the postsynaptic neuron, it must be of sufficient magnitude to bring the initial segment of the axon to threshold.

In contrast, inhibitory synapses are characterised by a hyperpolarising response ie the membrane potential becomes more negative and therefore less likely to generate an action potential. The mechanism behind this response is an opening of Cl^- or K^+ channel or both. The net result is a hyperpolarising response. This shift in membrane potential of the postsynaptic neuron is called an inhibitory postsynaptic potential (IPSP). It shares the same characteristics as the EPSP ie, it is a graded response which travels with decrement. Unlike action potentials neither the EPSP nor the IPSP possess a refractory period.

IPSP

SAQ 2.5

Make a list of four properties of both local potentials and action potentials to show how they differ from each other.

2.4.4 Synaptic integration

The situation described above ie a one-to-one synapse between a presynaptic neuron and a postsynaptic neuron is very simplistic. In reality a single dendrite or cell body may make synapses with hundreds of other neurons. Thus there exists the possibility that synapses can be sites of neuronal integration in terms of the rate at which local potentials are generated, their polarity, their magnitude and whether or not action potentials are eventually generated. Let us consider some of the ways in which synapses can process neural information. Two major types of integration are temporal and spatial summation (Figure 2.11).

The upper diagram in Figure 2.11 shows a neuron with two excitatory inputs (I and II), and an intracellular recording electrode to measure the membrane potential which is shown graphically in the lower diagram. Stimulation of either input I or input II causes a typical EPSP (A and B see lower figure). If we stimulate the neuron via input I and then before the potential has died away, we stimulate it again (C and D) we see a larger EPSP. This is produced by the process of temporal summation. This contrasts with the larger than normal EPSP observed in G, that results from summation of two different inputs occurring at the same time. This process is called spatial summation.

Π A third factor which influences the output of a neuron is where the input is in relation to the axon hillock. Why is this so?

The site of the input to the postsynaptic neuron is important because as you will remember the potentials generated are local, ie radiate with decrement from the site of origin. Therefore, if they are to generate an action potential in the axon they must travel to the initial segment of the axon and still be of sufficient magnitude to bring the membrane potential to threshold.

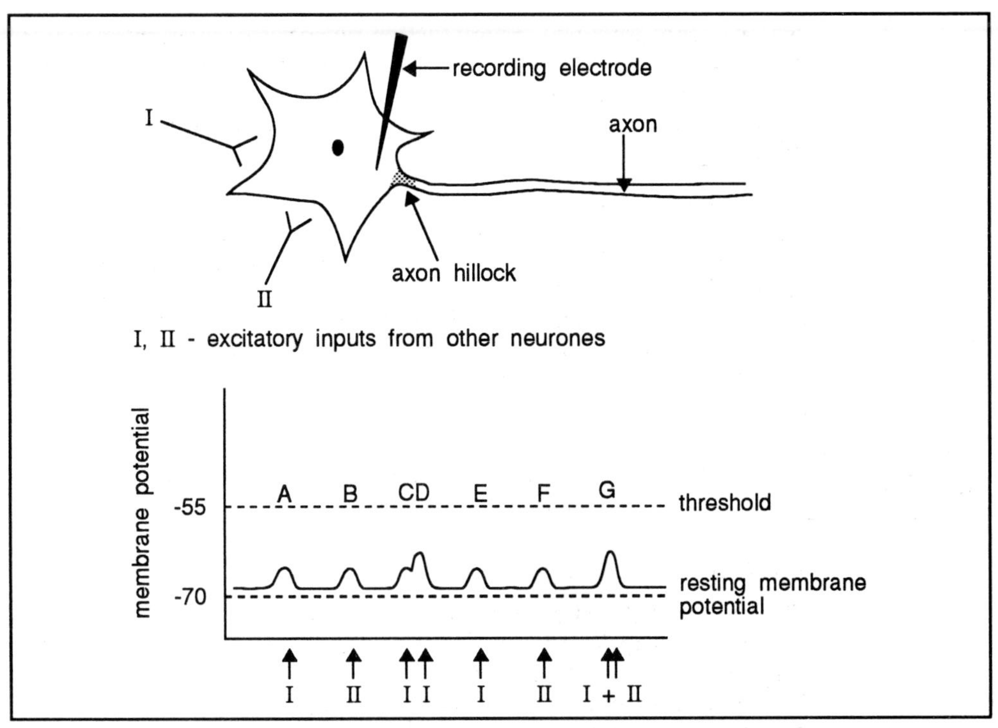

Figure 2.11 Diagram illustrating the concepts of temporal and spatial integration in a neuron. Redrawn from Vander AJ, Sherman, JH and Luciano, DS (1990) Human Physiology: The Mechanism of Body Function, McGraw Hill, New York.

SAQ 2.6

A neuron has 3 synaptic inputs all equidistant from the initial axon segment. Two are excitatory (A and B) and the third (C) inhibitory. When stimulated they produce changes in membrane potential of +5mV (A, B) or -5mV (C). Draw graphs to show the effect on resting membrane potential (-70 mV) of stimulating:

1) A alone.

2) Two impulses in A in quick succession.

3) C alone.

4) A and B together.

5) A, B and C together.

Explain your answer.

2.5 Sensory receptors

One way in which action potentials are generated in the nervous system is via stimulation of sensory receptors. The action potentials thus generated being conducted to the brain and spinal cord. In this section we will consider how sensory receptors function to keep us informed of what is happening in our environment. The sole function of sensory receptors is to convert, 'environmental energy', (heat, sound, light, pressure) into electrical energy (action potentials) which the nervous system can then process.

transduction

This conversion of energy is anatomically known as transduction. Anatomically, sensory receptors can be divided into two major groups. The first of these are receptors which are specialised endings of afferent (sensory) neurons. An example of this type are receptors which mediate the sensation of pain. Alternatively, sensory receptors may be specialised cells, which via the release of neurotransmitters, generate action potentials in an afferent neuron. Such an example are the hair cells of the inner ear which convert sound waves into action potentials. Regardless of the micro-anatomy of the sensory receptor, each receptor has a stimulus to which it responds more or less specifically. This is called the adequate stimulus. However, having said that, it should be borne in mind that nearly all receptors can be stimulated by several different forms of environmental energy. For example, the adequate stimulus of the eye is light. However, the eye can also be stimulated by pressure eg a punch in the eye. But even when stimulated like this it is still the sensation of light that is perceived. This illustrates another general principle of sensory systems, that is, any given receptor gives rise to only one sensation. Let us examine the stimulation of sensory receptors in more detail.

adequate stimulus

2.5.1 Sensory receptor function

Pacinian corpuscles

One of the best studied sensory receptors are Pacinian corpuscles (Figure 2.12). These are located in the skin and other tissues where they mediate the sensation of pressure.

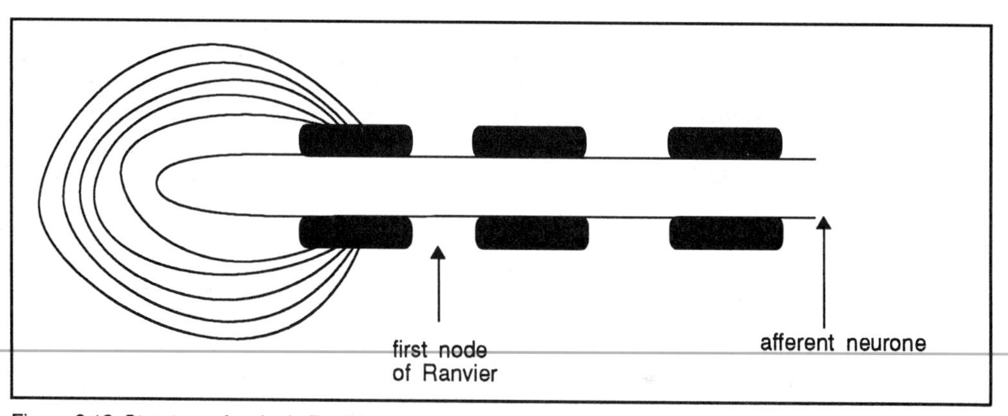

Figure 2.12 Structure of a single Pacinian corpuscle.

generator potential

Although we will confine our attention to Pacinian corpuscles, the general mechanism for the stimulation of sensory receptors is believed to be more or less similar for all receptor types. In all cases, the stimulus somehow alters the membrane permeability of the sensory receptor resulting in the production of local potential called the generator potential. The application of pressure to the Pacinian corpuscle causes the opening of ion channels in the membrane. This probably occurs simply by physical deformation of the membrane. The effect of opening these ion channels is the net movement of positive charge, probably Na^+ ions, into the afferent neurone. The generator potential produced in this way sets up local currents which flow to the first node of Ranvier of the afferent neuron. If the local currents are sufficient to bring the first node to threshold then an action potential is generated which then travels along the afferent neuron to the central nervous system.

Π Can you think what effect drugs that block Na^+ channels would have on sensory receptor function.

There are practical applications of interfering with sensory receptor function. Blockade of the Na^+ channels prevents the generation of action potentials. Therefore, even though the sensory receptor may be stimulated, the central nervous system is not aware of it. This is the basis of local anaesthesia where drugs eg Lidocaine are used to block ion channels in pain fibres.

2.5.2 Coding in sensory neurons

Information from sensory receptors is passed towards the central nervous system. The magnitude of the stimulus strength is coded for by the frequency with which action potentials pass along the afferent neuron. Several factors control the magnitude of the generator potential, and therefore the likelihood of the generation of an action potential.

These factors include:

- stimulus intensity;
- the rate of change of the stimulus intensity;
- summation of individual generator potentials;
- the degree of adaptation which occurs.

phasic adaptation

tonic adaptation

Adaptation is seen as the decrease in action potential frequency following the application of a constantly maintained stimulus. Adaption can either be phasic (fast) or tonic (slow). Taste receptors in the tongue show phasic adaptation. Applications of a substance to the tongue results in the production of a high frequency of action potentials in the afferent neuron which is connected to the sensory receptor. This is followed by a rapid decline or abolition in action potential frequency even through the substance is still in contact with the sensory receptor. On the other hand, thermoreceptors (temperature receptors) in the skin display tonic adaptation. In this case there is an initial high frequency of action potentials as the stimulus ie heat is applied. As the stimulus is maintained so the frequency of action potentials drops to a slightly reduced rate whilst the stimulus is maintained. Receptors displaying phasic adaptation are better suited to inform the central nervous system of whether the stimulus is present or not. Tonically adapting receptors are best suited to provide information regarding changes in the magnitude of a particular stimulus.

SAQ 2.7

List four ways in which the magnitude of the generator potential may be varied.

2.6 Spinal reflexes

reflex arcs

Having generated action potentials in sensory receptors, action potentials are transmitted along afferent neurons to the central nervous system. We will now consider how this is done and will look at some of the functions of the highest level of organisation of the nervous system ie the brain. In response to activity in the afferent nervous system there is usually some sort of response eg movement, which is produced by activity in the efferent or motor neurons. Not all the sensory information is processed by the brain. Some such information is dealt with at the level of the spinal cord. These responses are called spinal reflexes or reflex arcs and constitute the simplest level of functioning of the nervous system. An example of a spinal reflex is the withdrawal reflex. Thus, if our hands come into contact with a painful stimulus eg excessive heat, then our initial response is to withdraw the hand from the stimulus.

Π Can you think of any other spinal reflexes, how does your GP test your reflexes?

reflexes

Another example of a spinal reflex is the knee jerk response, evoked by stimulation of the tendon of the knee. Spinal reflexes may be characterised as either monosynaptic or polysynaptic (Figure 2.13).

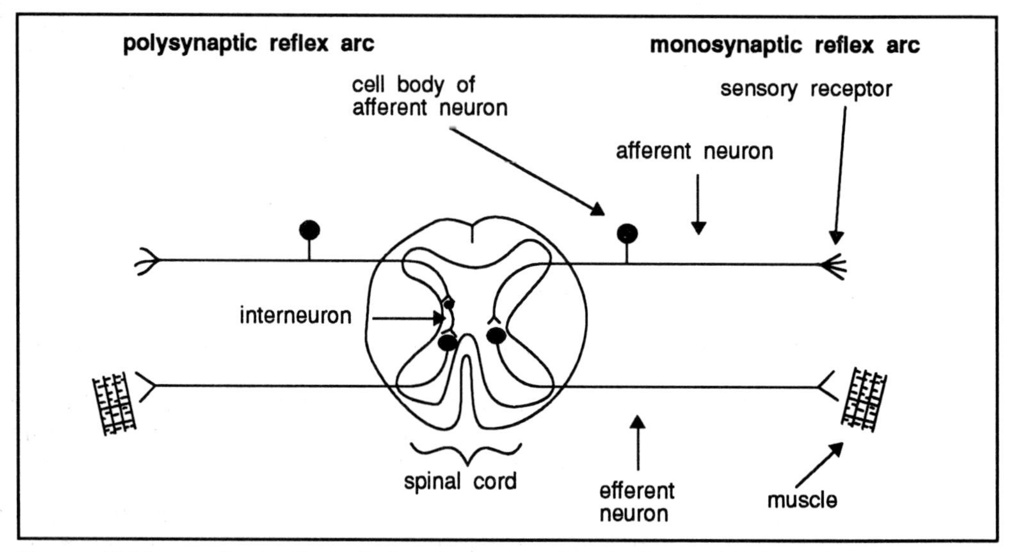

Figure 2.13 Diagram showing the basic plan of monosynaptic and polysynaptic reflexes. Note the presence of interneurons in the polysynaptic reflex arc. These are not present in monosynaptic reflexes.

As can be seen from Figure 2.13 a monosynaptic reflex is one in which the sensory neuron synapses directly with the motor neuron to produce a response. In a polysynaptic reflex, there are one or more of the so-called interneurons between the sensory neuron and the motorneuron.

SAQ 2.8

What reflexes do you think are the fastest, monosynaptic or polysynaptic? What are the sites of delay in the operation of a reflex?

As with most matters related to biological systems, everything is not as simple as it first seems. For example, it is possible for the brain to exaggerate or suppress these simple reflex activities.

Π Can you think of an everyday example where reflex activity is suppressed.

We have mentioned the withdrawal reflex. Your first response to picking up a hot object would be to drop it, ie remove your hand from a painful stimulus. What would your response be if the hot object you picked up and dropped contained your dinner which was spilt onto the floor?

A further complication of the simple withdrawal reflex is that not only does the afferent neuron influence the activity of efferent neurons, it will also relay information to the brain such that pain is perceived.

Neural pathways exist that send information from the peripheral neurons to the brain while others send information from the brain to the peripheral neurons. Thus when a sensory receptor in a peripheral tissue is stimulated this 'message' is sent to the spinal cord. Here other neurons are stimulated which pass the information to the brain. These neurons run along the length of the spinal cord. Nerve impulses can also be sent in the opposite direction. These pathways are called the ascending or descending pathways. The pathways are named such that the first part of their name signifies their origin and their latter part their destination eg the ascending spinothalamic tract originates in the spinal cord and terminates in a region of the brain called the thalamus. The precise nature of these pathways is very complex and beyond the scope of this chapter. However, the pathways carry more or less specific information and dysfunction may result in specific neurological abnormalities.

2.7 Autonomic system

Look back to Figure 2.1. You will see that the peripheral nervous system can be divided into the somatic nervous system which carries information from the central nervous system to skeletal muscle and the autonomic nervous system which carries information from the central nervous system to regulate the internal organs. In this section we will provide you with some more information about the autonomic system.

sympathetic and parasympathetic system

The autonomic system controls the activities of smooth muscles, cardiac muscle, endocrine glands and secretary glands. It regulates important functions such as respiration, circulation, digestion, excretion and hormone secretion. It is mainly composed of motor (efferent) nerves and can be divided into two subsystems, the sympathetic and parasympathetic systems. These work in balance with each other.

The sympathetic system prepares the body for action (flight or fight). The nerves of this system arise from the whole length of the spinal cord and act by releasing noradrenaline at synapsis. In contrast the parasympathetic system prepares the body for excretion, rest and reproduction. The major parasympathetic nerve (the vagus nerve) starts from the base of the brain.

Π Examine Figure 2.14 and list the organs the vagus nerve controls.

The other major nerve (the pelvic nerve) of the parasympathetic system arises from the lower part of the spinal cord. The nerves of the parasympathetic system use acetyl choline as the neurotransmitter.

Π Figure 2.14 shows the major effects of the parasympathetic and sympathetic systems. It would be a helpful form of revision to construct a table of the information shown in this figure. You will meet with these systems again in subsequent chapters.

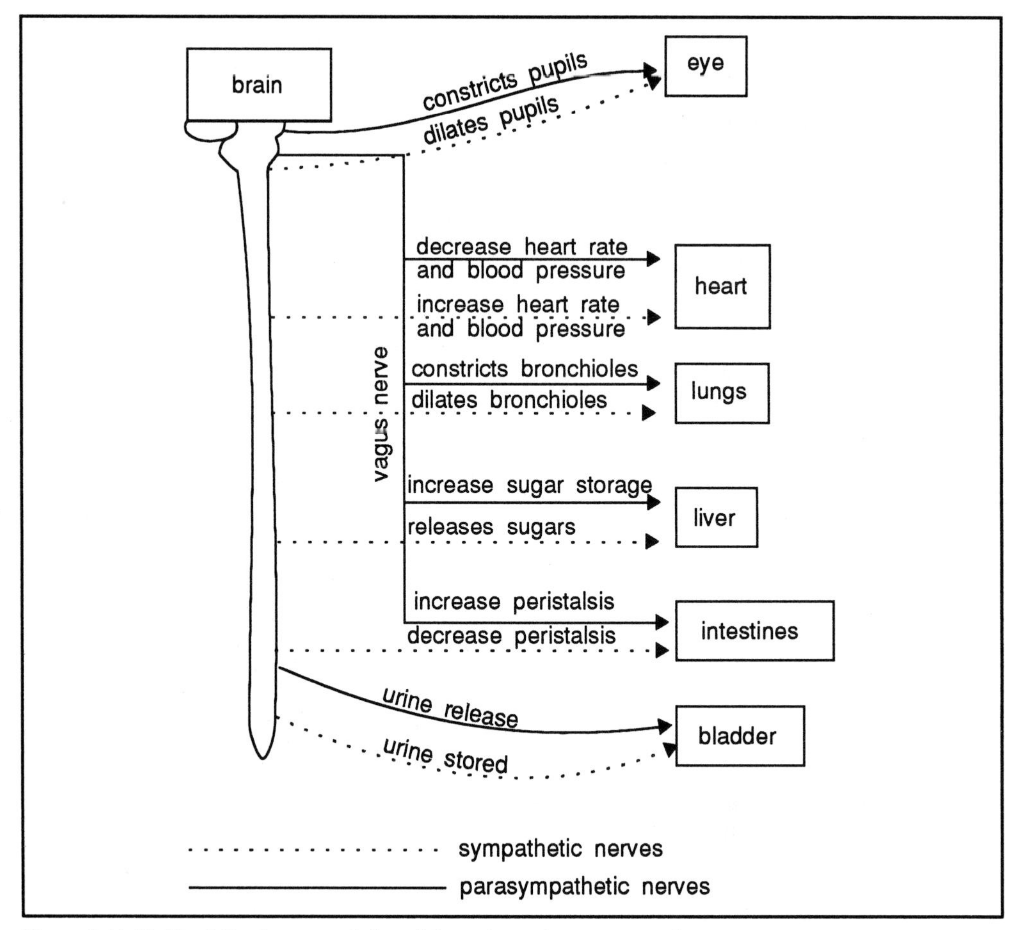

Figure 2.14 Highly stylised representation of the autonomic nervous system.

2.8 The brain

We started this chapter by looking at the activity of individual neurons. We have gradually built up through increasing levels of complexity until we have now arrived at the brain, the highest level of neuronal activity. Our knowledge of the functioning of the brain, although vastly increased over the last decade or so, is still far from complete.

Although this text is primarily concerned with the functional aspects of physiology, we need to have some idea about the anatomy of the organs involved in these functions. This is especially true of the brain. The brain is one of the largest organs of the body and, in an adult, weighs about 1.5kg. It has a very complex morphology and texts devoted to anatomy and neurology provide many details of this organ. Many terms have been used to describe various structures and, if you are interested in gaining further details of the anatomy of this organ, we refer you to appropriate texts in the 'Suggestions for further reading' section at the end of this text. Here we will provide a general description so that you have an overview of the structure of the brain. Figure 2.15 shows a cross section through the human brain. The brain can be divided into six major divisions. Starting from the top, and working downwards, these are:

- cerebrum; also called the forebrain
- diencephalon; also called the forebrain
- cerebellum; also called the hind brain
- midbrain; collectively referred to as the brain stem
- pons; collectively referred to as the brain stem
- medulla oblongata. collectively referred to as the brain stem

Let us examine each briefly in turn.

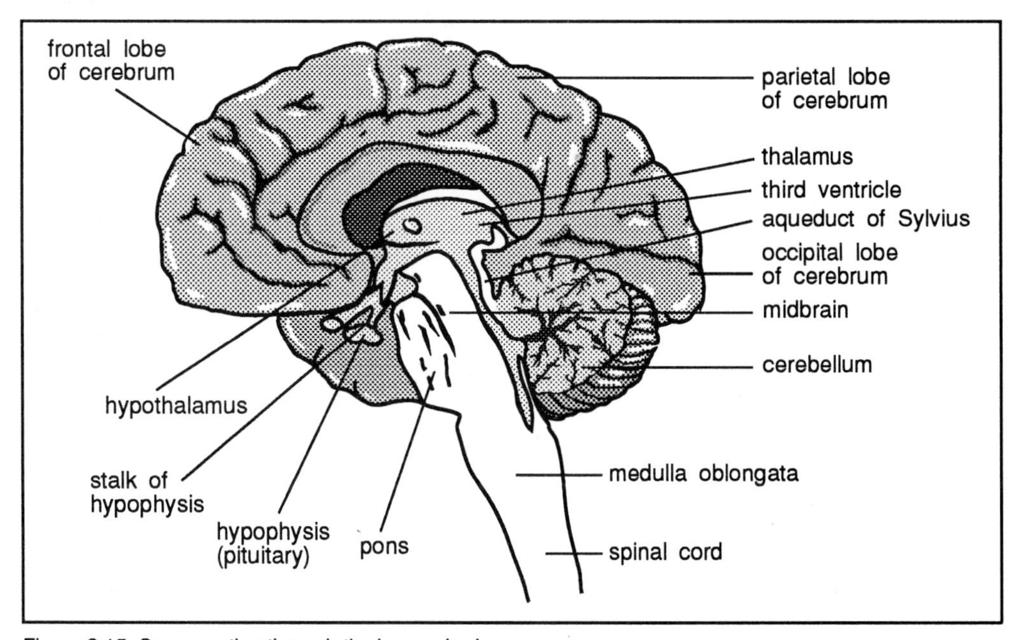

Figure 2.15 Cross section through the human brain.

2.8.1 Cerebrum

cerebral hemispheres

cerebral cortex

cerebral lobes

This is the largest division of the brain. It consists of two halves (cerebral hemispheres). The surface of the cerebral hemispheres is convoluted and is made up of six layers, the cerebral cortex. The cerebral cortex is only about 2-4 mm thick but contain millions of axons which are synapsed with many dendrites of other neurons. There are many groves (or fissures) within the convolutions. The major fissures (each of which have special names) divide each cerebral hemisphere into five lobes (frontal, parietal, termporal, occipital and insula). Beneath the cerebral cortex is a layer of white matter. The cerebrum is mainly concerned with the highest level of control including initiating, planning and programming movements.

2.8.2 Diencephalon

thalmus

hypothalamus

This part of the brain is found between the cerebrum and the midbrain. It consists of several structures around a structure called the third ventricle. Here we will focus onto two of its major components, the thalamus and hypothalamus.

The thalmus consists of two equal parts (right and left thalamus). Each is a rounded mass of grey matter. It plays a part in:

- recognition of pain, temperature, touch;
- transmission of nerve impulses from sensory organs to the cerebrum;
- the processing of nerve inputs responsible for emotions;
- arousal mechanisms;
- complex reflex movements.

The hypothalamus is found beneath the thalamus and consists of several structures; supraoptic nuclei, paraventricular nuclei, the stalk of the hypophysis (pituitary gland) and the neurohypophysis (posterior lobe of the pituitary gland).

neurohypophysis and anterior lobe of pituitary

The hypothalamus is, therefore, directly connected to the pituitary gland. We will learn more of this connection when we discuss the endocrine system. You should note, however, that the pituitary gland consists of two main lobes. The posterior lobe (or neurohypophysis) has many nerve fibres connected directly to centres in the hypothalamus whilst the anterior lobe has small blood capillaries which connect it directly with areas within the hypothalamus. Both nerve and blood connections pass along the pituitary stalk.

The hypothalamus plays a key role in the autonomic control system. It has direct connections with both the parasympathetic and sympathetic centres. It also functions in the processing and transmission of impulses from the cerebral cortex to the lower autonomic centres. We shall also learn in the next chapter that the hypothalamus plays a crucial part in the endocrine system and controls hormone secretion by a variety of glands. It also controls growth, plays a role in maintaining the waking state, influences appetite and in the maintenance of body temperature.

2.8.3 Cerebellum

This is found just below the posterior part of the cerebrum. Like the cerebrum it has a convoluted surface but the convolutions are much smaller. It performs three main functions all of which relate to the control of skeletal muscles. These are:

- maintenance of posture;
- in conjunction with the cerebral cortex, it produces skilled movements involving the co-ordinated action of several groups of muscles;
- the reduction of jerkiness and trembling to produce smooth, efficient and co-ordinated movements.

Brain stem

The medulla oblongata is the part of the brain that is attached to the spinal cord. It is composed of white matter and a mixture of white and grey matter called reticular formation. Within this reticular formation are various nuclei (clusters of neuron cell bodies). Some of these are connected to particular organs (hence cardiac nucleus, respiratory nucleus, vasomotor nucleus).

The pons is found just above the medulla. Many fibres run through the pons and into the cerebellum. The midbrain, as its name implies, forms the midsection of the brain.

vital centres

The brain stem performs reflex, sensory and motor functions. Thus the nuclei in the medulla contain a number of important reflex centres (for example, cardiac, respiratory, vasomotor). They serve to control heart action, respiration, blood vessel diameter and so on and are often called the vital centres.

You should realise that the brain is not only connected to the rest of the body via the brain stem and spinal cord. It also has nerve connections with the rest of the head and the major sense organs (eyes, ears, tongue, nose) and the muscles of the face.

Π We would now suggest you re-read this section about the structure of the brain and at the same time draw up a table of the information given. Use the following format. You will be able to add to this table as you read later chapters.

Division of the brain	Major structures	Functions
Cerebrum	Two hemispheres, divided into five lobes by fissures 6 layered cerebral cortex	

2.9 The organisation of the sensory and motor systems and the brain

We began this chapter by providing a brief overview of the basic organisation of mammalian nervous systems (see Figure 2.1). Now that you have some understanding of the various components of the nervous system, we can re-examine how the nervous system is organised and how it functions.

2.9.1 Secondary pathways and the organisation of the motor system

Basically, all sensory pathways tend to be organised in the same way. We can follow the nerve impulse from its point of origin to its reception in the central nervous system. We can represent this in the following way.

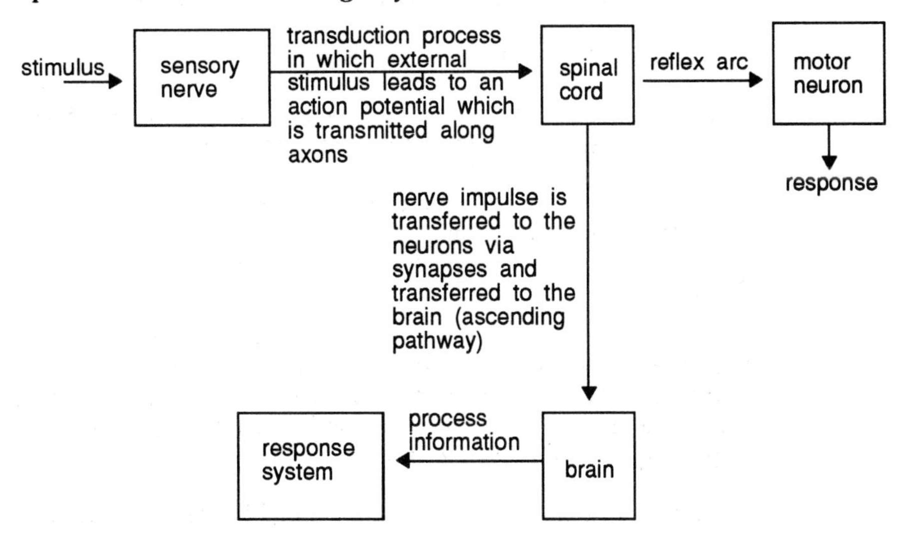

We will examine what happens in the brain a littlc latcr.

We have, of course, a wide variety of sensory systems.

Π Make a list of the sensory systems we pocess.

We anticipate that you would have included the senses of touch, taste, smell, sound and sight. You may have also included in your list such items as pain and temperature.

The nerve impulse is transmitted to the brain by the so called ascending pathway. In the brain the impulse is processed. The way it is processed depends upon the site of origin and the nature of the signal.

The signal is directed towards the appropriate part of the cerebral cortex in the cerebral hemispheres. Different regions of these hemispheres are involved with processing different functions (see Figure 2.16). We need not directly concern ourselves with which part of the brain is specifically concerned with which part of the body at this stage.

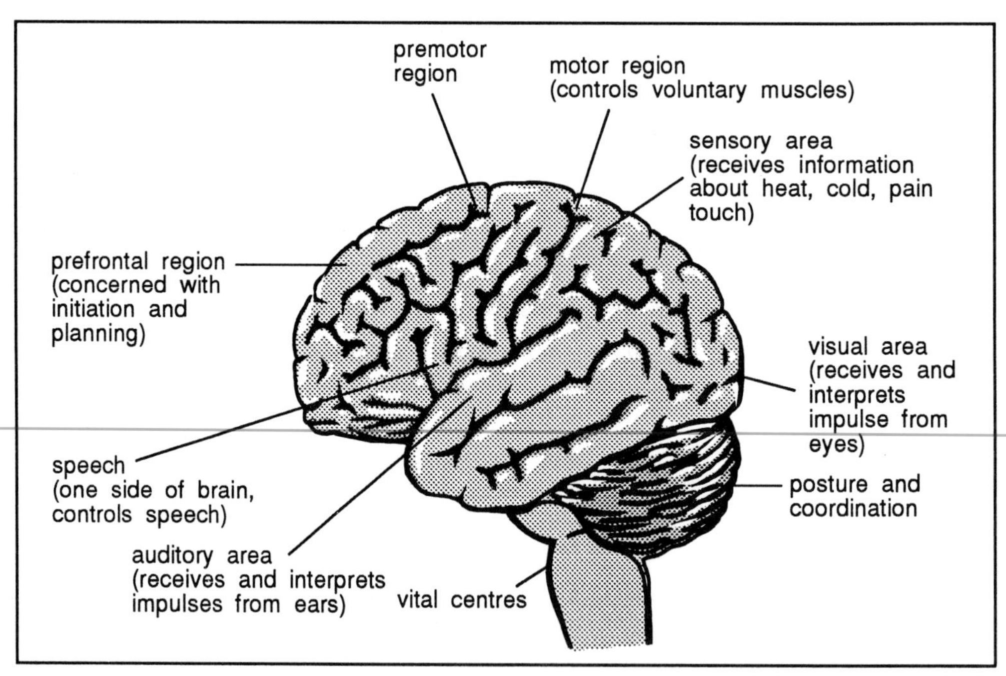

Figure 2.16 Localisation of the various functions in the human brain. Note particularly the prefrontal, premotor and motor regions as these have particular significance in the organisation of motor functions.

In addition to being processed the signal may be transferred to the thalamus and lead to an endocrine (hormonal) response (we will deal with the endocrine system in a later chapter). We can, therefore, represent the response to a sensory input as:

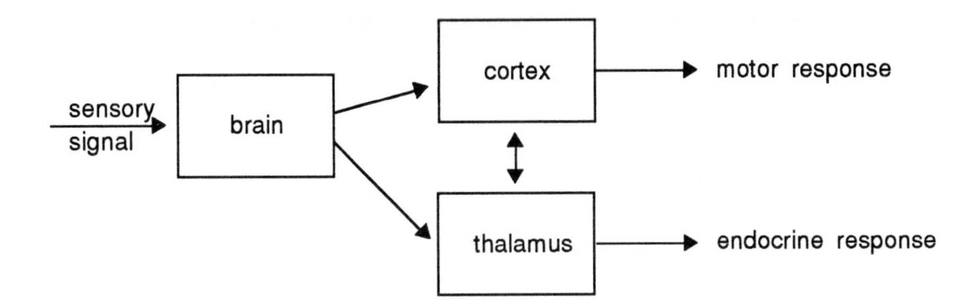

Let us focus on a motor response to a sensory signal. The motor systems are those areas of the nervous system that are primarily responsible for controlling an animal's movement. We have already examined one type of motor response the so called reflex reaction. We can perhaps call such a system a closed-loop. Here we would like to focus on the open-loops which contribute to 'voluntary' movement. These directly involve the brain.

We can regard the organisation of the motor system as being organised at five levels of control. We will call the highest level of control level 1 and the lowest level, level 5. These levels can be described in the following way:

Level 1 (highest level)

This is primarily organised in the cerebral hemispheres. This level is concerned with the initiation, planning and programming of movement other than those resulting from reflex actions. The areas responsible for this level of organisation are mainly the so called prefrontal and premotor cortex. This cortex has six layers of varying thickness (Table 2.3).

Layer	Content	Thickness
1	dendrites from cells in lower layers	few mm
2	pyramidal cells	200µm
3	pyramidal cells	200µm-5mm
4	granule cells	200µm
5	pyramidal cells	1mm

Table 2.3 Basic organisation of the cerebral cortex.

We can draw a stylised version of this in the following way:

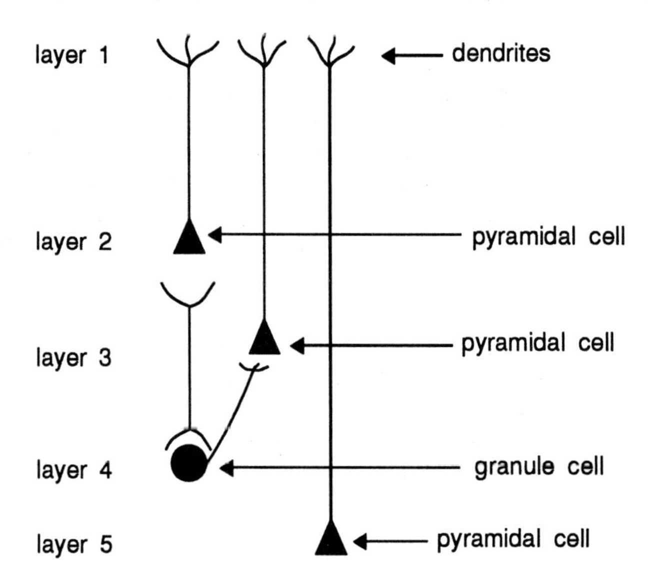

There is, of course, an output from this cortex via efferent neurons. Thus:

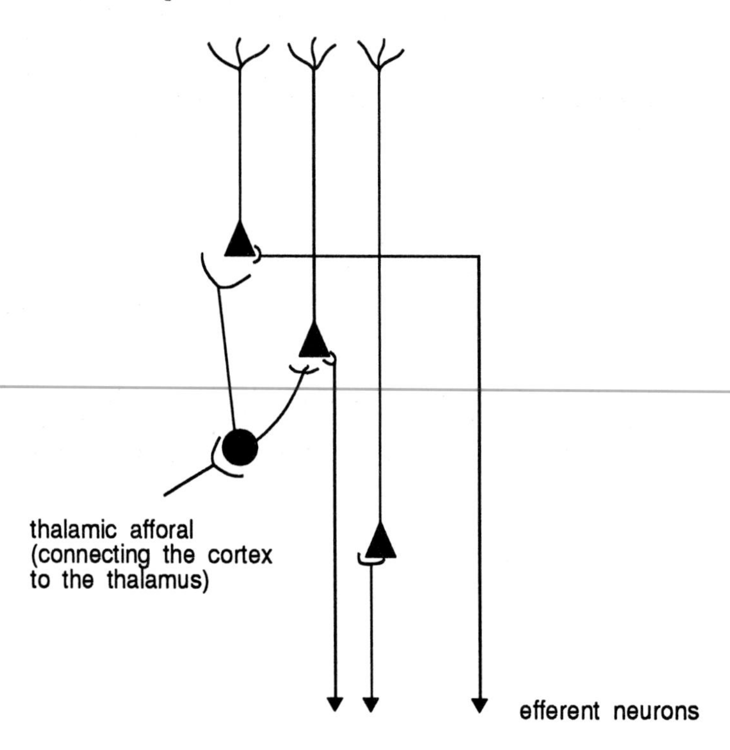

Note that the granule cells are connected to the thalmus via afferent neurons. Note also that our figure is very much simplified. The neurons are interconnected as a complex network.

Level 2

cerebellum

This level of organisation is concerned with the co-ordination of both voluntary and reflex movements with sensory guidance. For example balance involves co-ordinating movements and sensory information. It would for example be impossible for us to jump over a hurdle or to climb a flight of stairs if we could not co-ordinate movement with sensory (visual) information. Such co-ordination is conducted in the cerebellum. It is also the role of the cerebellum to act as a memory store of these movements.

Level 3

motor cortex

This level of organisation receives information from levels 1 and 2 either directly or via the thalmus. It may also receive information from the sensory systems directly. Its function is to convert these various inputs into motor responses. The area responsible for this is the so called motor cortex. The premotor cortex may also be involved. These areas transmit signals to the muscle via the spinal tract via the so called descending pathways.

Level 4

vital centres and central pattern generation

This level is mainly organised in the spinal cord and around the brain stem motor nuclei or vital centres. In these areas the interneurons form their own network. These networks can generate their own movements without additional nerve inputs. They are mainly responsible for automatic locomotive behaviour. These networks of interneurons are sometimes called central pattern generators (CPGs). Note, however, that CPGs may be modified by sensory signals.

Level 5 (lowest level)

This level consists of motor neurons and the muscles they innervate. Thus at this level the motor neurons conduct impulses from the spinal cord to the muscle. They synapse with muscles and the impulse is converted into muscle contraction. We will examine nerve: muscle interaction at the higher levels of motor control in Chapter 9.

We can represent what we have learnt about the brain and the organisation of the motor system in the following way. (It might be a good form of revision to draw your own version of this figure using the description given in the text).

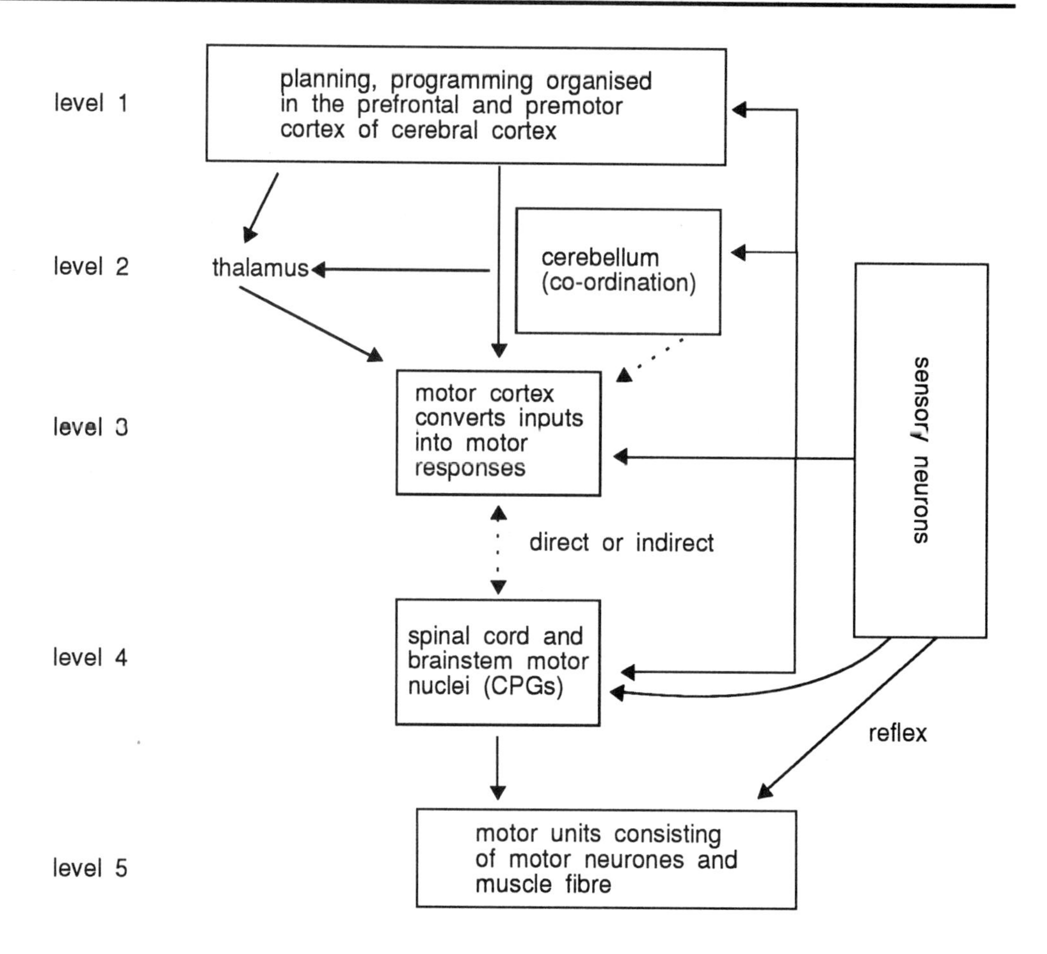

2.9.2 Concluding remarks

Our description of the organisation of the brain and the motor system given above is greatly simplifed. We have not for example discussed how emotions are generated nor have we responsed to such emotion. It has, however, not been our intention here to turn you into neuro-specialists. The topic is too large and complex to be able to achieve that in a single chapter. Nevertheless you should now have a good general appreciation of how the nervous system functions in animals. For detailed anatomical and functional descriptions of this system you are referred to the list of recommended further reading give at the end of the text.

SAQ 2.9

Select the appropriate word(s) from those provided.

1) The [ ] is the site of central pattern generators (premotor cortex, motor cortex, cerebellum, brain stem nuclei).

2) Transmission of sensory signals via interneurons to the brain are by [ ] pathways (ascending, radial, descending, reflex).

3) The cerebellum is involved with organisation of the motor system at [ ] (level 1, level 2, level 3, level 4, level 5).

SAQ 2.10

Assign the following functions to the appropriate part of the brain. (Choose from cerebrum, hypothalamus, cerebellum, medulla oblongata).

1) Maintenance of posture.

2) Heart beat rate.

3) Control of hormone secretion.

4) Peristalsis in the intestines.

5) Initiation and planning.

Summary and objectives

In this chapter we have introduced you to the nervous system. We examined neurons, first from a structural point of view for describing how they function especially in terms of transmission of action potentials. We also examined the junction between neurons (synapses). This leds us into a discussion of reflex arcs (loops) and the autonomic nervous system before we briefly discussed the structure and function of the brain. Our main point of focus in discussing the brain was the organisation of the motor system into a hierarchial system of organisation.

Now that you have completed this chapter you should be able to:

- describe the basic plan of the mammalian nervous system;
- describe the structure of a typical neuron;
- classify neurons according to their shape;
- list the different types of neuroglial cells and state their functions;
- describe the ionic basis of the resting membrane potentials and action potentials;
- explain how action potentials are transferred along the axon, stating the importance of myelin;
- describe the structure of a typical synapse;
- explain how neurotransmitters are released from neurons;
- explain how and what changes in membrane potential are caused by neurotransmitters interacting with neurons;
- explain how synapses may process and integrate neural activity;
- describe the basic structure and function of sensory receptors;
- describe the functioning of the reflex arc (loops);
- describe in outline the structure and principal functions of some regions of the brain.

The endocrine system

The endocrine system

3.1 Introduction

3.1.1 Homeostasis

Homeostasis is the term coined by an American physiologist, Walter Cannon to describe the principle of maintaining the constancy of the body's internal environment. This concept was first recognised by Claude Bernard in the eighteenth century, who realised that such constancy is essential for the normal functioning of the cells of which our bodies are composed. There are many different homeostatic mechanisms which function in the human body to control a wide variety of physiological parameters, eg temperature, pH, P_{O_2}, P_{CO_2} and others. Consequently there is a requirement to have an overriding control system in order to ensure that such mechanisms function in a co-ordinated manner to serve the needs of the body as a whole balanced organism.

The idea that such control could be brought about by chemicals being released into the blood was first suggested by Berthold in 1848 following experiments which showed that the development of sexual behaviour and male characteristics in cockerels was related to the chemical secretions of the testicular tissue when present in the animal.

hormones

Further work by Bayliss and Starling in the early twentieth century confirmed that co-ordination of organ function could occur by the secretion of chemical substances into the blood and it is these authors who coined the phrase hormone (from the Greek word meaning 'to excite').

In this chapter we shall be discussing the properties of this control system, usually called the endocrine system. The overall study of this system and its functions is called endocrinology. In fact the endocrine system works in conjunction with another control system of the body, that is, the nervous system. Although there are some fundamental functional differences between the two systems, research in the last fifty years has shown that there is a close functional relationship between them.

We will begin by briefly comparing the endocrine and nervous systems as systems of control and co-ordination. This will enable us to consider the nature of chemical communication between cells. We will then provide you with an overview of the endocrine system and explain the dominant position of the pituitary and hypothalamic glands in controlling the secretion of hormones. With this foundation we will then examine the structure and synthesis of the major hormones. This is followed by a description of their transport and metabolism. A major part of the chapter examines how hormones can bring about changes in cellular behaviour and metabolism. Here the approach is to explain the common mechanisms involved in hormone action especially through discussion of the roles of second messengers. We also examine the control of hormone secretions and some disorders which arise through the malfunction of the endocrine system. Although we have provided a list of the major hormones together with their general properties and functions (Table 3.3), we have not attempted to explain how they interact with each other in order to control particular physiological systems. This is best done within the context of these physiological systems themselves.

At this stage it is better to understand the general functions of hormones and the mechanisms by which they fulfill these functions.

3.1.2 Endocrine versus nervous system

glands

target cells

The endocrine system consists of a number of glands (endocrine glands) which secrete minute quantities of hormones (the chemical messengers) into the blood. These are carried to distant cells upon which they act (target cells) to bring about a particular physiological effect. The system is characterised by events which follow a slow and prolonged time course and which tend to be concerned with the metabolic functions of the tissues. Hormones are always regulatory in their action rather than being instrumental in initiating events.

This contrasts sharply with the nervous system which functions very rapidly, initiates relatively short term events and is concerned with the motor activities of the body.

The functional interrelationship between the nervous and the endocrine systems is apparent both in the influence of hormones on brain development and animal behaviour and also in the neural control of hormone secretion. The latter occurs through neurosecretion, a process by which specialised neurons not only are able to conduct action potentials along their axons but also can secrete hormones from the nerve endings. These so-called neurohormones are manufactured in the cell bodies and axonally transported to the terminal ending where they are stored.

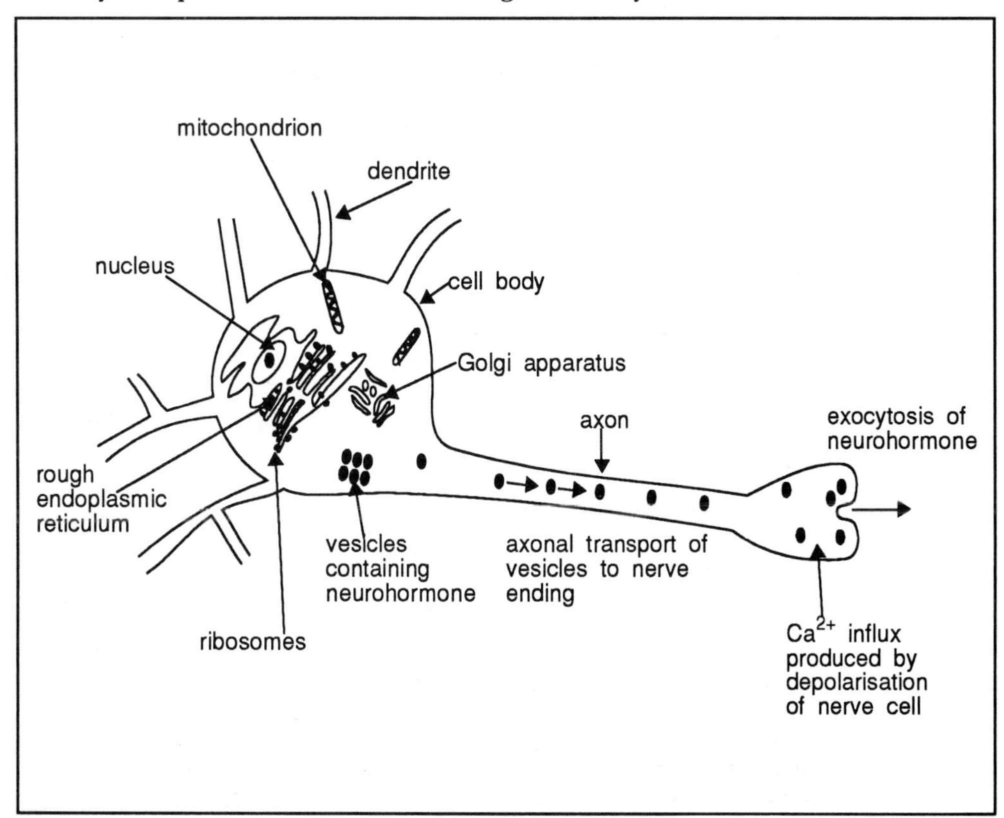

Figure 3.1 Diagram to illustrate a neurosecretory cell. The neurosecretory hormones are synthesised on the rough endoplasmic reticulum, processed by the Golgi apparatus and packaged into vesicles. These vesicles are transported down the axon to the secretory end (see also Sections 3.3 and 3.8.4).

Upon depolarisation of the neuron the stored neurohormones are secreted into the blood. Such neurosecretory cells are abundant in the hypothalamus where they function to control the secretion of most of the other hormones released in the body.

SAQ 3.1

Make a list of the differences between the endocrine and nervous control systems.

3.1.3 Other intercellular chemical messengers

Before moving on to consider some of the general properties of the endocrine system, first we must clarify the role of hormones with reference to other intercellular chemical messengers which exist to enable cells to communicate with each other. A hormone is considered to be a chemical messenger which is released into the blood by which it is transported to a distant target tissue where is exerts its influence. However, most nerve cells also communicate by using chemical messengers but, in contrast, these neurotransmitters are secreted into and diffuse across the synaptic cleft which separates the presynaptic and postsynaptic neurons. The neurotransmitters act within the immediate area from which they are released and specific enzymes are present to inactivate them.

neuro-transmitters

The neurosecretory cells mentioned previously appear to fall between these two extremes in that their chemical transmitters are released from a neuron but are transported by the blood to affect cells elsewhere in the body. These transmitters, therefore, are usually referred to as neurohormones.

neurohormones

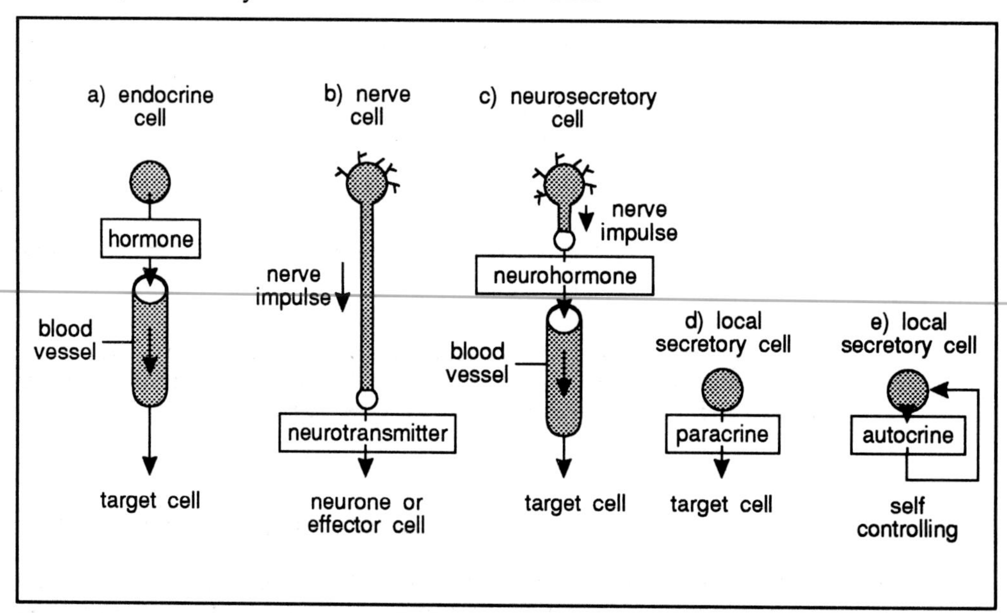

Figure 3.2 Diagram to show the different categories of chemical messengers: a) a hormone is secreted directly into the blood and carried to a distant target organ. b) A neurotransmitter is released from a nerve ending, as a result of excitation of that cell, into the synaptic cleft. c) A neurohormone is released from a nerve ending into the blood stream. d) A paracrine acts on cells in its immediate vicinity. e) An autocrine acts upon the cell which secreted it. Adapted from Figure 7.8 Vander, Sherman and Luciano (1990) Human Physiology: The Mechanism of Body Function, McGraw Hill, New York.

Other chemical transmitters affecting cells within the immediate area of their release are important in mediating local responses. Such substances have been defined as paracrines if they influence cells close by or autocrines when they affect only the cell from which they have been released (Figure 3.2).

However, it is very important to realise that some chemical messengers have been found to be released from a variety of different sites which do subserve many, if not all, of the above functions. Some chemical messengers may behave as a hormone when released from one site, but as a neurotransmitter on release from another.

An important group of intercellular messengers are the cytokines. These messenger molecules are especially important in the development of defence against infection. They are best considered within the context of the immune response and are discussed at length in the BIOTOL text 'Cellular Interactions and Immunobiology'.

eicosanoids

Finally there is the group of chemicals messengers (known as the eicosanoids) that exerts a wide variety of effects on many different tissues. Members of this group are prostacyclin, the thromboxanes and the leukotrienes. They are manufactured as a result of enzymatic breakdown of arachidonic acid derived from the cell membrane phospholipids upon activation of the cell and have been shown to act mainly as paracrines and autocrines. They are quickly metabolised once they have had their effect. However, there is some argument as to whether some of them may act hormonally via the blood to act on a target cell distant from their site of origin.

SAQ 3.2

Indicate which of the following statements are false. Give reasons for your answers.

1) A neurotransmitter substance is secreted directly into the bloodstream.

2) A paracrine messenger is one which has an effect on cells within the immediate locality of its site of release.

3) A neurohormone is released from an endocrine gland into the blood.

4) Chemical messengers may have dual roles both as a hormone and as a neurotransmitter.

Having briefly considered the different types of chemical messengers which exist in the body, we shall now move on to explore the properties of the endocrine system and its means of intercellular communication, the hormones.

3.2 An overview of the organisation of the endocrine system

secretory glands

The endocrine cells are usually grouped together into secretory glands. These glands may be distinctive separate structures (eg pituitary gland, thyroid gland) or may be part of an organ carrying out other functions (eg kidneys, pancreas or intestinal tract). Figure 3.3 shows the main hormone producing sites. Although a human body is used to demonstrate the major organism, the basic organ layout of domestic animals is entirely analogous. Note that the sexual organs have an endocrine function. This function is as important as their roles in reproduction. The hormones produced by the sexual organs can have a major impact on metabolic rates and growth, the levels of a wide variety of components (eg cholesterol) in blood and upon the morphological development of the rest of the animal.

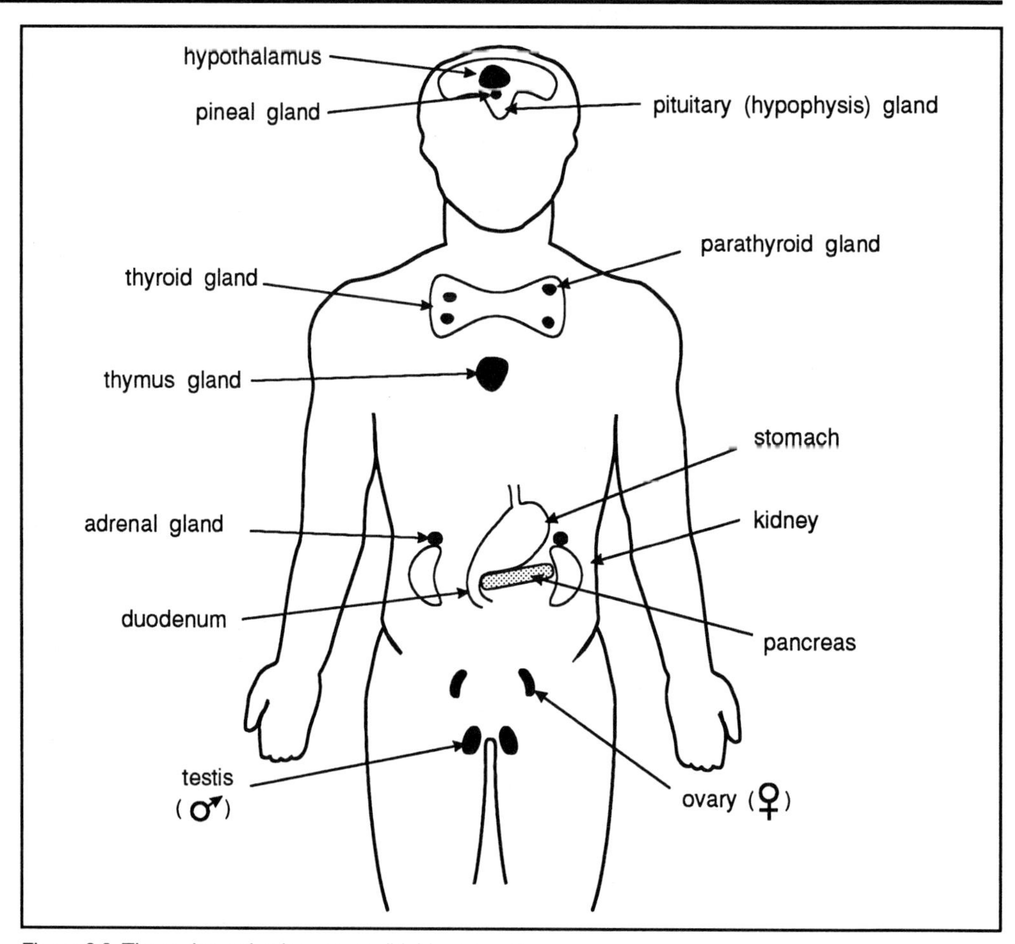

Figure 3.3 The major endocrine organs (highly stylised).

The implication behind Figure 3.3 is that the production of hormones is a function of specific groups of cells. There is also some specificity amongst the cells which respond to particular hormone action. This degree of specificity does however, vary. For example, muscle, liver and adipose (fat) cells all respond to adrenaline whilst thyroid cells are the only cells which are sensitive to thyroid stimulatory hormone (TSH). Note that some adipose cells cited above are also sensitive to insulin. The response of individual cells to a hormone may be dependent, therefore, not only on the amount of the hormone but also on the presence and concentration of other hormones. Whether or not a cell responds to a hormone depends upon whether it possess the appropriate receptors. Receptors are protein-based structures on or in the target cells which have the capability of binding tightly to a particular hormone.

receptors

The endocrine system is not just a collection of independent organs secreting hormones. There are several levels of organisation. A key organ is the hypothalamus. This receives a variety of stimuli both from the external environment and from within the body itself. The hypothalamus produces a variety of responses to these stimuli. These include production and secretion of a range of releasing hormones. These releasing hormones are passed into the blood supply that runs along the stalk of the hypophysis to the anterior lobe of the hypothysis (pituitary gland). There the releasing hormones stimulate target cells to release a variety of hormones (especially so called trophic hormones) which are secreted into the general circulatory system. These hormones then circulate around the body until they reach their target organs. The main target organs of the hormones released by the anterior lobe of the pituitary are themselves hormone producers. The hormones produced at this stage exert their effect on target tissues. We have represented this cascade in Figure 3.4.

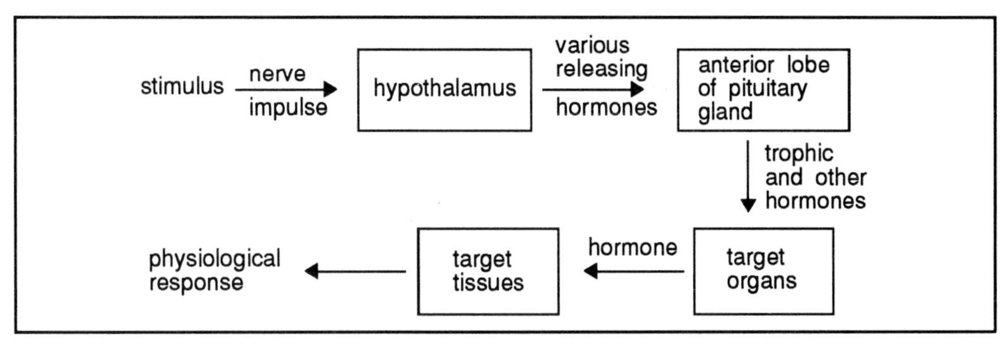

Figure 3.4 An outline scheme showing the cascade of signals beginning with the stimulus and ending with the physiological response.

In Figure 3.5, we have given some examples of this type of cascade. Let us just follow one of these cascades by looking at the production of cortisol by the adrenal cortex in response to an appropriate stimulus (for example traumatic stress). The stimulus is received by the hypothalamus as a nerve impulse and the hypothalamus is stimulated to release adrenocorticotrophic hormone releasing hormone (ACTH-RH) into the blood vessels in the pituitary stalk. It moves down to the anterior lobe of the pituitary where it stimulates target cells to release adrenocorticotrophic hormone (ACTH) into the general blood stream. On reaching its target tissue (the adrenal cortex), this target tissue is stimulated to release cortisol into the blood stream. The cortisol then influences a number of body tissues to bring about a physiological response.

Π Use the example of cortisol to follow the cascades which lead to the secretion of the hormones drawn in circles in Figure 3.5.

You should realise that the levels of hormones need to be carefully controlled. In the general description given so far we have described how the secretion of a hormone may be stimulated. It can also be reduced either by being metabolised or excreted via the kidneys. Many hormones whilst their own secretion. For example, cortisol inhibits the secretion of ACTH-RH and ACTH and thus indirectly inhibits its own secretion. We will examine this type of control (so called negative feedback control) in more detail later. Some other negative feedback loops are shown in Figure 3.5. It would be worthwhile making yourself familiar with Figure 3.5.

Before we leave this overview you should realise that the pituitary gland is a multilobed structure. The front (anterior) lobe is glandular and produces many releasing hormones. The rear (posterior) lobe contains many neurosecretory cells. Between these two is the pars intermedia.

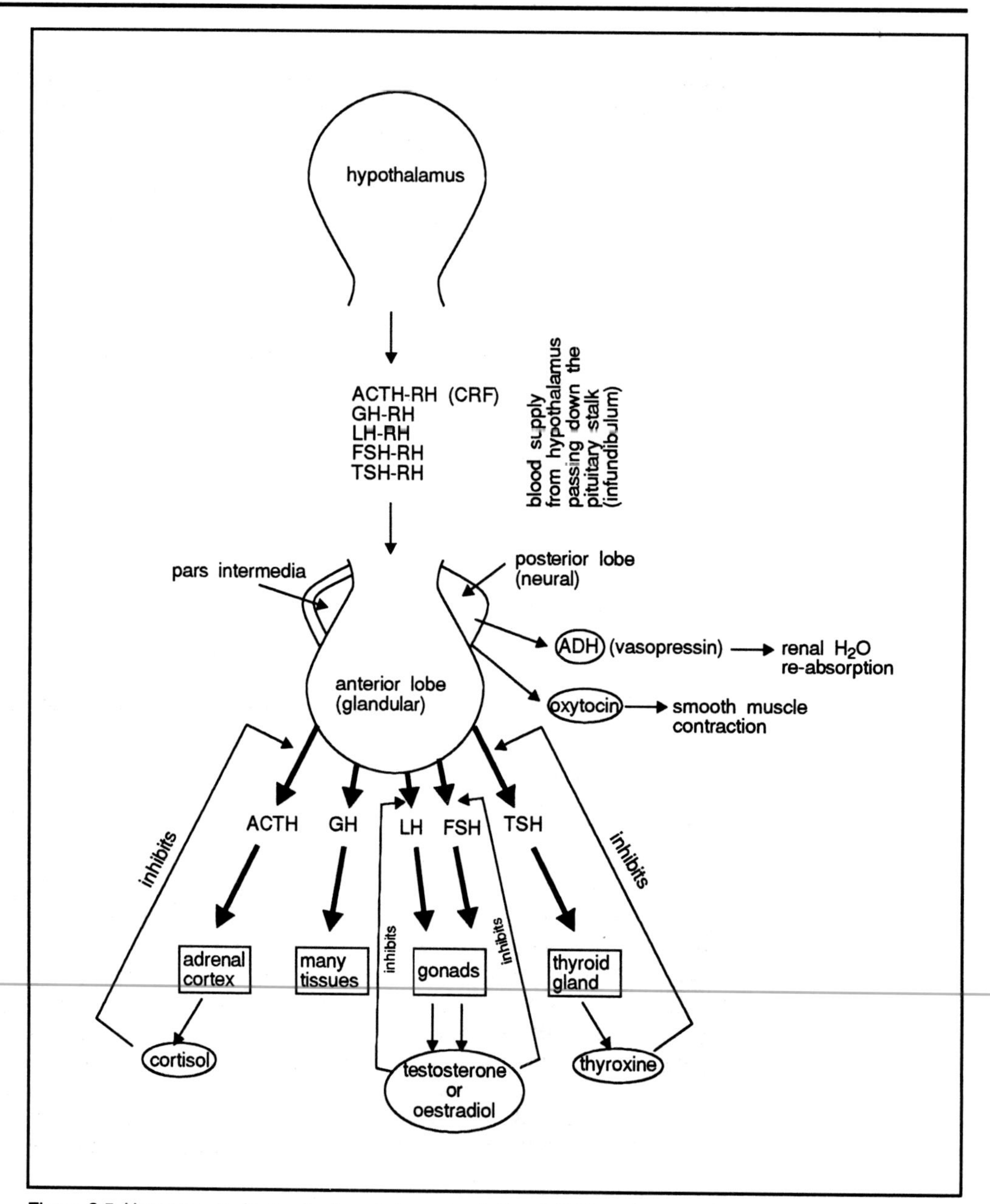

Figure 3.5 Hormone production by the pituitary gland. Key: ACTH = adrenocorticotrophic hormone, GH = growth hormone, LH = luteinising hormone, FSH = follicle stimulating hormone, TSH = thyroid stimulating hormone, GH-RH, ACTH-RH etc = releasing hormones, ADH = antidiuretic hormone.

The picture that should have emerged from this section is that the animal body contains many organs (or glands) capable of producing hormones and a kind of cascade of signals take place. In this the hypothalamus produces hormone releasing-hormones which stimulate the anterior lobe of the pituitary to release hormones. These hormones, in turn, activate target organs. The effects on the target organs are that they are stimulated to mature and to begin releasing the hormones. By the time you have completed this chapter you will have much more experience of this cascade of signals and the control of this cascade.

Before we examine each of the major organs of the endocrine system and the hormones they produce, it is worthwhile examining some underpinning features of hormone action. In the next section we will briefly examine the structure and synthesis of hormones before exploring the mechanisms of hormone action. Finally we will briefly review the endocrine organs and the hormones they produce. The intention is not to provide details of the molecular and biochemical consequences of hormone action but to provide a working knowledge of the sites of production and general physiological function of the major hormones.

3.3 The structure and synthesis of hormones

Hormones can be classified into three main groups according to their chemical structure:

- polypeptide or protein hormones eg insulin, vasopressin (antidiurctic hormone);
- steroid hormones, eg cortisol, testosterone;
- tyrosine based hormones, eg thyroxine.

3.3.1 Polypeptide and protein hormones

Use Figure 3.6 to help you follow the description of polypeptide and hormone synthesis and secretion given below.

polypeptide hormones

These hormones consist of a chain of amino acids varying in length from small peptides (eg oxytocin and antidiuretic hormone) to large polypeptides (eg insulin and growth hormone). Their synthesis begins on the ribosomes of the endocrine cell as a large inactive protein molecule known as a pre-prohormone. The molecular structure of this parent hormone is modified in the endoplasmic reticulum to form a smaller prohormone which is then packaged by the Golgi apparatus into a membrane bound vesicle containing a specific proteolytic enzyme. Further breakdown of the molecule occurs inside the vesicles to produce the smaller active hormone molecule and other peptide chains. This occurs whilst the vesicles migrate to the cell periphery where they are stored. Secretion of the hormone takes place by exocytosis, a process by which the contents of the vesicles are released into the extracellular fluid. This secretion is triggered by a rise in the concentration of calcium ions in the cytosol caused by opening of voltage sensitive calcium channels in the cell membrane or by release of calcium from intracellular stores. These two events, in turn, are brought about by the stimuli which excite that particular endocrine cell.

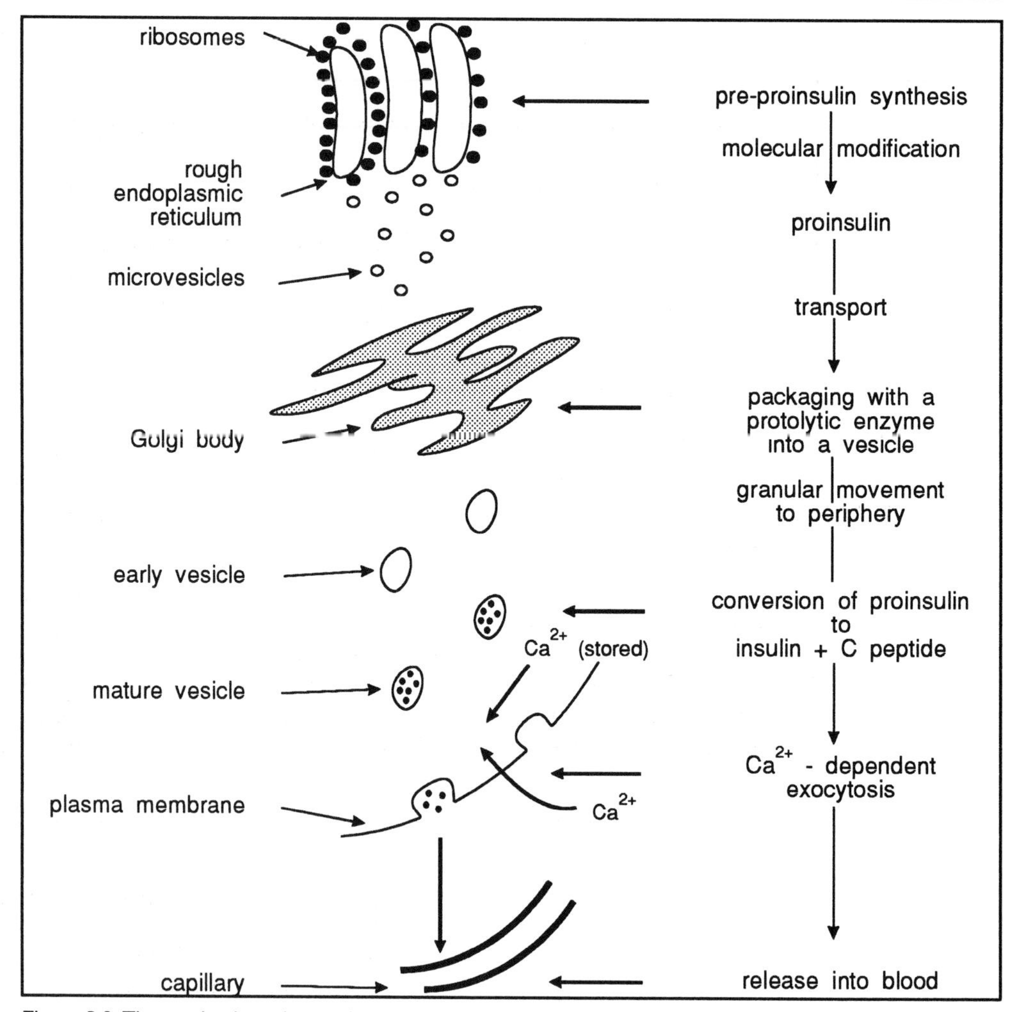

Figure 3.6 The synthesis and secretion of insulin (an example of a peptide hormone).

SAQ 3.3

Compare and contrast the release of peptide hormones from endocrine glands and the release of neurotransmitters from nerve endings (refer to Chapter 2). Do this in the form of a table.

SAQ 3.4

Place these phrases into the correct sequence to describe the synthesis and release of a typical peptide hormone.

1) Influx of calcium causes release of hormone by exocytosis.

2) The prohormone is packaged into vesicles by the Golgi apparatus.

3) The secretory vesicles migrate to the periphery of the cell.

4) The final breakdown of the hormone from its inactive to active form by a protease enzyme occurs in the vesicles.

5) The pre-prohormone is formed on the ribosomes.

6) Modification of the molecular structure of the parent molecule takes place in the rough endoplasmic reticulum.

3.3.2 Steroid hormones

These hormones are all derived from cholesterol which is either delivered to the cell from the blood, manufactured metabolically or stored in the cell in lipid droplets. The cholesterol is first converted to pregnenolone by the enzyme cholesterol desmolase. This occurs in the mitochondria of the steroid producing cells (Figure 3.7). The second stage takes place in the abundant smooth endoplasmic reticulum, characteristics of these cells, where the pregnenolone is converted to 17α-hydroxyprogesterone or progesterone.

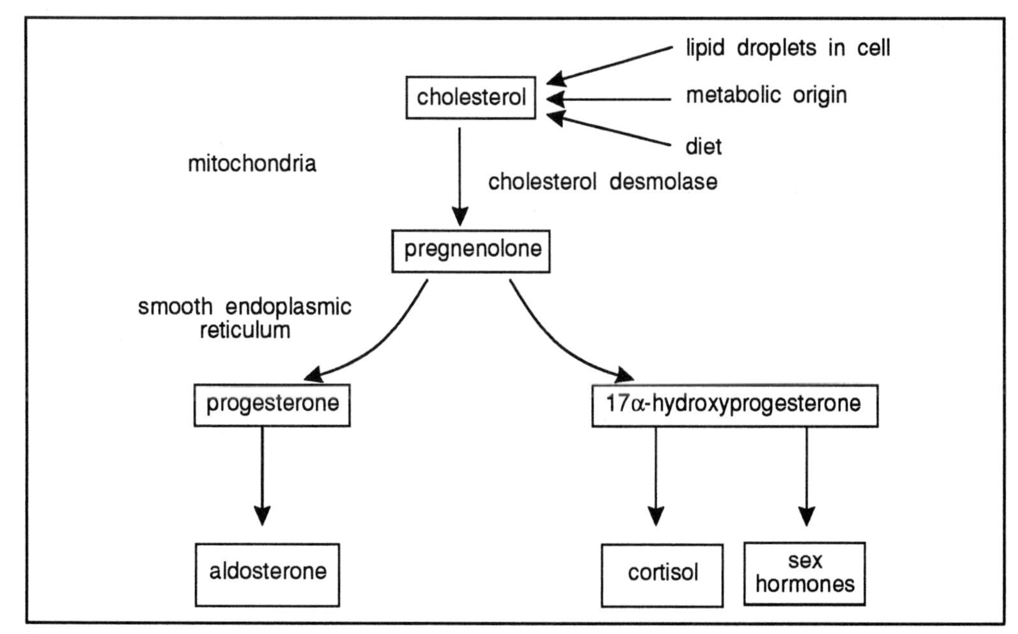

Figure 3.7 A simplified flow diagram of the synthesis of the steroid hormones (many intermediate steps have been omitted).

The subsequent stages of synthesis depend upon the presence or absence or specific enzymes which vary from tissue to tissue. The steroid hormone which is finally produced, therefore, depends upon the specific enzyme systems present in the endocrine cell eg cortisol is produced in the fascicular and reticular zones of the adrenal cortex which contain the enzyme 17α-hydroxylase whereas aldosterone is manufactured in the glomerular zone of the adrenal which lacks this enzyme.

Π You might find it helpful to redraw Figure 3.7 for yourself but this time insert the relevant chemical structures. The structures you should use are:

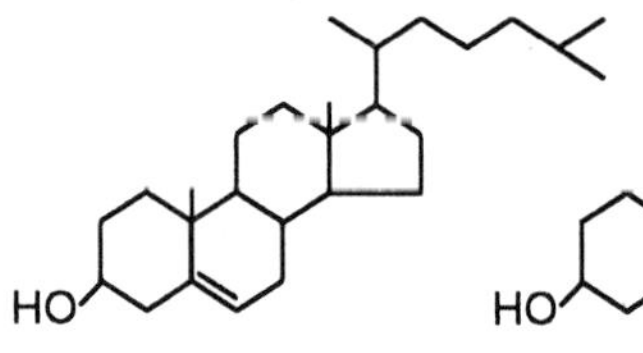

cholesterol 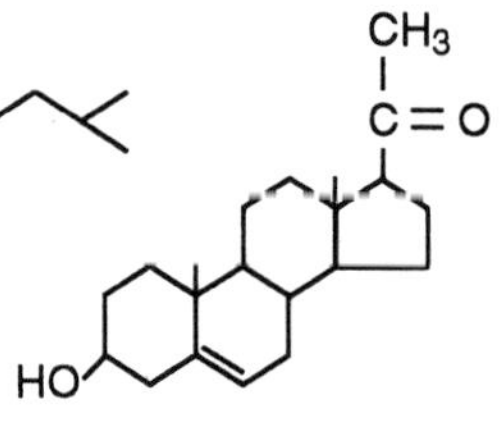 pregnenolone 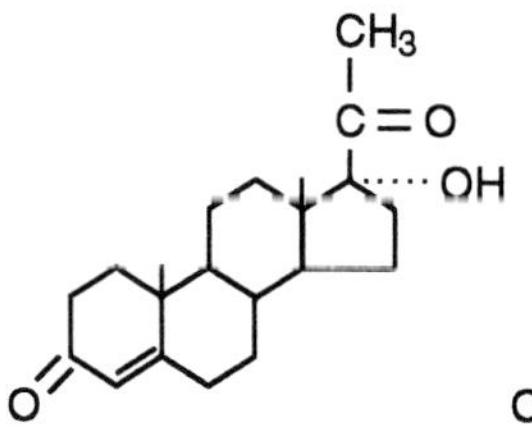 17α hydroxyprogesterone 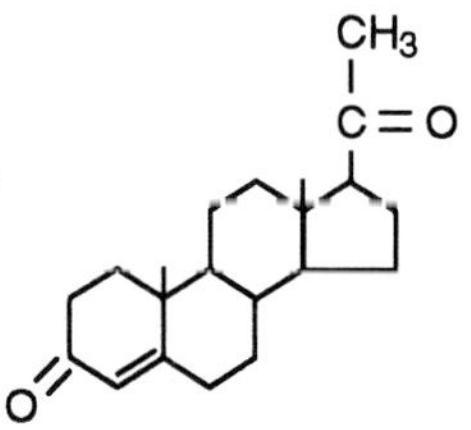 progesterone

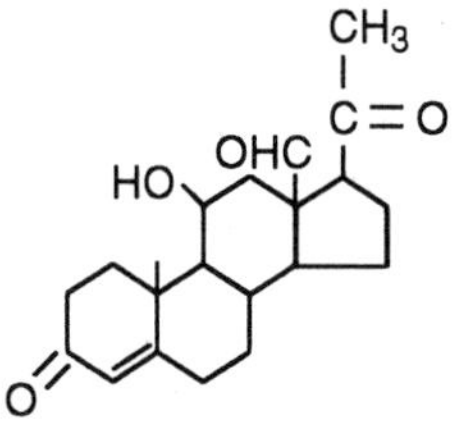

aldosterone 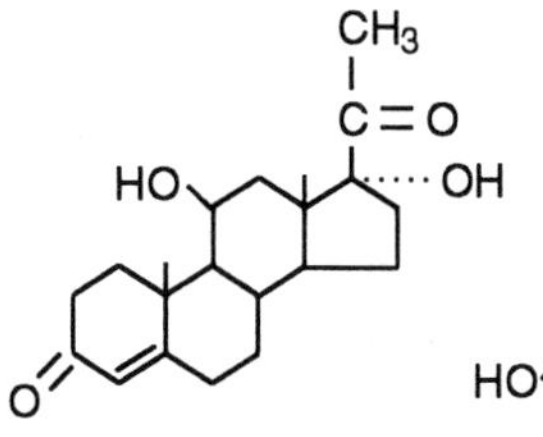 cortisol 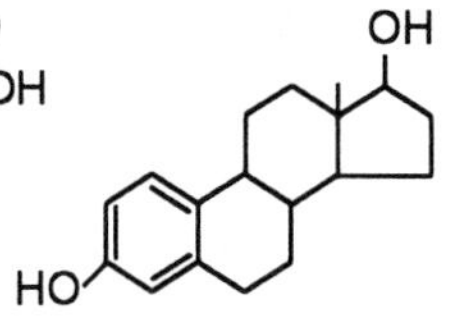 oestradiol 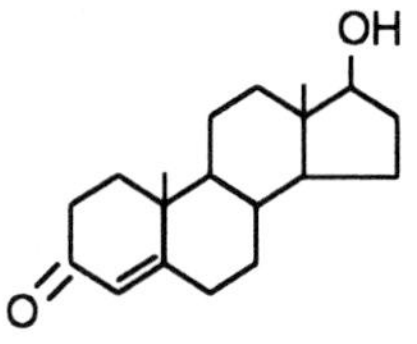 testosterone

Unlike the peptide hormones, the steroid hormones are not stored in the cell. Once synthesised, they are rapidly secreted; because of their lipid soluble structure they diffuse across the cell membrane into the extracellular fluid and into the blood.

3.3.3 Tyrosine based hormones

These hormones are derived from the amino acid tyrosine and include dopamine, adrenaline, noradrenaline (from the adrenal gland) and thyroxine and triiodothyronine (from the thyroid gland).

production of catecholamine

The synthesis of the catecholamines (noradrenaline and adrenaline) occurs by enzymatic modification of dopamine which is formed from the tyrosine molecule. This takes place in the chromaffin tissue of the body (the adrenal medulla) as well as in sympathetic adrenergic nerve endings. The presence of a specific enzyme controls the final conversion of noradrenaline to adrenaline. As more of this enzyme is present in the adrenal medulla than in the adrenergic nerve endings, the adrenal medulla secretes more adrenaline than noradrenaline. In contrast, the nerve endings secrete mostly noradrenaline (Figure 3.8).

The hormones are stored in granules with a carrier protein called chromagranin and their release is thought to be by exocytosis in a manner analogous to that involved in the secretion of peptide hormones.

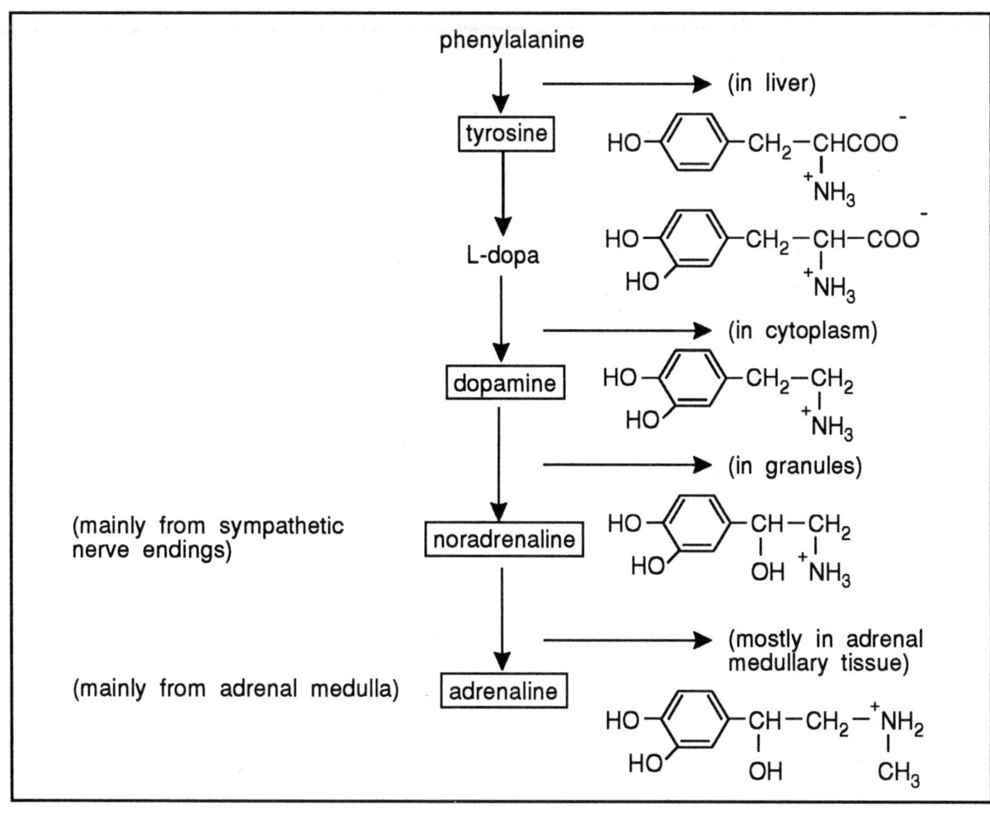

Figure 3.8 The synthesis of the catecholamines. L-dopa = L 3,4 dihydroxyphenylalanine.

site of thyroxine production

The thyroid gland synthesises triiodothyronine (T_3) and thyroxine (T_4) by the iodination of tyrosine. This occurs in the thyroid follicles where the tyrosine residue of a protein, thyroglobulin (produced by the follicular cells) is iodinated (the addition of iodine). The monoiodotyrosine and diiodotyrosine thus formed are condensed to form either of the two hormones. The release of these hormones is complex and results in the hormones being split from the thyroglobulin and transported from the colloidal centre of the follicles through the follicular cells into the blood. Figure 3.9 is a simplified diagram illustrating these events.

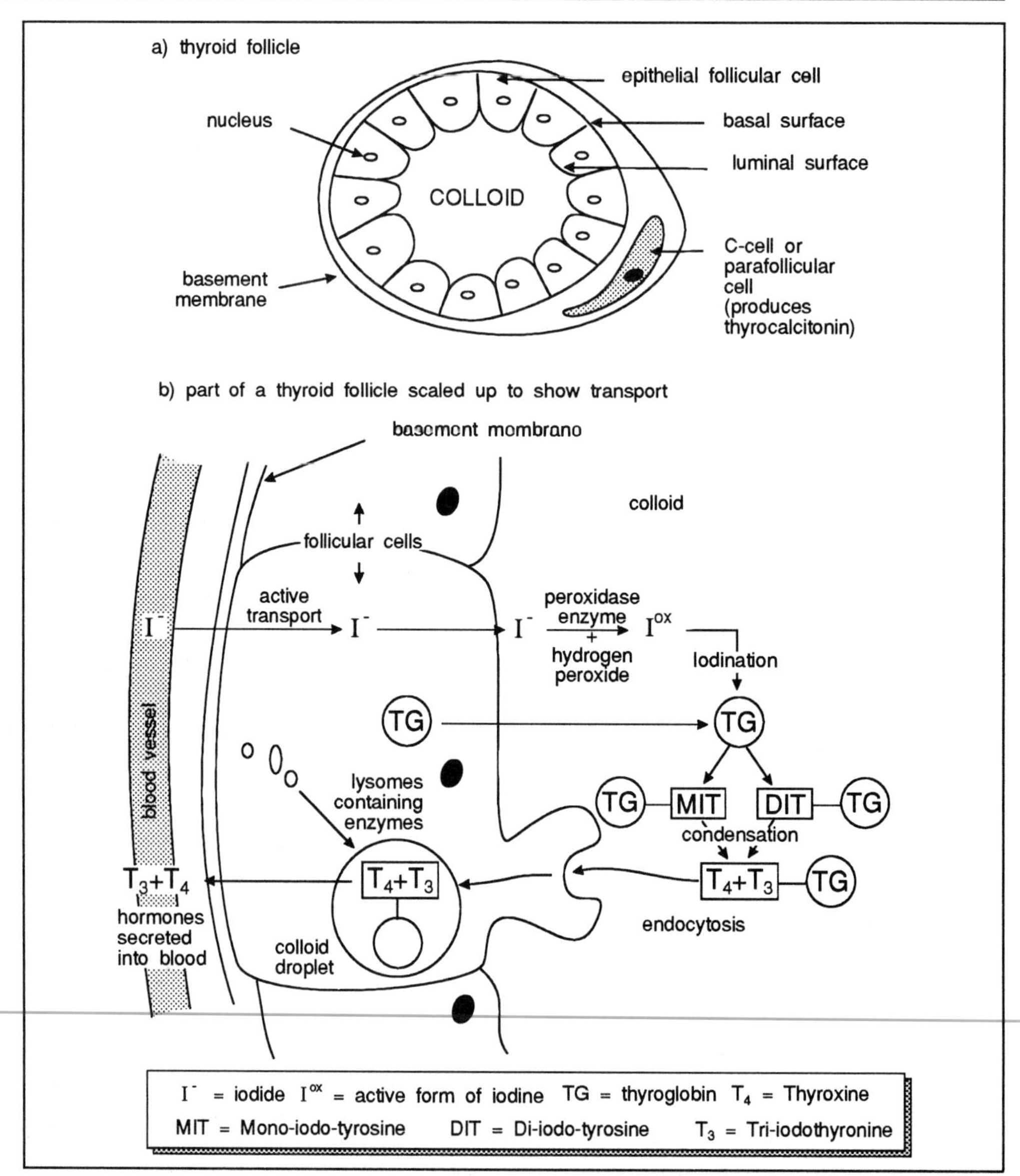

Figure 3.9 a) Diagram to show a typical follicle. b) A brief summary of the synthesis and release of thyroid hormones. Colloid is a gelatinous substance containing the protein thyroglobulin (made and secreted by the follicular cells) and the thyroid hormones in all stages of synthesis. The oldest hormone is stored in the centre of the colloid; the most recently synthesised is at the periphery of the lumen ie near the luminal surface of the follicular cells. The thyroid gland is composed of hundreds of these follicles interspersed with many blood capillaries.

The structures of the intermediates are:

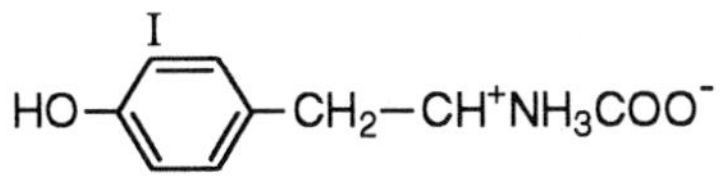

HO–C₆H₃I–CH₂–CH⁺NH₃COO⁻

monoiodotyrosine (MIT)

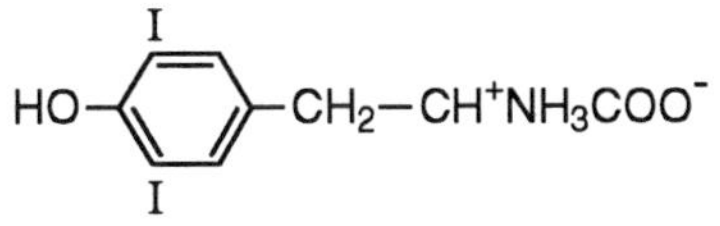

HO–C₆H₂I₂–CH₂–CH⁺NH₃COO⁻

diiodotyrosine (DIT)

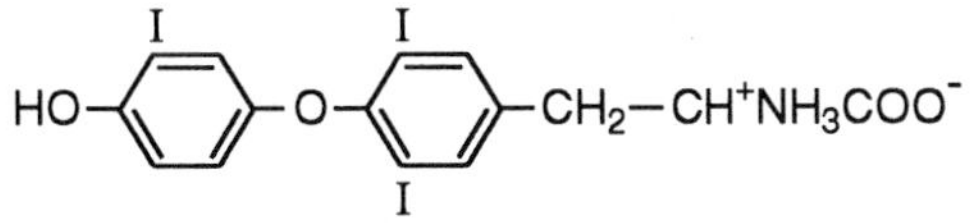

HO–C₆H₃I–O–C₆H₂I₂–CH₂–CH⁺NH₃COO⁻

triiodothyronine (T_3)

HO–C₆H₂I₂–O–C₆H₂I₂–CH₂–CH⁺NH₃COO⁻

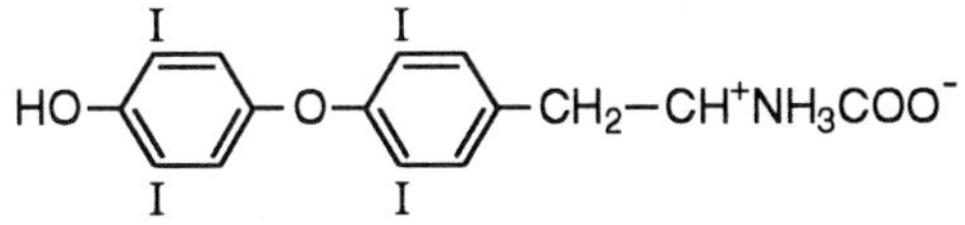

thyroxine (T_4)

SAQ 3.5

Indicate which of the following statements are false. Give reasons for your answers.

1) Steroid hormones are all derived from cholesterol.

2) Steroid hormones are stored in the cells in which they are synthesised.

3) Peptide hormones are synthesised from large, inactive parent molecules called pre-prohormones.

4) Calcium ions are essential for the release of all types of hormones.

5) The molecular structure of both thyroid hormones and catecholamines is based on tyrosine.

3.4 Transport of hormones in the blood

water solubility and hormone transport

The peptide hormones and catecholamines are water soluble molecules and they are carried, mostly, in the blood dissolved in the plasma. The steroid and thyroid hormones are bound to plasma proteins because of their relative water insolubility. The binding occurs either to specific carrier proteins such as cortisol binding globulin or thyroid binding globulin or generally to the plasma albumins. Such bindings prevents both renal filtration of the hormones and their enzymatic degradation of hormones. The carrier proteins serve as a reservoir for those hormones controlling their slow release. The latter point can be understood by reference to Figure 3.10 in which you can see that, although mostly bound to proteins, some hormone remains free (unbound) in the plasma and it is this fraction which is physiologically active causing the response in the target cell.

The unbound portion of circulating hormone is always in equilibrium with the bound fraction. Once the free hormone interacts with the receptor on the target cell, (the receptor has higher affinity for the hormone than does the plasma protein) then the equilibrium is disturbed and more of the bound hormone is released (dissociated) from the plasma proteins.

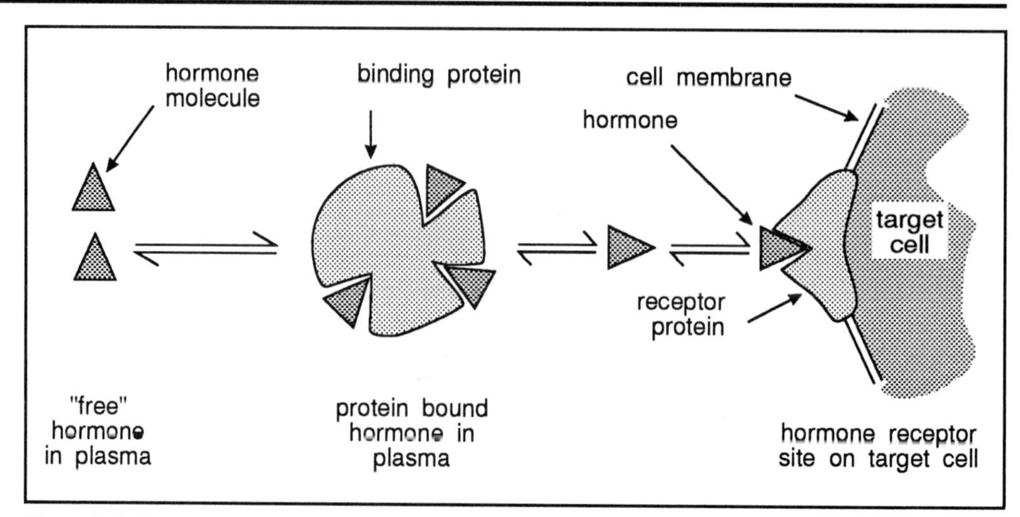

Figure 3.10 The equilibrium between free and protein bound hormone in the plasma and receptor bound hormone on the target cell. Hormone receptor protein has greater affinity for the hormone than does the plasma binding protein. Thus hormone moves to receptor. Note that without the receptor site on the target cell, the target cell would not respond to the hormone.

3.5 Metabolism and excretion of hormones

sites of hormone inactivation

The time a hormone remains in the bloodstream varies from a few minutes (eg antidiuretic hormone) to several days or weeks (eg thyroid hormones). This variability depends upon how the hormone is specifically metabolised and excreted. The liver and kidney are the main routes for the removal of hormones but other sites such as the blood and the target tissues are also involved. Often the hormones is enzymatically degraded in the liver (bound hormones) or target tissue (free hormones) before being excreted in the urine.

An important point to note is that the metabolism of some hormones after their secretion is necessary to convert them into their active form. In other words, these hormones are secreted into the blood in an inactive form which is later metabolised to the active form capable of bringing about the physiological responses of these hormones. This activation may occur in the blood during its passage through a particular tissue (eg angiotensin I is converted to angiotensin II in the lung) or in the target tissue (eg T_4 to T_3). Figure 3.11 summarises the fate of hormones once they have been secreted.

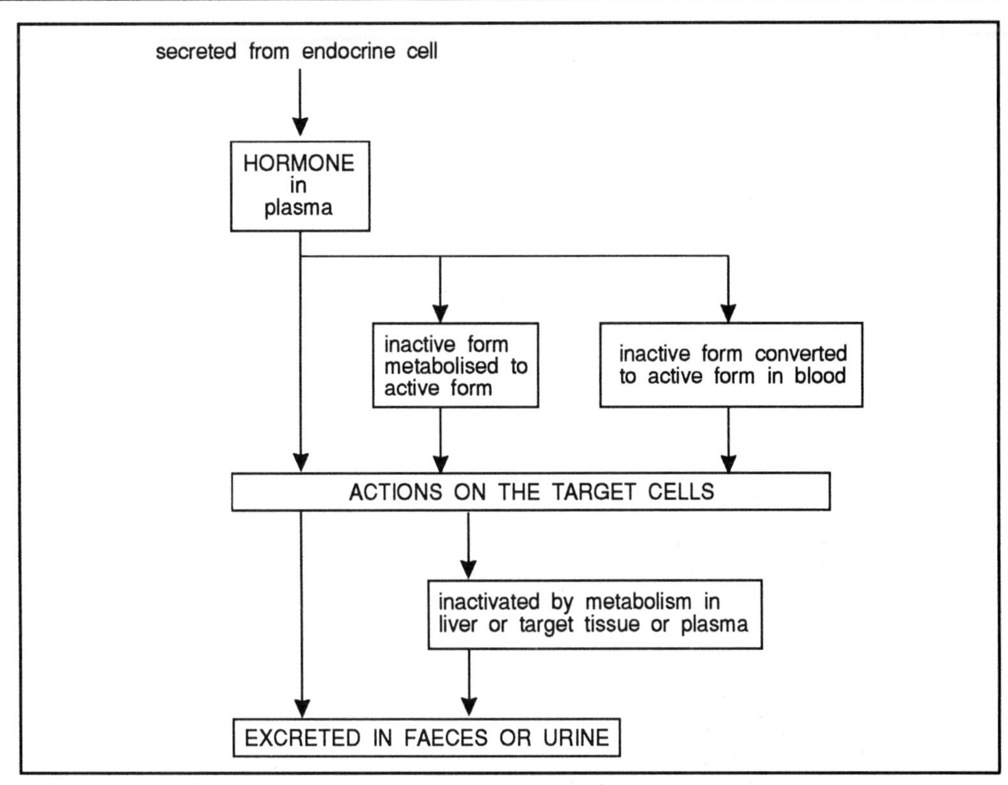

Figure 3.11 A flow diagram to summarise the possible fates of hormones following their secretion.

Π List three factors which determine the concentration of a hormone in the blood at any one time.

The rate of secretion, the rate at which the hormone is broken down and the rate of excretion of the hormone will all control the concentration of that hormone in the blood at any given time.

3.6 The mode of action of hormones

3.6.1 Receptors

Hormones are carried to virtually all tissues in the body by way of the bloodstream. However, not all tissues respond to all of the hormones. The response of the tissues is highly specific and involves only the target cells responding to a particular hormone. This ability to respond depends on the presence on, or in, the target cells of receptor molecules which are specific for a given hormone. Thus the specificity of the target cell depends upon the presence of these receptors. That is to say, if a cell has a receptor for a particular hormone it will respond to that hormone. If it does not possess a receptor, then, although the hormone may be transported to that cell, the cell will not respond.

Π Why do you think thyroid gland cells will respond to thyroid stimulating hormone whereas ovarian cells will not?

Ovarian cells do not pocess a receptor for thyroid stimulating hormone whereas thyroid cell do.

importance of receptor molecules

The receptor molecules are proteins which exist either in the cell membrane or in the cytosol or nucleus of the target cell. They function on a 'lock and key' basis in that the molecular shape of the receptors exactly matches the shape of the specific hormone molecule. Once binding occurs, the receptor hormone complex acts as a 'switch' to initiate events inside the cell leading to the cell's response (Figure 3.12).

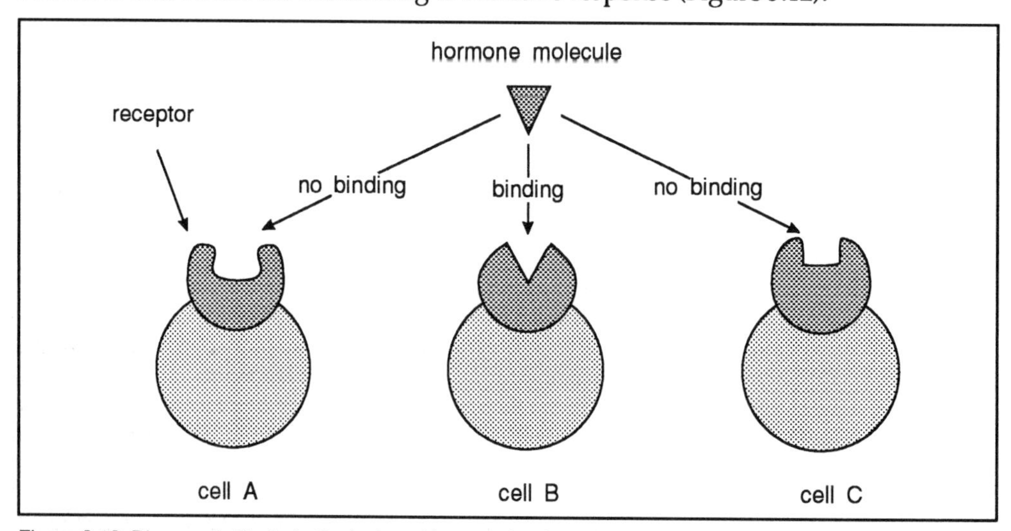

Figure 3.12 Diagram to illustrate the lock and key mechanism of hormone/receptor binding.

A cell may have more than one receptor type which may respond to either the same hormone or to different hormones. This means that a cell can give different responses to the same hormone (where the different receptors are acting as switches for different cellular events) or the cell may respond to more than one hormone. An example of the former behaviour is illustrated in Figure 3.16 where you can see that adrenaline produces two different responses from smooth muscle depending on whether an α-receptor or a β-receptor is involved.

3.6.2 Regulation of receptors

The actual number of a particular receptor type that exists on a target cell and the affinity of those receptors for their specific hormone are under physiological regulation. If a cell is exposed for a prolonged time to a high concentration of a hormone, then the number of receptors on the cell for that hormone is decreased. This is known as down-regulation. It has the effect of decreasing the cell's responsiveness to the hormone over a period of time, therefore, in effect, acting as a local negative feedback control.

down regulation

up regulation

Conversely, changes in the opposite direction, known as up-regulation, occur when a cell synthesises more receptors to a particular hormone in response to a prolonged exposure to low concentration of that hormone. This change increases the sensitivity of the tissue to that specific hormone. Physiologically, down-regulation of receptors is found to occur much more frequently than up-regulation.

The increase or decrease in receptor numbers is possible because there is a continuous turnover of the receptor molecules within a cell as a result of both receptor degradation and synthesis of the constituent proteins. The changes may occur as a result of either normal physiological control or also as a result of pathological processes. We shall be considering the latter, briefly, in a later section.

SAQ 3.6

Which of the following possibilities is the correct cause of down-regulation of hormone receptors?

1) A receptor is exposed for a short time to high plasma levels of its specific hormone.

2) A receptor is exposed for a prolonged time to low plasma levels of its specific hormone.

3) A receptor is exposed to high plasma levels of its specific hormone for a prolonged length of time.

4) A receptor is exposed to high levels of any hormone in the plasma.

3.6.3 Receptor interaction

permissiveness and hormones

A final property of receptor molecules to be mentioned is the one which underlies the phenomenon of permissiveness whereby the presence of one hormone is essential for, or potentiates, the response of the cell to a second hormone. This characteristic depends upon the fact that a hormone is able, not only to influence the number of its own specific receptors on a cell, but also the number of receptors for other hormones. This change may be either an increase or a decrease in the number of receptors for the second hormone. The phenomenon is illustrated in Figure 3.13. When a hormone A results in an increase in the number of receptors to a hormone B then hormone A is said to be permissive to hormone B. Often only small amounts of the permissive hormone are required in order for the full effect of the hormone B to be achieved. The example in Figure 3.13 shows that small amounts of the permissive thyroid hormones are required for the full effect of adrenaline upon adipose (fatty) tissue.

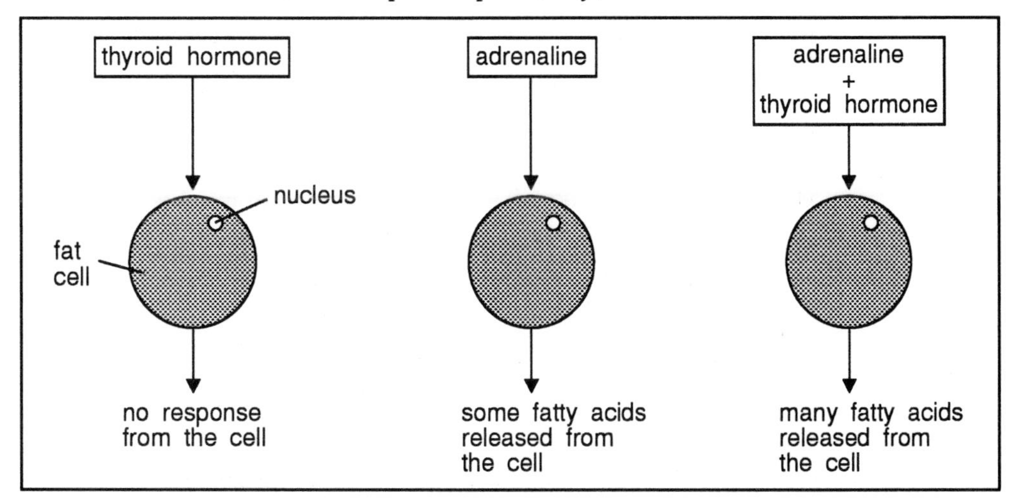

Figure 3.13 The permissive action of thyroid hormone on the effects of adrenaline on adipose tissue. Adapted from Vander, Sherman and Luciano (1990) Human Physiology: The Mechanisms of Body Function, Figure 10.7, McGraw Hill, New York.

3.7 Hormone receptor binding (signal transduction mechanisms)

3.7.1 Peptide hormones

second messengers and signal amplification

For many hormones the intracellular events triggered by their binding to the plasma membrane receptors are mediated by intracellular molecules known as second messengers. The binding hormone is called the first messenger. Such systems are utilised by those hormones unable to penetrate the cell membrane because of their large size or their lipid insolubility. As one molecule of hormone can lead to the production of hundreds of molecules of second messenger, it is also important as a means of amplifying the original signal such that only minute amounts of hormones are required for physiological action. The most commonly known second messengers are cyclic AMP, cyclic GMP, calcium ions, inositol triphosphate and diacylglycerol. We shall consider these systems in turn.

3.7.2 Secondary messengers and protein kinases

Cyclic 3′, 5′ adenosine monophosphate (cyclic AMP or, simply cAMP) is produced in the cytosol from adenosine triphosphate (ATP). It is used as a second messenger by a great number and variety of hormones including thyroid stimulating hormone, adrenaline, antidiuretic hormone and luteinising hormone.

phosphorylation

The binding of these hormones to their receptor molecules on the surface of the cell membrane results in an increase in the cytosolic concentration of the cAMP. How does this happen? The process begins with the binding of a hormone to the receptor which causes a conformational change (a change in the molecular shape) in the receptor molecule which enables it to combine with a membrane regulatory protein called a G-protein. This G-protein, which can be either an excitatory (stimulatory) Gs-protein or an inhibitory Gi-protein, in turn, changes the activity of the membrane bound enzyme adenylate cyclase whose function is to catalyse the conversion of ATP to cAMP in the cytosol. Hence, the hormone binding to the receptor results in a rise in the cytosolic concentration of cAMP. The cAMP is a second messenger triggering a series of intracellular events which bring about the cell's ultimate response to the hormone.

G-proteins

protein kinases A-kinases

These events are mediated by protein kinases, present in the cytosol, which depend upon cAMP for their activation (cAMP dependent protein kinases or A-kinases). These kinases phosphorylate other proteins, often enzymes, within the cell. Such phosphorylations alter the activity of these proteins and thereby bring about the required changes in the cell's behaviour. The great number of different protein kinases in the cytosol allows for a variety of responses to occur in the cells which use cAMP as a second messenger.

The effect of the cAMP is terminated by a cellular enzyme, called phosphodiesterase, which catalyses the breakdown of the second messenger to inactive non-cyclic AMP (Figure 3.14).

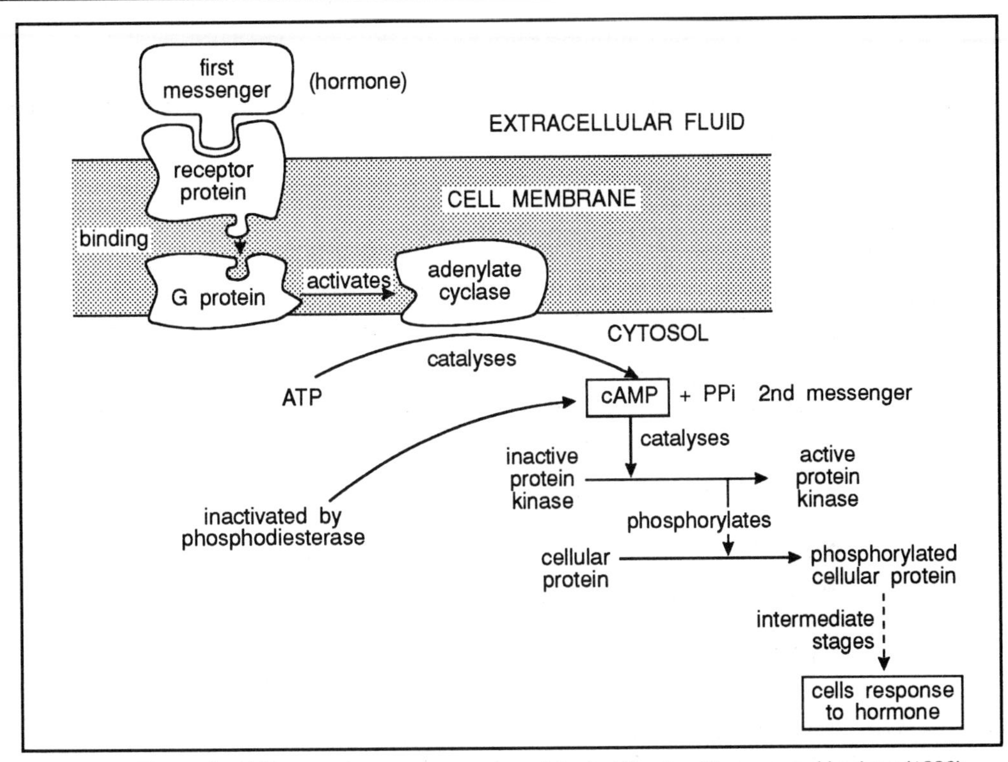

Figure 3.14 The cyclic AMP second messenger system. Adapted Vander, Sherman and Luciano (1990), Human Physiology: The Mechanism of Body Function, Figure 7.17, McGraw Hill, New York.

We should note, also, that the concentration of cAMP can be decreased, rather than increased, by some hormones. In these cases the hormones bind to receptors which combine with different regulatory proteins (Gi-proteins) that inhibit, rather than activate, adenylate cyclase. This results in a fall in the concentration of cAMP and, therefore, in a reduction of the extent of the phosphorylation of key proteins in the cell.

adenylate cyclase

dependant protein kinases

Cyclic GMP. Cyclic 3′,5′-guanosine monophosphate is another cyclic nucleotide which serves as a second messenger in a system analogous to that of the cAMP. Cyclic GMP activates cGMP dependant protein kinases which phosphorylate different cellular proteins from those activated by cAMP dependant kinases and bring about alternative cellular responses. This system is not as widely used as the cAMP second messenger system.

Π A hormone uses cAMP as its signal transduction mechanism but can elicit 5 different responses from its target cells. A drug is able to block only one of these responses. Explain whether the drug is blocking the action of the receptor, the G-protein, adenylate cyclase, cAMP or none of these stages in the second messenger pathway.

As all of these stages are common to all the responses produced by the hormone, then the drug cannot be blocking any of the above.

SAQ 3.7

Put the following phrases into the correct order to explain the sequence of events in the cyclic AMP second messenger system.

1) The cAMP is inactivated by phosphodiesterase.

2) The first messenger binds to the specific membrane receptor.

3) The activated protein kinase phosphorylates proteins involved with the cell's response to the hormones.

4) The regulatory protein (Gs) activates the membrane bound enzyme adenylate cyclase.

5) Adenylate cyclase catalyses the conversion of ATP to cAMP, resulting in an increase in the cytosolic concentration of cAMP.

6) The receptor/hormone complex binds to a membrane regulatory protein called a Gs-protein.

7) Cyclic AMP activates specific protein kinases in the cytosol.

Calcium ions and inositol triphosphate

Calcium ions (Ca^{++}) are very important second messengers. Calcium ions function in a great number and variety of cellular responses to hormones and other chemical messengers. In common with cAMP, the key to calcium activity as a second messenger is a rise in the intracellular cytosolic concentration of the ion as a result of the hormone binding to its receptor at the cell membrane. The rise in cytosolic concentration of calcium ions may occur in several ways. Firstly, the binding of the hormone to the receptor may open up ligand, or receptor operated calcium channels in the cell membrane allowing calcium to diffuse into the cell down its concentration gradient. The calcium channels open up either because the receptor protein forms part of the channel and the conformational change which occurs on binding at the receptor causes the channel to open, or because the receptor protein links up with a G-protein which, in turn, opens the channel up (see Figure 3.15). It should be noted also that some calcium channels can be opened or closed by electrical events, rather than by chemical events, at the cell surface. These calcium channels are described as voltage gated channels.

calcium channels

Secondly, the cytosolic concentration of calcium ions can be increased by the release of calcium from intracellular stores. This occurs as a result of a different sequence of events which involves one of the phospholipid constituents of the cell membrane. In this sequence, the bound receptor activates a membrane bound enzyme called phospholipase C. The coupling of the receptor/hormone complex to this enzyme is mediated by another G-protein different from that involved in the cAMP pathway. The action of the phospholipase C is to catalyse the breakdown of a membrane phospholipid, phosphatidylinositol 4,5-bisphosphate (PIP_2) to inositol triphosphate (IP_3) and diacylglycerol (DAG). The IP_3 enters the cytosol where it causes the movement of calcium ions from intracellular stores. The mechanism of this removal is, as yet, unknown, but evidence suggests the involvement of regulatory proteins within the membranes of the intracellular structures usually the endoplasmic reticulum.

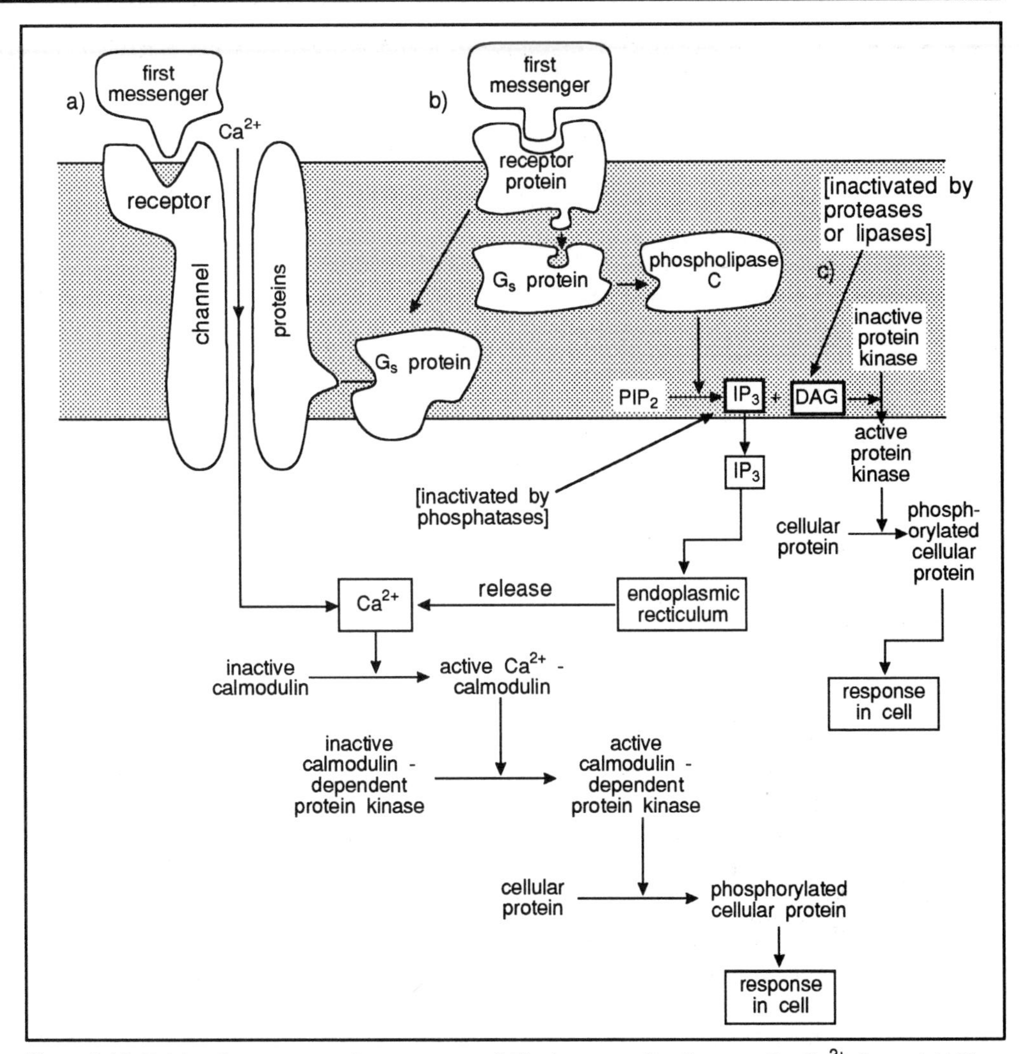

Figure 3.15 Calcium ions as second messengers. a) The hormone directly opens the Ca^{2+} channel. b) The hormone indirectly opens the Ca^{2+} channel via a G_s protein. c) The cascade leading to the release of intracellular stored Ca^{2+} (see text for details). Adapted from Vander, Sherman and Luciano (1990), Human Physiology: The Mechanism of Body Function, Figures 7.19/7.20, McGraw Hill, New York.

The effect of IP_3 is terminated by the action of specific cytosolic phosphatases which remove the phosphate groups leaving free inositol to be reincorporated into the cell membrane (Figure 3.15).

Thus two mechanisms exist to increase the concentration of calcium ions in the cytosol: a) opening up calcium membrane channels to allow entry of calcium ions from the extracellular fluid, b) the release of calcium ions from intracellular stores. However, it has been found that they often work in conjunction with each other rather than functioning exclusively; the two mechanisms may even be initiated by the same hormone.

SAQ 3.8

A hormone's response in a cell is mediated via a rise in the concentration of intracellular calcium ions. If a drug which blocked all the membrane receptor operated channels for calcium were given, would it abolish the response of the cell completely? Explain your answer.

So, how do the raised intracellular calcium levels lead to the eventual responses of the cell to a hormone? The first step in this process is that the calcium binds, specifically, to calcium binding proteins present in the cytosol. The most important of these is calmodulin. Calmodulin then activates or inhibits various calmodulin dependent protein kinases which, in turn, lead to phosphorylation of proteins involved in the ultimate response of the cell to the hormone.

calmodulin

Other proteins similar to calmodulin function in the same way to mediate the second messenger actions of calcium. In some instances calcium ions may have a direct effect on the cytosolic proteins involved in the cell's response without the mediation of specific calcium binding proteins.

Table 3.1 summarises the ways in which calcium acts as a second messenger.

The increase in calcium ion concentration in the cytosol can occur in two ways:

- opening up of plasma membrane channels for calcium;
- release of calcium from intracellular stores mediated by inositol triphosphate.

The calcium ions then act as second messengers in one of three ways:

- they bind to calmodulin, which then activate a calcium calmodulin dependent protein kinase whose function is to phosphorylate cellular proteins and, thereby, bring about the cellular responses;
- calcium binds with other calcium binding proteins other than calmodulin. The events are analogous to that above;
- calcium binds to the cellular (response) proteins directly without any mediators.

Table 3.1 Role of calcium ions acting as second messengers.

Recent evidence has shown that inositol triphosphate (IP_3) can be further phosphorylated to form inositol tetraphosphate (IP_4). This molecule appears to act as an intracellular messenger to open membrane calcium channels which allow calcium ions to enter the cell and either replenish the calcium stores depleted by IP_3, or perhaps to enable calcium redistribution within the cell.

Diacylglycerol (DAG)

The final second messenger to be considered is the molecule diacylglycerol produced during the breakdown of PIP_2 by phospholipase C in the following reaction.

```
                      O                                          O
                      ||                                         ||
      O     CH2—O—C—R'                           O     CH2—O—C—R'
      ||     |                                   ||     |
R"—C—O—CH                          ——————►  R"—C—O—CH
             |                                          |
            CH2—O—P—O—inositol 4,5                     OH     +   inositol 1,4,5
                       bisphosphate                                triphosphate
         PIP2                                          DAG         IP3
```

The DAG (a lipid) remains in the cell membrane where it activates a specific protein kinase, called protein kinase C, which is capable of phosphorylating a large number of other proteins in the cell. It appears to be very important in the down-regulation of membrane protein receptors, as it phosphorylates these receptors thus rendering them ineffective. (Cyclic AMP dependent protein kinases involved in down-regulation seem to function in a similar manner).

The DAG is inactivated either by specific enzymes and reincorporated into the cell membrane or by a lipase to liberate arachidonic acid. This latter molecule is the precursor of the eicosanoids (prostaglandins), a further groups of intracellular messenger molecules.

synergism

It is important to emphasise that a hormone initiating events via IP_3 will simultaneously release DAG and, therefore, will activate protein kinase C. There is evidence in many cases that the molecules act synergistically ie they are both required in order to initiate the cell's response to a hormone. The use of various pharmacological agents has shown that if either of the molecules is inactivated, then the cell's response to the hormone may be blocked.

Not only is there a functional link between IP_3 and DAG but also there exist important interactions between all the different second messengers. They frequently function together, sometimes causing effects which are in opposition to each other, other times the changes are in the same direction. Figure 3.16 shows how effects of adrenaline on smooth muscle are mediated by two different signal transduction mechanisms resulting in opposing responses from the muscle.

Receptors acting as protein kinases

So far we have described four molecules which can activate protein kinases, cAMP, cGMP, calcium activated calmodulin and DAG. To finish this section we must consider one further mechanism by which the first messenger can trigger the activation of protein kinases. This mechanism involves the receptor itself. When the first messenger binds to the receptor, the receptor then becomes the active protein kinase which phosphorylates either itself or other cytosolic or membrane proteins. This group of protein kinases are referred to as tyrosine kinases because the site of phosphorylation is the tyrosine portion of the target protein. An example of a hormone which is thought to function in this manner is insulin.

tyrosine kinases

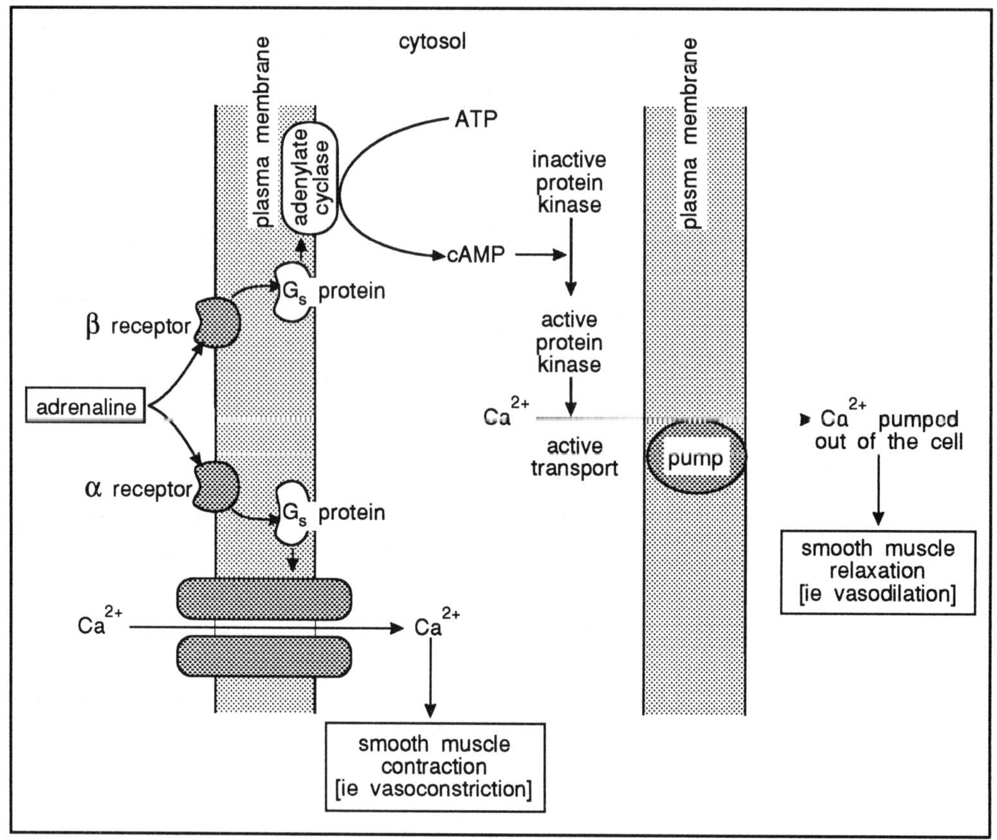

Figure 3.16 The effects of adrenaline on smooth muscle cells mediated by α and β receptors. Adrenaline binding to α receptors open Ca^{2+} channels allowing Ca^{2+} to enter the cell. This leads to smooth muscle contraction. Adrenaline binding to β receptors cause Ca^{2+} to be pumped out of the cells leading to smooth muscle relaxation. This diagram illustrates how a cell can have two different receptors for the same hormone, each mediating a different cellular response. Adapted from Vander, Sherman and Luciano (1990), Human Physiology: The Mechanism of Body Function, Figure 7.21, McGraw Hill, New York.

In summary a particular receptor type is often associated with a particular signal transduction mechanism in a variety of cells. The different events triggered by this mechanism result in various physiological and biochemical responses within the cell.

SAQ 3.9

Which of the following statements is incorrect. Give reasons for your answers.

1) IP_3 acts to increase cytosolic concentration of calcium ions by opening up membrane channels for calcium.

2) DAG activates protein kinase C.

3) Phosphodiesterase inactivates both cAMP and DAG.

4) The same regulatory G-protein mediates transduction of all signals across cell membranes.

5) A hormone initiating the activation of the IP_3 pathway will simultaneously cause an increase in the concentration of DAG in the cell membrane.

SAQ 3.10

You are presented with the problem of identifying the second messenger systems involved in three different responses (A, B, C) of a particular cell type to a hormone. You are given four pharmacological agents to manipulate the pathways:

1) Forskolin, which activates adenylate cyclase.

2) A calcium ionophore, which causes an increase in the concentration of calcium ions in the cytosol.

3) Theophylline, which inhibits phosphodiesterase.

4) Imidazole, which activates phosphodiesterase.

Your results are summarised in the table below (results are expressed in percent cell response):

Agent	A	B	C
forskolin	100	0	25
ionophore	0	100	25
theophylline	100	0	25
imidazole	0	0	0
theophylline + ionophore	100	100	100

Table showing results or pharmacological manipulation of cellular responses A, B and C.

From this data try to deduce which second messenger system is the mediator in each response.

3.7.3 Steroid and thyroid hormones

steroid hormones

Steroid and thyroid hormones share a signal transduction mechanism which is different from that of the peptide hormones. These hormone molecules are able to penetrate the cell membrane; steroids because of their lipid solubility and thyroid hormones because they are transported into the cell.

translocation

The specific receptors for these hormones are found in the cytosol or in the nucleus. In the case of the cytosolic receptors, the hormone molecules bind to the receptor and the complex is translocated into the nucleus of the cell (Figure 3.17). Once inside the nucleus the receptor hormone complex binds to the chromatin, stimulating the production of specific mRNA. The mRNA then enters the cytosol to act as a template for the synthesis of cellular proteins involved in the cell's response to the hormone. These events follow a longer time course than the second messenger systems such that steroid and thyroid hormones are usually characterised by a typical lag period of 45 minutes before a response in the cell is apparent. This contrasts with the virtually instantaneous effects of the other second messenger systems.

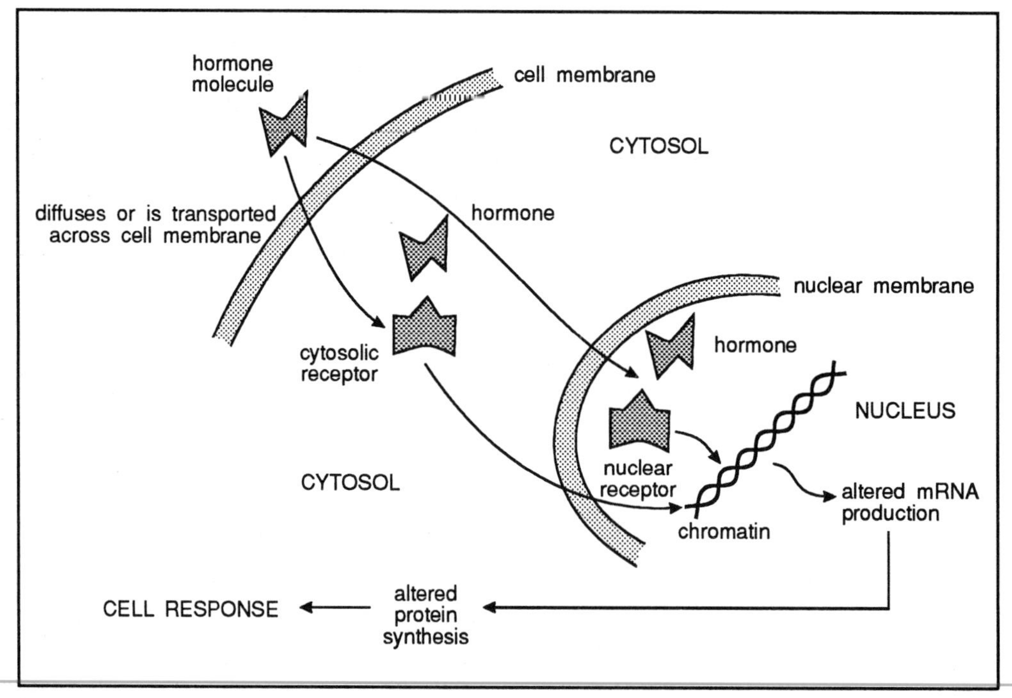

Figure 3.17 Cellular mechanisms of action of steroid and thyroid hormones.

Π Actinomycin-D is an antibiotic which blocks RNA synthesis. What affect would this have on the action of a steroid hormone?

Of course, it would block the action of the hormone.

3.8 Control of hormone secretion

3.8.1 Rhythm

circadian, monthly and season rhythm

Many hormones are not secreted continuously. Instead, their fluctuating plasma concentrations follow a variety of rhythmical patterns. Amongst others, these include the occurrence of short bursts of output every few hours (growth hormone), a 24 hour cyclical variation (circadian rhythm eg cortisol), a monthly rhythm (gonadal hormones) or seasonal variations in output (pineal hormones).

Often, such rhythms result from the influence of external factors such as sleep, light/dark cycles, environmental temperature and physical or mental shock, which are mediated by the nervous system.

The secretion of hormones is usually homeostatically controlled so that there is neither oversecretion nor undersecretion of the hormone as related to the needs of the body. If the regulation is defective then pathological states often ensue.

3.8.2 Feedback control

As with other homeostatic mechanisms in the body, hormone output is controlled by negative feedback. In such a feedback a change in a given parameter results in a response in the opposite direction in order to return that parameter to its original value. With some hormones there is a simple feedback control that involves the blood levels of an inorganic ion or organic substance which the hormone controls. For example, insulin, which is controlled by the level of glucose in the blood; parathyroid hormone, is controlled by the calcium levels in the blood (see Figure 3.18). In the case of insulin, a rise in blood glucose stimulates the output of insulin from the pancreas, the insulin then acts to encourage the cellular uptake of the glucose which has the effect of lowering the levels in the blood. The decreased blood glucose level then remove the original stimulus to the insulin output, therefore, insulin output falls.

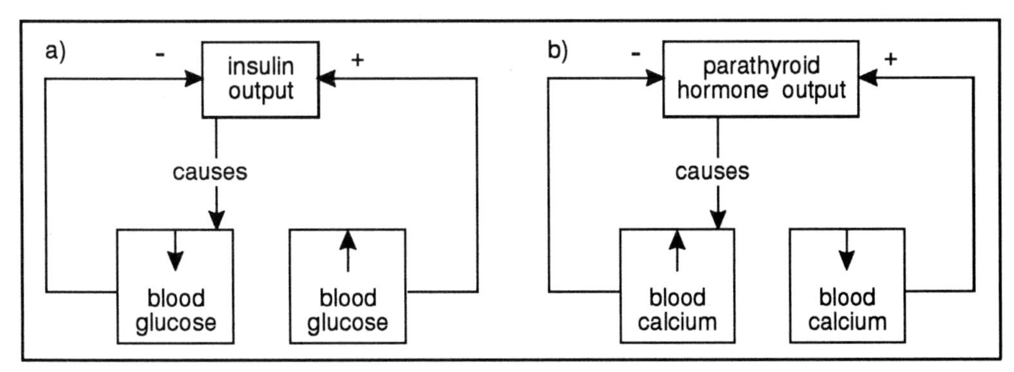

Figure 3.18 Negative feedback control of a) blood glucose on insulin output. b) plasma calcium levels on parathyroid hormone output.

In a few instances, the hormone output is controlled by a positive feedback system whereby the controlling input increases the output even further, rather than reducing it. This is the case for both the output of oxytocin in parturition and the ovulatory surge in the output of luteinising hormone. During parturition oxytocin is released as the foetus descends into the birth canal. The oxytocin causes uterine contractions which push the foetus further down, causing more output of oxytocin. This continues until the birth of the infant is complete (see Chapter 10).

menstrual cycle

In the early menstrual cycle, the rising levels of oestrogen exert a negative feedback on the output of luteinising hormone (LH) from the pituitary gland. However, once oestrogen levels reach a certain level in the plasma which is maintained for a given length of time, then this feedback suddenly becomes a positive feedback leading to a surge in the output of the LH and the release of the ovum from the ovary (ovulation). The mechanism for the sudden reversal of the feedback is, as yet, unknown.

3.8.3 Nervous control

The output of other hormones is controlled, not by the circulating level of a specific compund in the blood, but by the nervous system. This can be a direct effect of neurotransmitters released from nerve endings supplying the gland, as seen in the output of adrenaline from the adrenal gland which is stimulated by the excitation of sympathetic nerves. Some hormones are released directly from nerve endings by the process of neurosecretion. The two well known examples of this are antidiuretic hormone (ADH) and oxytocin, both of which are neurosecreted from the posterior pituitary as a result of excitation of hypothalamic neurosecretory cells whose axons terminate there.

The nervous system also influences the endocrine output via the hypothalamus and anterior pituitary gland which are important for the control of the majority of hormones in the body. Before considering this system in more detail, we should mention one or two other points on the control of hormone output. Firstly, that hormones are capable of influencing the output of each other and, secondly, that the output of a given hormone is often a result of more than one of these controlling inputs. In many instances it is a combination of blood parameters, nervous input and plasma levels of other hormones which influence the final output of a given hormone.

3.8.4 The hypothalamus and pituitary

The key to the function of the hypothalamo-pituitary influence on endocrine output is the process of neurosecretion. As described at the beginning of this chapter neurosecretion is the manufacture and secretion of peptide hormones by specialised neurons, called neurosecretory cells. Such neurons exist in two main areas of the hypothalamus, from which axons originate and which in turn influence the pituitary gland.

components of the pituitary gland

The pituitary gland (hypophysis) lies at the base of the brain is connected to the hypothalamus by the pituitary stalk (the infundibulum). It is divided into two main sections, the anterior or glandular portion (the adenohypophysis) and the posterior or neural part (the neurohypophysis). In some animals there is a marked third section, the intermediate lobe (pars intermedia), but this is very small in humans.

The posterior pituitary is a downgrowth of the hypothalamus and is composed of the long axons and axon terminals of the neurosecretory cells situated in two distinct areas of the anterior hypothalamus (Figure 3.19). As described in the last section, the hormones ADH and oxytocin are manufactured in these cells, transported down the axons to be stored in vesicles in the axon terminals. The hormones are released by exocytosis as a result of excitation of these nerve cells by a variety of synaptic inputs. The process is summarised in Figure 3.19.

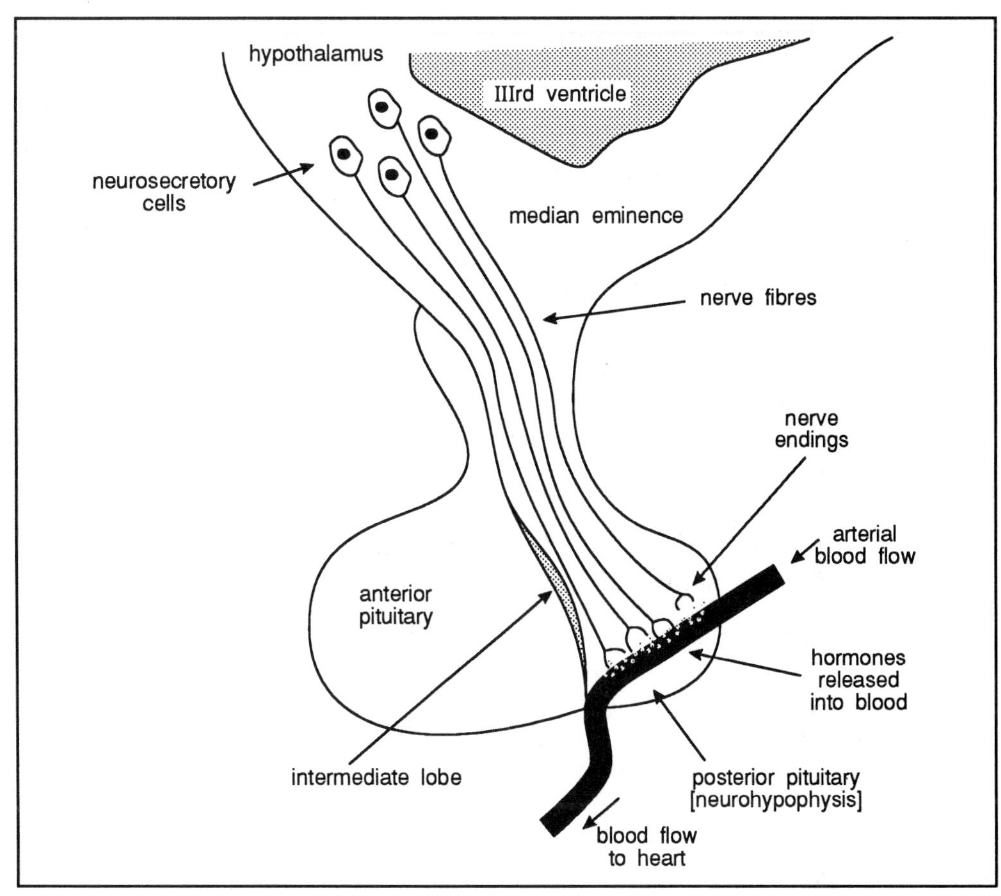

Figure 3.19 Diagram to show the posterior lobe of the pituitary and its nervous connections to the hypothalamus.

The anterior pituitary, however, is glandular tissue developed as an outgrowth of the embryonic foregut. It is connected to the hypothalamus by a plexus of blood capillaries (the primary plexus) which drains into the hypophysial portal vessels passing down through the infundibulum to the anterior pituitary (see Figure 3.20a). Neurosecretory cells from another, less well differentiated, area of the hypothalamus send short axons to terminate in the region of the primary plexus at the top of the pituitary stalk. These cells produce a series of chemical messengers called hypothalamic releasing or regulatory hormones or factors. The latter distinction depends upon whether the chemical structure of the molecule is known (hormone) or not (factor). These factors may be either excitatory or inhibitory. They are released into the capillaries of the primary plexus from which they are transported via the portal vessels to the glandular cells of the anterior pituitary (Figure 3.20b).

regulatory hormones and factors

This direct blood route to the adenohypophysis avoids the hormones being diluted or metabolised as they would be if they were secreted into the general circulation for distribution by the heart. When the hypothalamic releasing factors arrive in the anterior pituitary, they cause the gland cells to release their own stored hormones. There are six different anterior pituitary hormones, each synthesised by different types of cells. These hormones, in turn, are transported to their target endocrine glands where they stimulate the secretion of the target gland hormones. These hormones, then bring about a response from their specific target tissue.

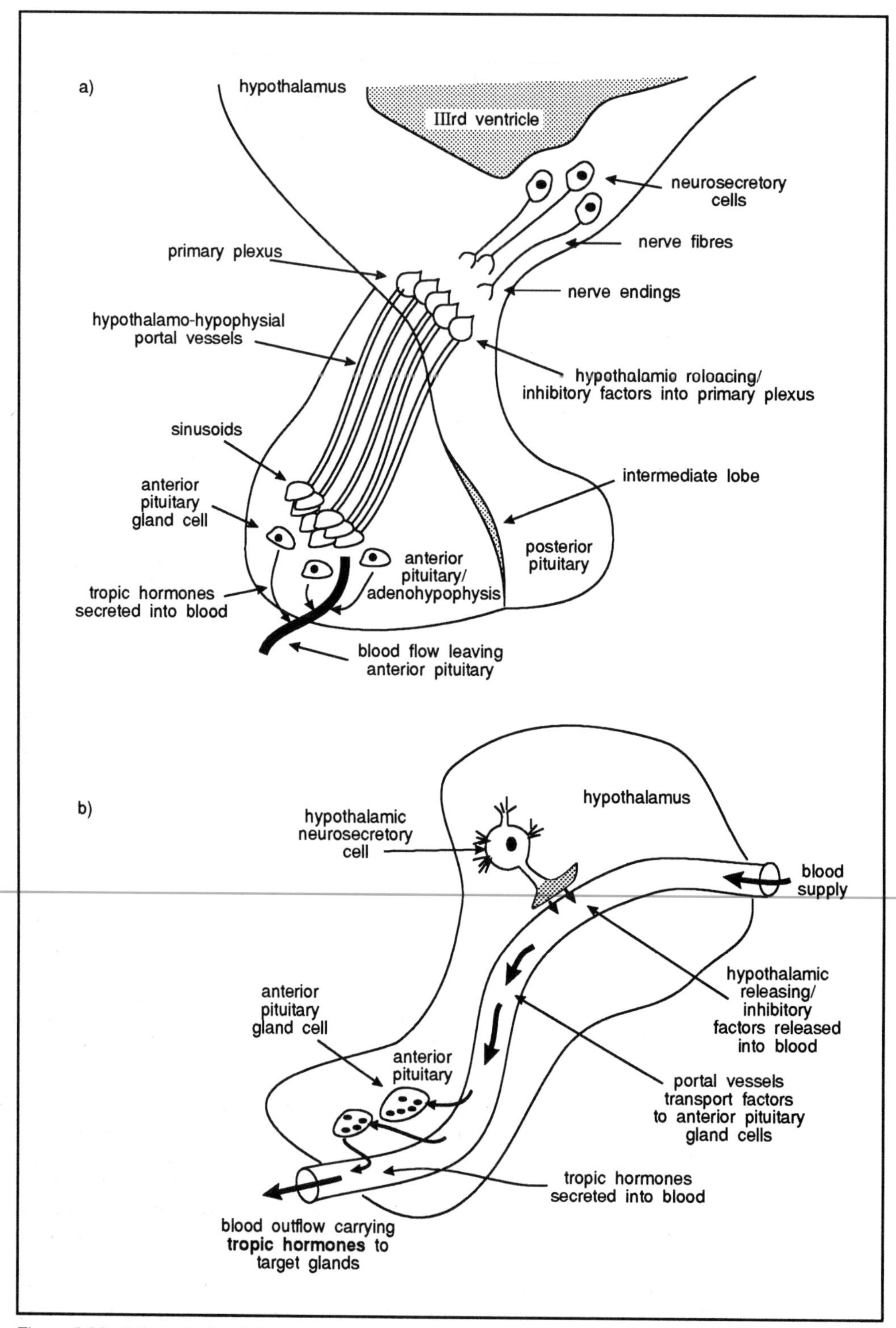

Figure 3.20 a) the anterior pituitary and the blood vessel connections with the hypothalamus. b) Mechanism of influence of the hypothalamic releasing/inhibitory factors onto the anterior pituitary.

Π You should be able to list 5 of the hormones produced by the anterior pituitary lobe. Have a go and check your answer by examining Figure 3.5. The only one we did not include is prolactin. This hormone influences milk production by mammary glands and we will discuss this in Chapter 10. You might like to add prolactin to Figure 3.5.

The anterior pituitary hormones are often referred to as tropic (or trophic) hormones as they function to make other endocrine glands secrete their hormones.

The sequence of events therefore is summarised as:

the secretion of a hypothalamic releasing or inhibiting factor/hormone which controls the secretion of the anterior pituitary tropic hormone which in turn controls the output of another hormone from a target endocrine gland and this in turn brings about a response from its target tissue.

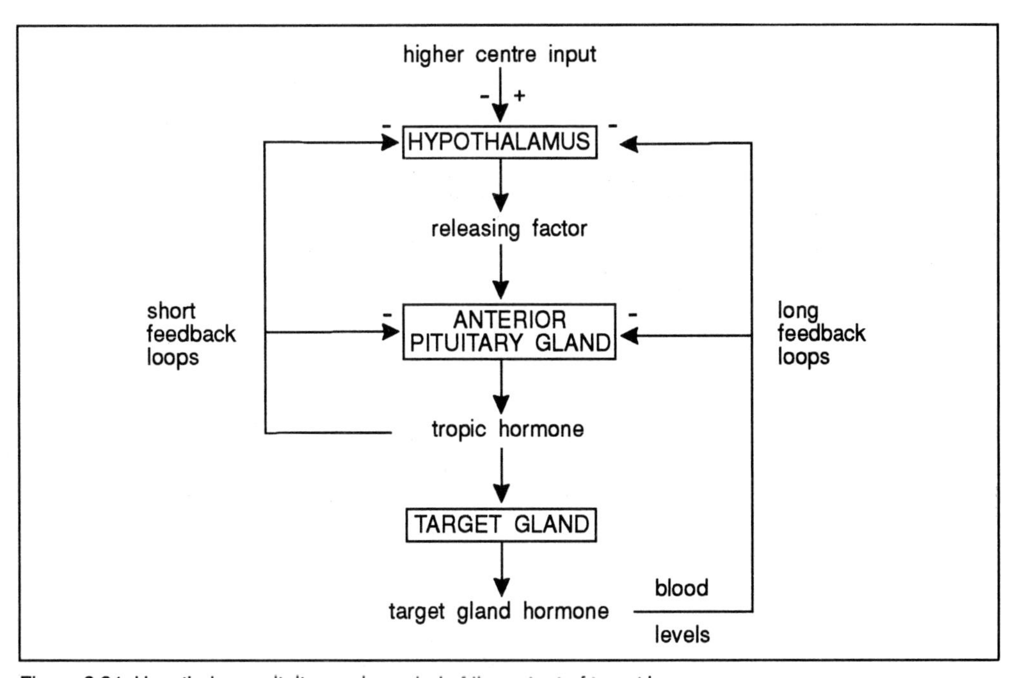

Figure 3.21 Hypothalamo-pituitary axis control of the output of target hormones.

The whole cascade of events is controlled by various negative feedback loops and from higher centre inputs to the hypothalamic neurosecretory cells (Figure 3.21). The main feedback is from the level of the target hormone in the blood to the pituitary cells producing the tropic hormone. If the level of the target hormone becomes too high, the output of the tropic hormone will be reduced, resulting in less stimulation of the target endocrine gland and, therefore, in a decrease in the output of the target hormone.

This feedback loop is also thought to extend to the hypothalamus to reduce the output of the releasing factor, which in turn further decreases the output of the tropic hormones. (Such a feedback control will also function for the inhibitory factors). These feedback effects of the target hormone onto the pituitary and hypothalamus are described as long loop negative feedback controls.

feedback effects of the target hormone onto the pituitary and hypothalamus are described as long loop negative feedback controls.

Some of the pituitary hormones do not have a specific target endocrine gland whose output can regulate their secretion. Examples are prolactin which affects mammary tissue and growth hormone which affects all tissues in the body. The negative feedback in these cases is via a short loop feedback in which the pituitary hormone itself, exerts a negative feedback onto the output of the hypothalamic neurosecretory cells. The hormones regulate the output of their own releasing factors. Such short loop feedbacks may also exist for the other pituitary hormones, thus adding a further refinement to the control of their output. In addition, recent evidence indicates that there may be some interaction between the various pituitary hormones and their feedback loops. It is thought that plasma levels of one hormone may influence the sensitivity of several pituitary cells to their hypothalamic releasing factors in addition to those regulating their own output.

SAQ 3.11

Draw a flow diagram to summarise the control of output of thyroid hormones from the hypothalamus and pituitary. The hypothalamic releasing factor is called thyroid releasing hormone and the pituitary tropic hormone is called thyroid stimulating hormone.

3.9 The various hormones of the body and their functions

The general properties of some human hormones are summarised in Table 3.3. The table includes their chemical classification, site of production, control and their major physiological function. There is an enormus amount of information in this table so read it carefully. Do not worry if you cannot remember all of these hormones immediately. As you learn more about physiology and the various functional systems such as nutrition and reproduction that operate in animals, you will meet with many of these hormones again and they will become more familiar to you.

Hormone	Site of production	Type	Control	Function
GHIF (somatostatin)	hypothalamus	peptide	GH levels	Control of GH levels
GHIH	hypothalamus	peptide	GH levels	Control of GH levels
TRF	hypothalamus	peptide	T_4 & T_3 levels	TSH output
GnRF	hypothalamus	peptide	gonadal function	LH/FSH output
CRH	hypothalamus	peptide	cortisol/ACTH levels	ACTH output
PIF/PRF	hypothalamus	peptide	foetal develop-ment	prolactin output
Growth hormone	anterior pituitary	peptide	GHIF/GHRF	growth
Prolactin	anterior pituitary	peptide	PIF levels	milk production
TSH	anterior pituitary	peptide	T_4 & T_3 levels	$\uparrow T_4$ & T_3 levels

Table 3.2 Summary of the general properties of the major hormones.

ACTH	anterior pituitary	peptide	cortisol/TSH levels	↑cortisol output
LH/FSH	anterior pituitary	peptide	relaxin/inhibin levels	ovulation follicle developments and sperm development
ADH	hypothalamus/ posterior pituitary	peptide	blood osmotic pressure	↑water reabsorption
Oxytocin	hypothalamus	peptide	suckling/stretch of birth canal	milk ejection/ uterine contraction
Cortisol	adrenal cortex	steroid	ACTH levels	metabolism/stress responses immune system
Aldosterone	adrenal cortex	steroid	renin/angiotensin	↑sodium reabsorption
Adrenaline	adrenal medulla	tyrosine based	sympathetic nervous system	stress/metabolism
Testosterone	testis	steroid	GnRH/LH/FSH	male 2° sexual characteristics/ spermatogenesis
Oestrogen/ Progesterone	ovary	steroid	GnRH/LH/FSH	female 2° sexual characteristics/ oogenesis
Throid hormones	thyroid	tyrosine based	TRF/TSH	metabolism/ growth maturation
Calcitonin	thyroid parafollicular cells	tyrosine based	↑calcium levels in blood	↓calcium levels in blood
Parathyroid hormone	parathyroid gland	peptide	↓calcium levels in blood	↑calcium levels in blood
Insulin	pancreas β-cells	peptide	↑blood glucose levels	↓blood glucose levels
Glucagon	pancreas α-cells	peptide	↓blood glucose levels	↑blood glucose levels
Somatostatin	pancreas γ-cells	peptide	insulin/glucagon levels	feedback control of insulin and glucagon output
Pancreatic polypeptide	pancreas	peptide	?	?
Gastrin/secretin GIP somatostatin	gastro-intestinal	peptide	digestive products pH	control of digestion/motility
Thymosin	thymus	peptide	?	immune system
Melatonin	pineal	peptide	?	body rhythms/ sexual maturity
ANF	cardiac atria	peptide	↑blood volume	↑sodium excretion
Inhibin	ovary/testis	peptide	↑FSH levels	FSH levels

Table 3.2 continued

Abbreviations
ACTH Adrenocorticotrophic hormone; ADH Antidiurectic hormone; ANF Atrial natiuretic factor; CRH Corticotrophin releasing hormone; FSH Follicle stimulating hormone; GH Growth hormone; GHIH Growth hormone inhibiting hormone; GHRH Growth hormone releasing hormone; GIP Gastric inhibitory peptide or glucose dependant insulinotropic hormone; GnRF Gonadotrophin releasing factor; LH Luteinising hormone; PIF Prolactin inhibitory factor; PRF Prolactin releasing factor; T4 Thyroxine; T3 Triiodothyronine; TRF Thyroid releasing factor; TSH Thyroid stimulating hormone. (Adapted from Vander, Sherman and Luciano (1990), Human Physiology: The Mechanisms of Body Function, Table 10.1, McGraw Hill, New York.

3.10 Endocrine disorders

In conclusion we will consider some of the pathological conditions associated with alterations in the blood levels of some of the hormones. These conditions often arise as a result of either:

- malfunction in the homeostatic control of the hormone output;
- an alteration in the sensitivity of the receptors;
- a pathological change in the tissue in which they are synthesised and from which they are released.

3.10.1 Hyposecretion

Where the blood levels of a hormone are found to be less than normal for a given situation, this is described as hyposecretion of the hormone. The lack of hormone is called primary hyposecretion if it results from the endocrine gland being unable to produce enough of the hormone. This may occur as a result of such factors as disease, toxic chemicals, lack of essential constituents or genetic absence of specific enzymes required in the synthesis of the hormone. One well known example of primary hyposecretion is Type I diabetes mellitus. In this disease there is a lack of β cells in the pancreas whose function is to produce insulin to control the levels of glucose in the blood. The lack of insulin results in the blood having high levels of glucose which is associated with the symptoms of polyuria (increased urine production), polydipsia and polyphagia (excessive drinking and eating respectively). If untreated, the high glucose levels eventually cause coma and death. The disease, which is also known as juvenile onset diabetes as it is commonly developed in people younger than twenty, is treated by intramuscular injections of insulin as required. The treatment of the disease is by the use of genetically engineered human insulin or by insulin extracted from animal tissue.

diabetes mellitus

The hyposecretion of a hormone can also result from lack of stimulation of the endocrine gland by its pituitary tropic hormone. This is called secondary hyposecretion. This may also be caused by a lack of hypothalamic releasing factor. In order to be able to make a diagnostic distinction between a primary and a secondary hyposecretory state, the physician needs to measure not only the blood levels of the target hormone, but also the levels of the relevant tropic hormone.

Π Hypothyroidism can be diagnosed by finding a low level of thyroid hormones in the blood. In order to decide whether this is of a primary or a secondary origin, the blood level of which other hormone needs to be measured? If it were a secondary hypothyroidism what changes in the levels of these two hormones would you expect?

A low level of TSH (thyroid stimulating hormone) accompanied by low levels of thyroid hormones indicates secondary hypothyroidism. If it were primary, the levels of TSH would be higher than normal because the negative feedback effect of the thyroid hormones onto the pituitary would be lessened.

3.10.2 Hypersecretion

Oversecretion of a hormone is termed hypersecretion. This also may be a primary disease (where the gland is secreting too much of the hormone) or a secondary disease, in which case there is overstimulation of the endocrine gland by excessive amounts of the tropic hormone. When thyroid hormones are oversecreted, the condition is known as thyrotoxicosis. The primary disease (Grave's disease) is due to a diseased thyroid gland and is characterised by goitre (enlarged gland), exopthalmos (protruding eyeballs) and general physical, mental and metabolic overexcitability. Treatment of the disease may involve surgical removal of all or part of the thyroid gland (thyroidectomy), administration of radioactive iodine (which selectively destroys cells into which it is taken up for synthesis of the hormone) or by treatment with antithyroid drugs which interfere with the synthesis of the hormone.

Grave's disease

Overproduction of cortisol from the adrenal cortex is called Cushing's disease. It is due to hyperplasia (overgrowth) of the adrenal cortex caused by overstimulation by ACTH (adrenocorticotropic hormone) from the anterior pituitary. It is a secondary hypersecretory state produced as a consequence of pituitary ACTH, rather than adrenal disease.

Cushing's disease

3.10.3 Hyposensitivity

Finally, we will consider a third way in which endocrine disease can be brought about and that is by a change in the number or sensitivity of the receptors which mediate the cell's response to a particular hormone. This dysfunction will occur even though normal blood levels of the hormone are present. In some instances, there is a genetic lack of receptors for a particular hormone eg certain men lack receptors for dihydrotestosterone so that the target cells do not respond to testosterone. This means that some of the male sexual characteristics do not develop despite normal amounts of the hormone being present.

In other diseases, the receptors can bind the hormone, but the ensuing cellular events do not occur due to the absence of specific enzymes used in the pathway.

A third type of hyporesponsiveness is seen when the sensitivity of receptors to a particular hormone is altered. This is thought to be the basis of Type II diabetes or maturity onset diabetes. This is diabetes which develops in late middle age, in overweight individuals. They have high levels of blood glucose but normal insulin levels. The receptors, therefore, would appear to have lost their sensitivity to the hormone. One possible explanation for this is, that the continual exposure of the receptors to high levels of insulin (produced because of the high sugar intake) results in down regulation of the receptors. Consequently the insulin produced eventually has progressively less effect on decreasing the blood glucose levels. Successful treatment of the disease is usually accomplished by a reduction in the caloric intake of these people.

Summary and objectives

In this chapter we have examined the endocrine system and drawn attention to its relationship with the nervous system. Our approach has been a mechanistic one in which we have described the strucure, synthesis, transport and metabolism of hormones and explained how they can mediate changes within cells as a consequence of binding to specific receptors. We have shown how the response of cells to exogenous hormones is mediated by a series of second messengers. We also described in general terms how the production of hormones is controlled and the medical consequences of endocrine dysfunction.

Now that you have completed this chapter you should be able to:

- list the difference between the endocrine and the nervous systems;
- describe the wide range of chemical messengers used to communicate between cells;
- know the location of the main endocrine tissues in the body;
- distinguish between the three classes of hormones and explain how they are synthesised;
- describe the transport in the blood and the metabolism of hormones;
- understand the different modes of action of hormones at the cellular level including receptors and their regulation, the different second messenger systems involving cAMP, cGMP, calcium ions, IP_3 and DAG, tyrosine kinases and nuclear receptor mechanisms for steroid and thyroid hormones;
- describe the different output patterns of hormone secretion;
- understand the feedback control of hormone output;
- understand the hypothalamic and pituitary control of hormone output.

Digestion and absorption

Digestion and absorption

4.1 Introduction

The functions of the digestive system and of the organs that make up the gastrointestinal tract and associated structures are to break down food both physically and chemically, and to absorb nutrients. This chapter will describe these functions in more detail and provide some answers to the questions; 'How is food absorbed?', 'How is digestion controlled?', 'What are the controls of food intake?' and 'Why are problems of the gastrointestinal tract so common?'

We begin by giving a brief overview of the components of the alimentary tract.

4.2 Components and functions of the digestive system

If you were requested to design a chemical factory which could extract products from a mixture of raw materials, you might start by making a list of properties the system should have, eg a crushing system.

Π Try to approach the digestive system from this angle and list the properties that would be desirable for the digestive system.

Among the more obvious are:

- a system for crushing and grinding;
- a means of moving material through the system;
- a means of solubilising the wanted products;
- a means of separating wanted from the unwanted products.

Other desirable properties include the ability to renew worn out parts, to control the processes according to demand for the products, and to protect the system against either invasion by micro-organisms or damage by toxins. The digestive system has all of these properties.

Figure 4.1 shows the main components of the digestive system. Food entering the mouth is chewed and moistened by saliva. This saliva contains mucus and a digestive enzyme, amylase, which is secreted by the salivary glands. A lipase is also produced by lingual glands beneath the tongue. On swallowing, the food passes down the oesophagus undergoing waves of contraction (peristalsis), pushing the food towards the stomach. The lower oesophageal sphincter relaxes to allow food to enter the stomach, but at other times remains constricted to prevent reflux of food back into the oesophagus.

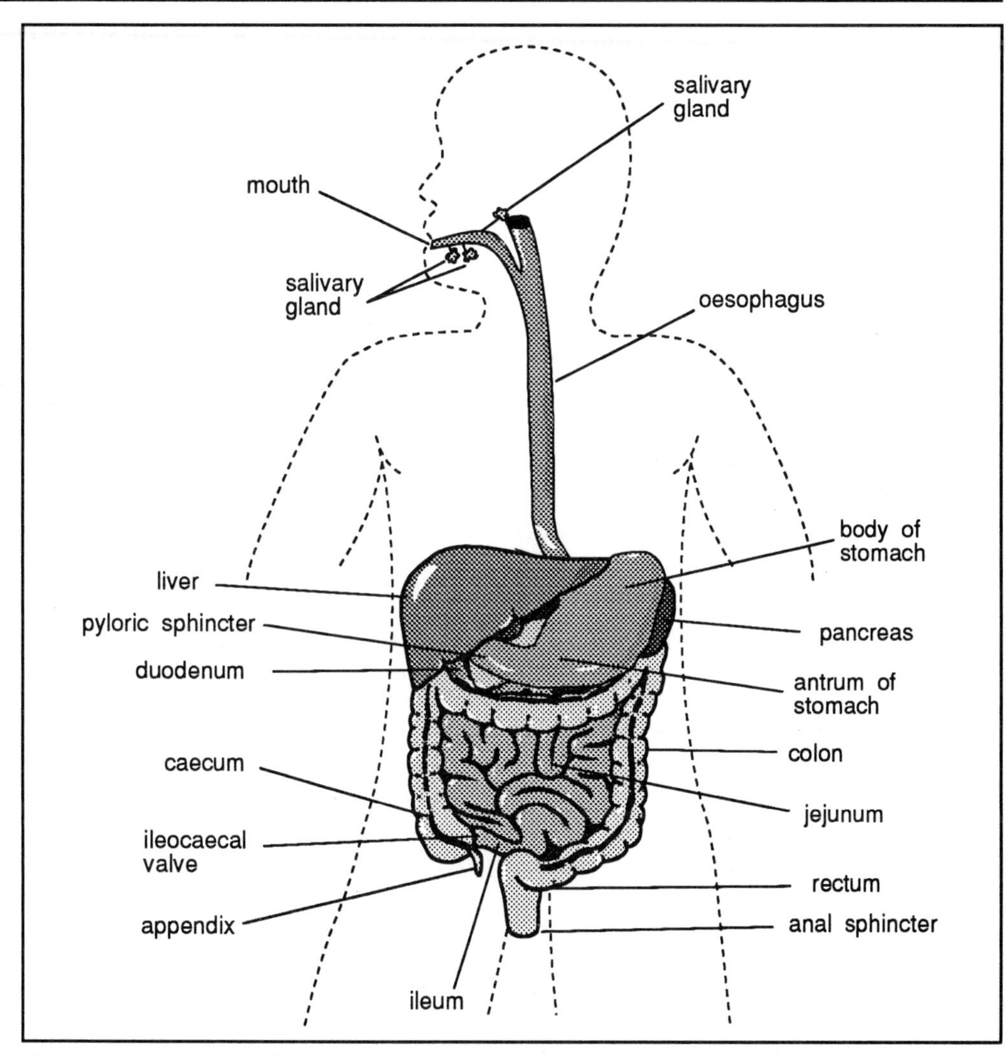

Figure 4.1 The gastrointestinal tract and associated organs.

stomach

chyme

duodenum

The stomach is a site for storing food and initiating protein digestion. The larger part of the stomach, the body, is chiefly concerned with storage of the food and is the major site for secretion of gastric juice. Gastric juice contains a proteolytic enzyme and HCl. Some mixing of the food occurs here, but most occurs as food moves into the antrum of the stomach resulting in the liquefaction of the food into the chyme. This chyme is allowed to enter the duodenum by periodic relaxation of the pyloric sphincter. It might be thought that all the food would move at the same rate, once it was mixed in the stomach, but actually fat moves relatively slowly, and also large particles are retained in the stomach until the digestion of other matter further down the gastrointestinal tract has been completed. Little chemical digestion of the food occurs in the stomach, except for some protein hydrolysis and a little carbohydrate breakdown. Although some fat-soluble drugs may be absorbed, of the nutrients only water and ethanol are taken up by the stomach wall. It is thus possible to survive without a stomach, but with less efficient digestion and absorption.

jejunum
ileum

Brunner's gland

caecum
colon

Absorption of nutrients begins in the duodenum and continues in the upper (jejunum) and lower (ileum) parts of the small intestine. Chyme entering the duodenum has a low (acid) pH because of the gastric juice it contains. The duodenum is protected from the acidic material by alkaline mucus secretions from the Brunner's gland at the beginning of the duodenum. In addition secretions from the gallbladder associated with the liver and pancreas (containing detergent-like bile and an alkaline mixture of digestive enzymes, respectively) also enter the duodenum. Further digestive enzymes are secreted by the various glands of the small intestine. Waves of muscle contractions (peristalsis) keep the chyme moving slowly onwards, but also there are segmenting contractions which occur frequently and these divide the intestine into small sections and mix and churn the chyme. Eventually, the chyme is moved to the ileocaecal valve where is stays until another meal stimulates the opening of the valve, so that it can pass into the caecum, and then through the colon. Water and electrolytes are progressively removed so that the liquid chyme becomes the semi-solid faeces. Onward movement is slow, and the contents of the colon are mixed by contractions of longitudinal and circular muscles. Some fermentation of undigested carbohydrate material may occur in the colon, leading to the formation of volatile fatty acids which are readily absorbed. Several vitamins are synthesised by bacterial action and are absorbed.

Faecal excretion occurs when faeces move into the rectum, stimulating the relaxation of the anal sphincters. The external anal sphincter is under voluntary control and when this is relaxed, voluntary contractions of the muscles of the abdominal wall and a descending diaphragm increase the intra-abdominal pressure to expel the faeces.

Π It may be helpful to you to redraw Figure 4.2 (perhaps in a more stylistic way) and write on the additional information given in the text.

The sort of drawing we envisage you could produce is:

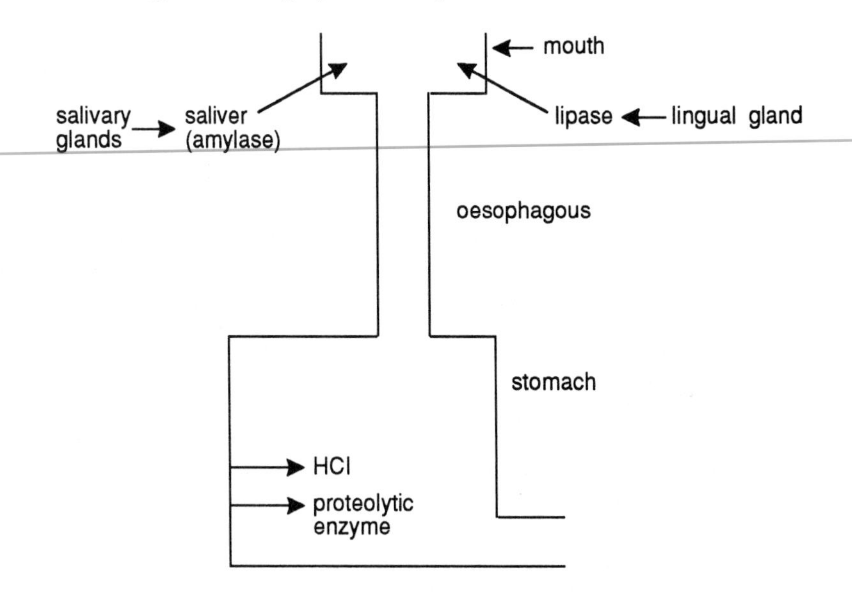

By working down the alimentary tract in this way, you will generate a better understanding of its general layout. Keep the figure you have drawn because you can add further information to it.

This text is not primarily concerned with anatomical details. Nevertheless, it is important for you to realise that the structure of the alimentary tract is greatly adapted to perform its various functions. For example you should be able to predict that the oesphagous is surrounded by rings of muscle tissues which can sequentially contract to drive the food towards the stomach. It will not however have adsorbative or secreting cells as no digestive enzymes are released in this part of the tract nor is there any adsorption of digested nutrients. To illustrate the adaption of the alimentary tract to fulfil its function we will explore in more detail the cellular structure of the small intestines in the next section.

4.2.1 The organisation of the alimentary tract at the cellular level - the small intestines

Π Given the kinds of functions that are mentioned in the previous section, what kinds of tissues would you expect to find forming the structure of the small intestine?

The functions include secretion, movement along the tube, and absorption. You might expect, therefore, to find glandular epithelium (secretion), muscular tissue (movement) and lining epithelium (absorption and lining the tube). In addition, there would have to be connective tissue (to hold other tissues together) and nervous tissue (for control of the system). Figure 4.2 is a schematic diagram of a section of the small intestine.

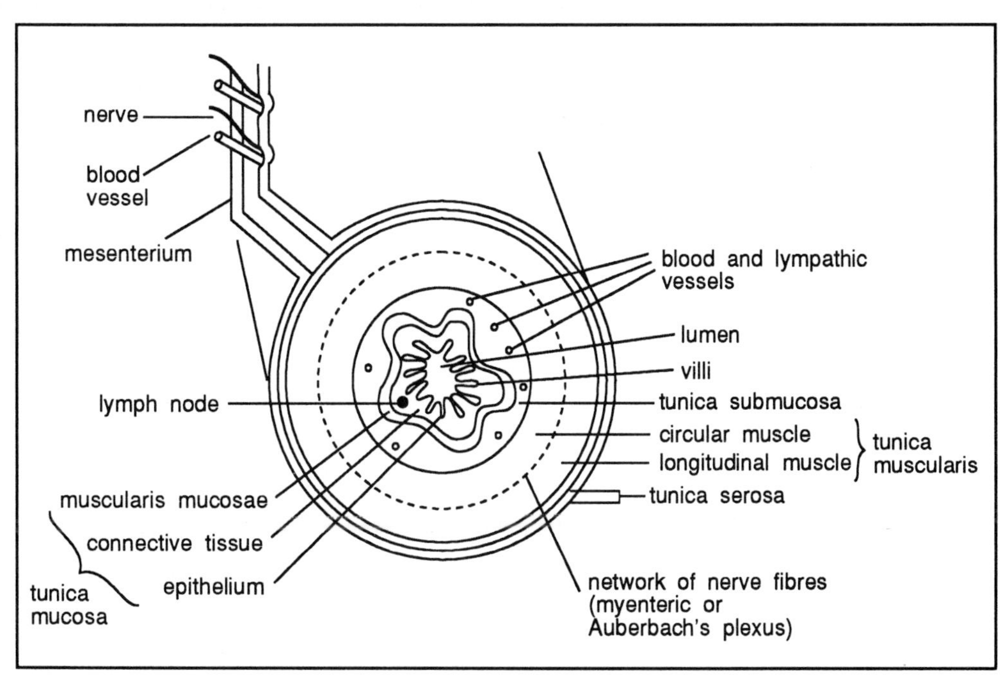

Figure 4.2 Section of small intestine.

A layer of connective tissue (tunica serosa) forms the outermost wall. Inside are two layers of muscle, the outer longitudinal and the inner circular. Between the two muscle layers lies a network of nerve fibres, the myenteric (or Auberbach's) plexus. Inside the circular muscle layer is the submucosal layer (tunica submocosa) containing glands, lymphatic vessels, blood capillaries and the submucosal (or Meissner's) nerve plexus.

Within this is a very thin layer of smooth muscle, the muscularis mucosae and finally the mucosa, a layer of epithelium covering connective tissue. The same general pattern is found throughout the gastrointestinal tract.

In the small intestine the structure of the epithelium is modified to increase the area available for absorption (Figure 4.3). The mucosal surface forms finger-like projections (villi) each of which (a villus) is covered with epithelial cells but contains blood and lymph vessels.

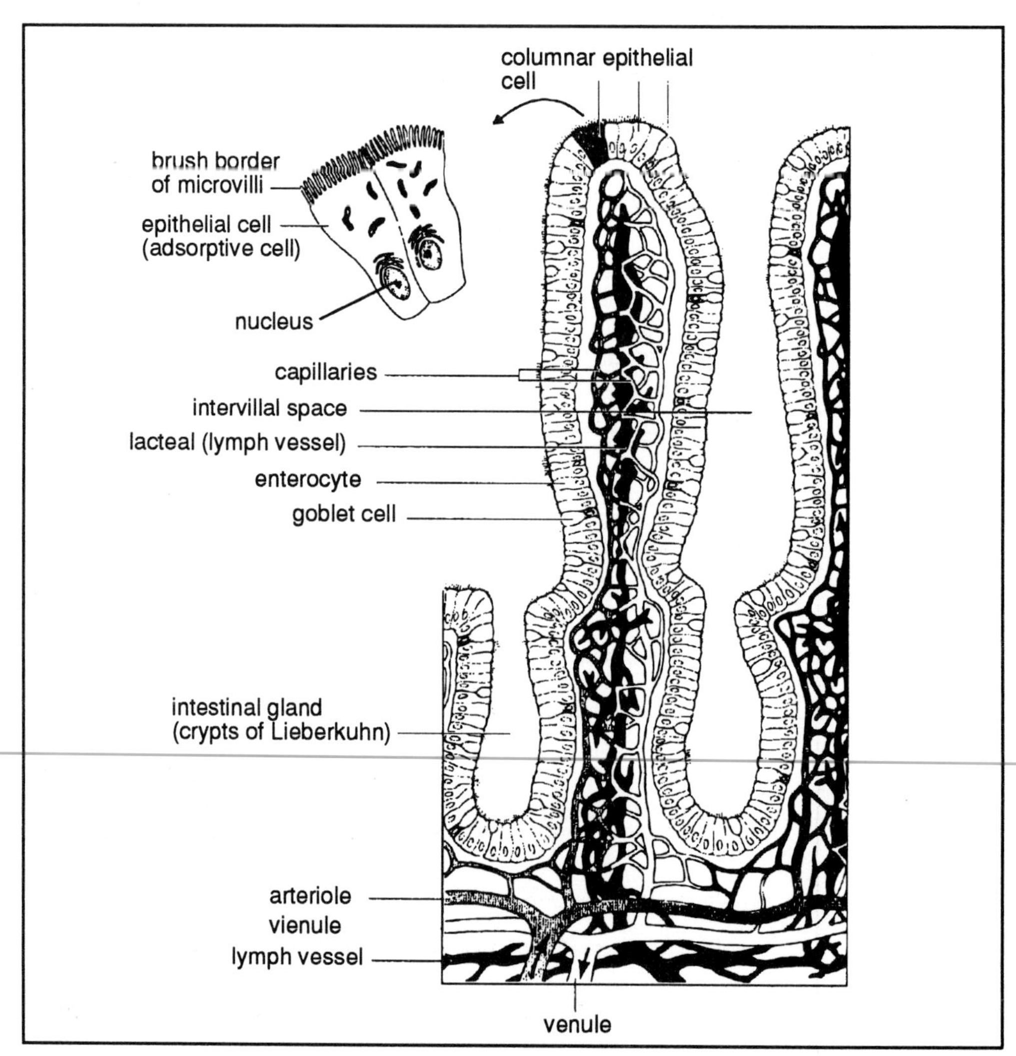

Figure 4.3 Diagram of an intestinal villus showing general appearance. Note particularly the arrangement of lymph and blood vessels and the structure of the epithelial cells.

Each absorptive epithelial cell carries on its lumenal surface in contact with the chyme numerous minute projections of the cell membrane called microvilli. These projections give the epithelial cells what is called their brush border.

goblet cells

Simple tubular glands (intestinal glands, the crypts of Lieberkuhn) open at the surface between the villi. In addition to absorptive cells there are also mucus producing cells in the villi known as goblet cells. The epithelial cells lining the small intestine have a short lifespan. New cells are formed by division of the cells in the crypts. These replacement cells then move up the villus shaft to reach the tip within about 48 hours, and are shortly afterwards shed into the lumen. Note the rich supply of blood and lymph vessel needed to carry blood and lymph into the villi where they can adsorb nutrients collected from the lumen by the epithelial cells. Note also the mucosal and basal membranes of the epithelial cells. We will be discussing their importance in adsorption in later sections.

SAQ 4.1

To examine your knowledge of the functions of the organs of the digestive system.

From the list below, complete the following table, writing in the numbers of the appropriate responses. Note, numbers may be used more than once, and more than one number may be used for each organ.

Organ	Function
Salivary gland	
Oesophagus	
Stomach	
Small intestine	
Large intestine	
Liver	
Pancreas	

1) To provide a passage between the mouth and stomach.
2) To store food temporarily.
3) To carry out a significant amount of chemical breakdown of food.
4) To carry out a significant amount of mechanical breakdown of food.
5) To secrete digestive enzymes.
6) To secrete fluids to change the pH of the chyme.
7) To secrete mucus to lubricate and protect the epithelium.
8) To secrete detergent like chemicals.
9) To absorb carbohydrates, lipids and amino acids.
10) To absorb water and electrolytes.

Now that we have an overview of the alimentary tract, let us turn our attention to the two main processes that take place in this tract; digestion and adsorption.

4.3 Digestion and digestive enzymes

Digestion, absorption and secretion are intimately linked processes. Digestion refers to the process by which food is broken down into molecules which can be taken up by the body (absorbed). For digestion to occur many cell products have to be secreted into the lumen of the gastrointestinal tract. These include hydrochloric acid, and diverse enzymes capable of hydrolysing numerous types of large molecules into smaller ones. A problem which arises is that these enzymes which break down food are potentially capable of breaking down the cells that line the gastrointestinal tract including the cells that produce them. All parts of the gastrointestinal tract, therefore, secrete a viscous coat of mucus which helps protect them.

saliva

The saliva, produced by the salivary glands, contains the enzyme amylase (ptyalin) which acts on starch. Some break down of starch occurs while the food is stored and mixed in the stomach, though after some time the amylase is inactivated by the low pH.

parietal cells

chief cells

In the stomach, the so called oxyntic (parietal) cells secrete HCl, causing the pH of the chyme to fall as low as 2. This acid itself probably catalyses some hydrolysis, but will also help to kill bacteria and to denature proteins. The chief (peptic) cells secrete pepsinogen an inactive precursor enzyme. This can be converted into its active proteolytic form, pepsin, by either contact with acid, or by other pepsin molecules. The chief cells are protected from autodigestion because they secrete the inactive form of the enzyme (zymogen).

pancreatic juice

The pancreatic juice secreted by the pancreas contains a wide range of enzymes. The protein degrading enzymes are generally secreted as zymogens (trypsinogen, chymotrypsinogen, and procarboxypeptidase). Trypsinogen is converted to the active trypsin either by an enterokinase secreted by intestinal cells, or by other trypsin molecules. Trypsin can also activate the zymogens of chymotrypsin and carboxypeptidase. Other enzymes include pancreatic lipase, colipase and phospholipase, (which hydrolyse fats), amylase (which hydrolyse starch) and nucleases which hydrolyse various nucleic acids.

The pancreatic secretion also contains bicarbonate to neutralise the acidity of the chyme.

intestinal

Intestinal juice secreted by cells in the small intestine contains little enzyme activity except for that contributed by cells that slough off the villi. The absorptive cells of the small intestine contain aminopeptidases dipeptidases, nucleotidase, nucleosidases and disaccharidases located in their brush borders and dipeptidases in their cytoplasm.

The roles of these enzymes are given in Table 4.1.

Enzyme	Substrate	Products
aminopeptidase	polypeptides	smaller peptides, amino acids
dipeptidase	dipeptides	amino acids
nucleotidases	nucleotides	nucleosides + H_3PO_4
nucleosidases	nucleosides	purines, pyrimidines pentoses
disaccharidases	sucrose	glucose + fructose
	maltose	glucose
	lactose	galactose + glucose

Table 4.1 Roles of digestive enzymes, their substrates and products.

Another important secretion which is released into the duodenum is bile. Bile is produced by the liver. It contains no enzymes but contains water, bile salts, phospholipids and cholesterol, plus various electrolytes, conjugated bilirubin (a haemoglobin derivative) and a variety of waste products. Bile salts are made by the oxidation of cholesterol to the bile acids chenodeoxycholate or cholate, which are then conjugated with glycine or taurine to form salts. Bile salts are powerful detergents and are able to promote the formation of mixed micelles of lipids effectively solubilising them so that they are prepared to be absorbed by the cells of the small intestine. Mixed micelles are a mixture of lipids and bile salts which are stable in aqueous suspensions. Stability is achieved by the hydrophilic parts of the bile salts extending towards the aqueous solvent. We can represent these structures in the following way.

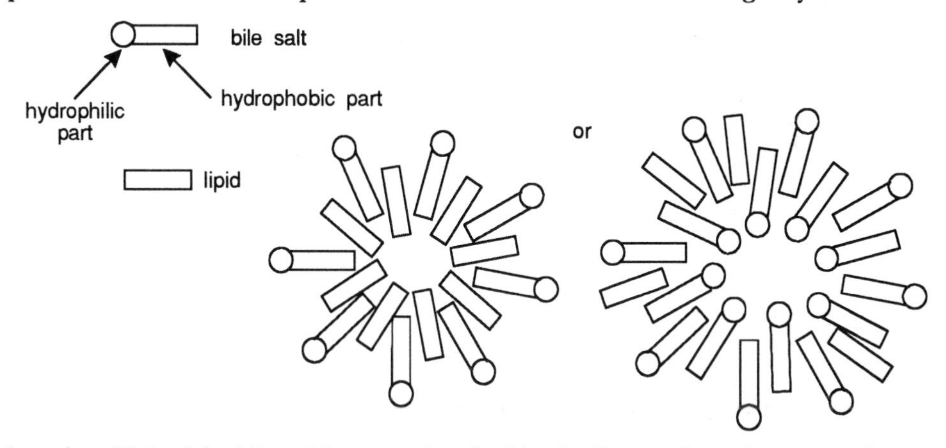

enterohepatic circulation

More than 90% of the bile acids are reabsorbed in the ileum where they enter the portal vein and are thus transported back to the liver where they are recycled. This recycling process is called the enterohepatic circulation.

Cholesterol is also recirculated to some extent. Normally about one half of that secreted with the bile is absorbed and passes into the lymphatic system together with triglyceride. The remainder is excreted in the faeces. This excretion together with the loss of some bile acid, is the only method for the removal of cholesterol from the body (approx 1g/24 hours).

Π Go back to the figure you drew in section 4.2 and add to it the various digestive enzymes and other agents that are secreted into the tract that we have described in this section. You will find this a good way to remember these various agents used in digestion.

SAQ 4.2

Complete the following table by filling in the blanks with the name of the appropriate enzyme, substrate or product.

Enzyme	Substrate	Product
1) Salivary amylase		branched oligosaccharides
2) Pepsin		polypeptides
3) Trypsin	Protein and polypeptides	
4) Chymotrypsin	Protein and polypeptides	
5) Carboxypeptidase	Protein and polypeptide	
6) Ribonuclease		nucleotides
7) Pancreatic amylase		maltose & dextrose
8) Lipase & colipase		monoglycerides, fatty acids & glycerol
9) Aminopeptidase	Polypeptides	
10) Dipeptidase	Dipeptides	
11) Sucrase	Sucrose	
12)	Lactose	galactose & glucose

4.4 Absorption

Figure 4.4 shows the sites of absorption of the major classes of nutrients. Absorption of the nutrients into the epithelial cells may be achieved by simple diffusion, facilitated diffusion or active transport.

Π From your knowledge of cell membranes, which types of molecules would you expect to be absorbed by each of these processes?

You probably recall that water passes freely across cell membranes according to the osmotic pressure. Small lipid molecules also cross membranes easily, and will tend to move from the side with the higher to that with the lower concentration. However, for many molecules we need to actively take up nutrients. Thus for many of the products of digestion, facilitated and active transport mechanisms are operative. Active transport allows us to take up nutrients against a concentration gradient but such a process costs energy. Facilitated diffusion is faster than free diffusion but does not consume energy and does not allow compounds to be taken up against a concentration gradient. (A detail description of the various transport mechanisms is given in the BIOTOL texts 'Biosynthesis and the Intergration of Cell Metabolism' and 'Infrastructure and Activities of Cells').

Examine Figure 4.4 carefully it contains a lot of information. So far we have described the overall arrangement of the alimentary tract and described in outline the processes of digestion and adsorption. Now let us follow the fate of particular nutrients with particular emphasis on the way in which they are absorbed.

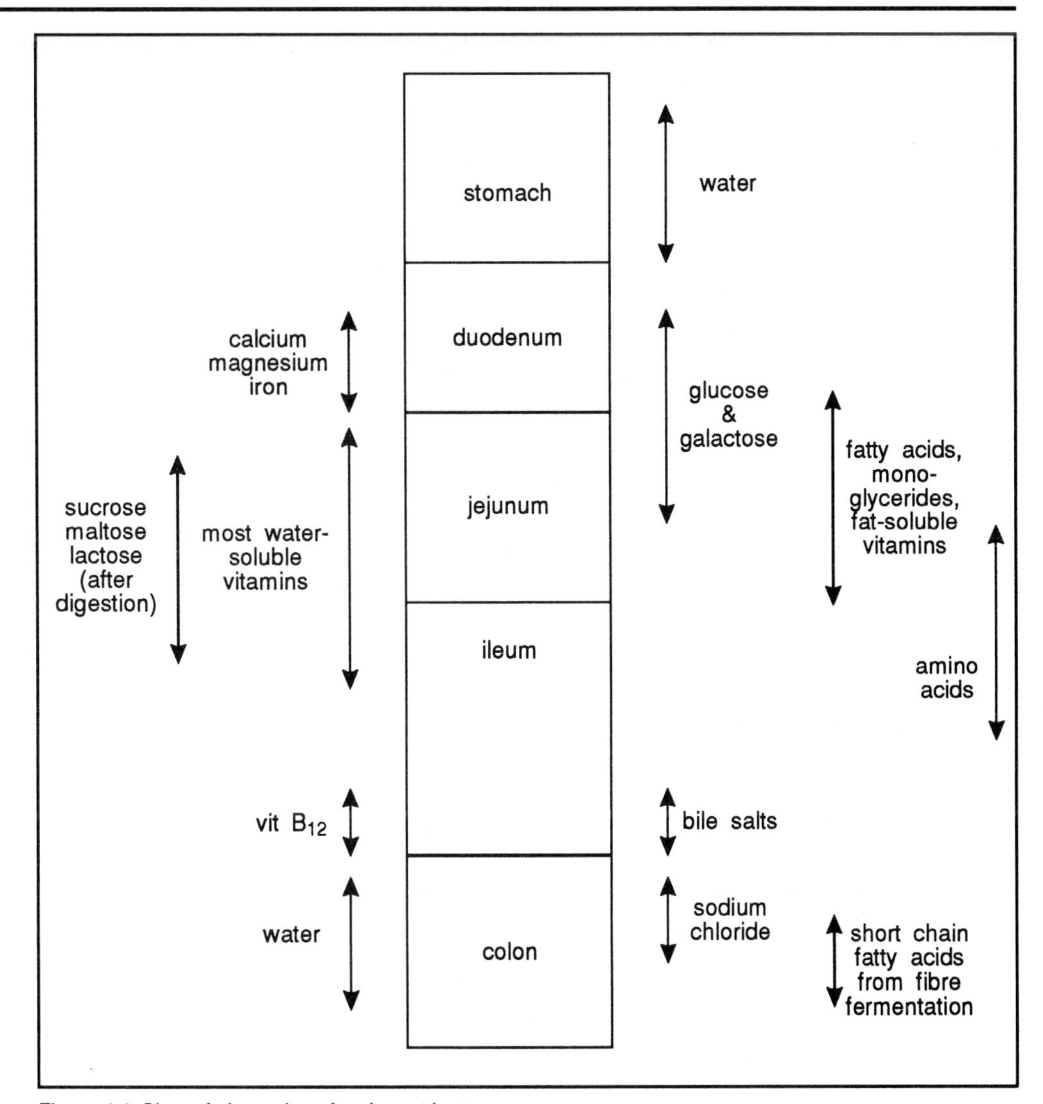

Figure 4.4 Sites of absorption of major nutrients

4.4.1 The digestion and absorption of carbohydrates

So far we have learnt that complex carbohydrates such as starch and glycogen are hydrolysed by a variety of enzymes to release short oligosaccharides, disaccharides and monosaccharides. This mainly takes place in the upper parts of the alimentary tract beginning with the introduction of amylase in the saliver. This mixture of partially hydrolysed polysaccharides are further hydrolysed by additional hydrolytic enzymes. We have represented this hydrolysis of carbohydrates simply by the conversion of starch to α dextrins (oligosaccharides), maltotroise and maltose in Figure 4.5. A detailed description of the biochemistry of this process is more appropriately dealt with in the context of biochemistry and is given in the BIOTOL text 'Principles of Cell Energetics'. Also shown in Figure 4.5 are the two disaccharides lactose and sucrose. You will note that these compounds are in the lumen of the alimentary tract. The key question is how do these compounds become transferred to the bloodstream.

fructose by facilitated diffusion

First you should realise that the brush borders produce significant levels of hydrolytic enzymes which further hydrolyse these carbohydrates present in the lumen to produce a series of monosaccharides (see Figure 4.5). These monosaccharides are all water soluble. These water soluble products are transported into and out of the intestinal cells by two main mechanisms, facilitated diffusion and by active transport. In facilitated transport, the molecule binds with a carrier which speeds up its transport across membranes. Fructose for example combines with such a carrier. Since fructose is readily metabolised its concentration in blood is low and the fructose is able to diffuse into the bloodstream.

In contrast, glucose is absorbed by a mechanism that indirectly requires active transport. A transport protein with two binding sites, one for glucose (or galactose) and one for Na^+ is loaded at the lumen side of the membrane and off loads the glucose and Na^+ into the epithelial cell. The Na^+ is pumped out of the cell via the basal side of the cell, using a Na^+-K^+-ATPase. This keeps the intracellular concentration of Na^+ low, and there is thus a steep concentration gradient for Na^+ between the lumen and the cell interior. Thus, Na^+ is moving down its concentration gradient at the same time that glucose is moving up its gradient. The glucose inside the cell is able to move across the basal membrane using facilitated diffusion, as the concentration in the cell is higher than that in the blood (Figure 4.5).

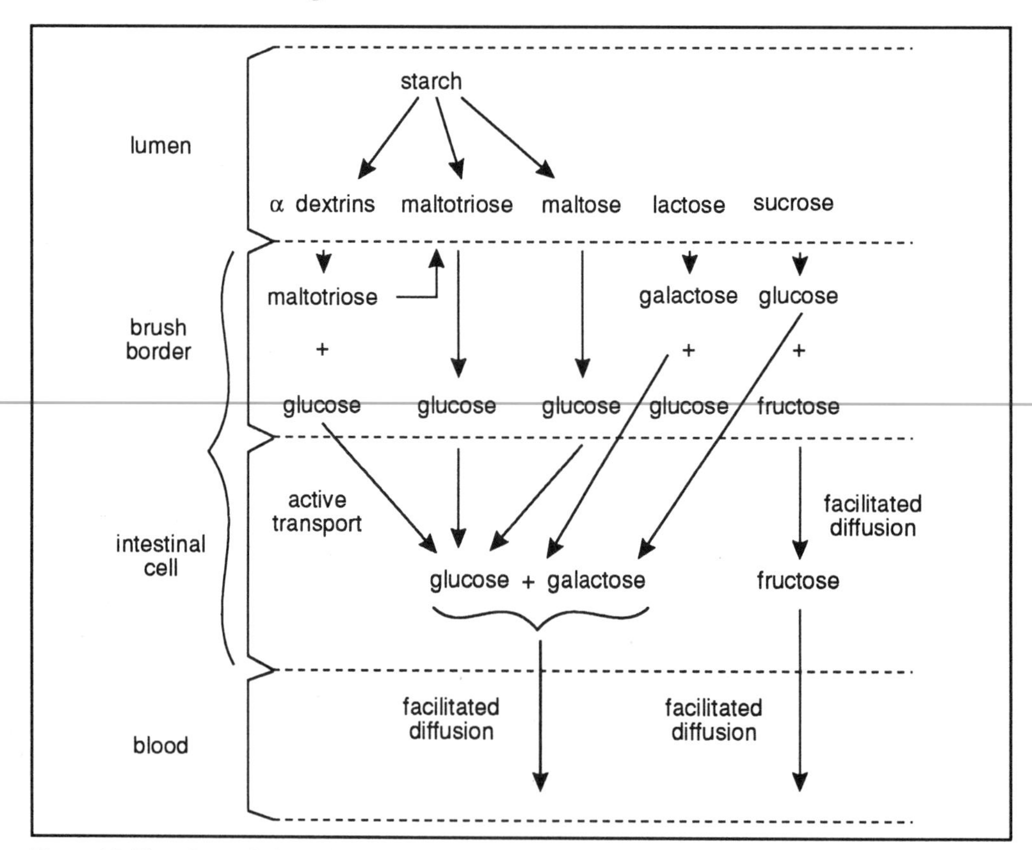

Figure 4.5 Digestion and absorption of carbohydrates (see text for further details).

4.4.2 The digestion and absorption of proteins and amino acids

endocytosis
exocytosis

Some proteins or polypeptides appear to be able to pass into the bloodstream without being completely hydrolysed. This is especially true in infants. It is thought that this occurs mainly by endocytous (pinotosis) in which the cell membrane invaginates to form vesicles entrapping the peptides. These are subsequently ejected by exocytosis on the opposite membrane. We can represent this process as:

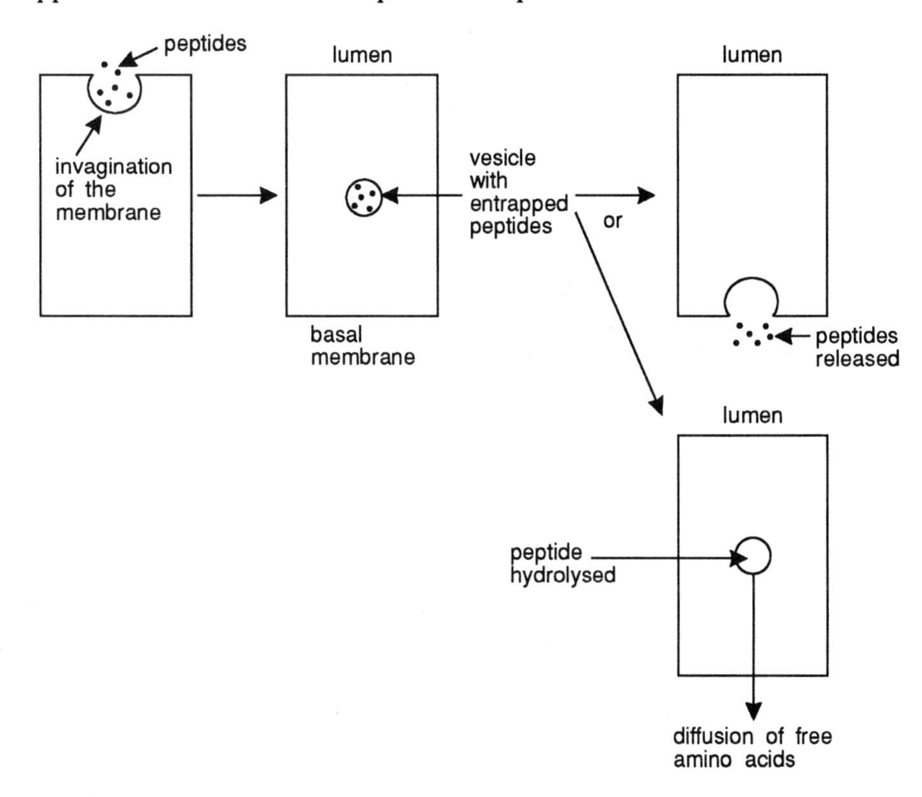

Note that the processes of endo- and exocytosis are described in the BIOTOL text 'Infrastructure and Activities of Cells'.

Let us however focus onto more mature animals. We have learnt that the proteins are hydrolysed partially in the stomach by pepsin and possibly by HCl catalysed hydrolysis. The bulk of the hydrolysis, however, takes place in the small intestines under the influence of pancreas-derived enzymes (trypsin, chymotrypsin and carboxypeptidase). The transfer of the products of this hydrolysis (peptides and amino acids) are transported into the brush borders. In the brush borders, the aminopeptidases and dipeptidases may further hydrolyse the peptides. The amino acids in the brush borders are then transported into the intestinal cells by facilitated diffusion using a range of carriers. These carriers have overlapping specificities. The following carrier types have been demonstrated.

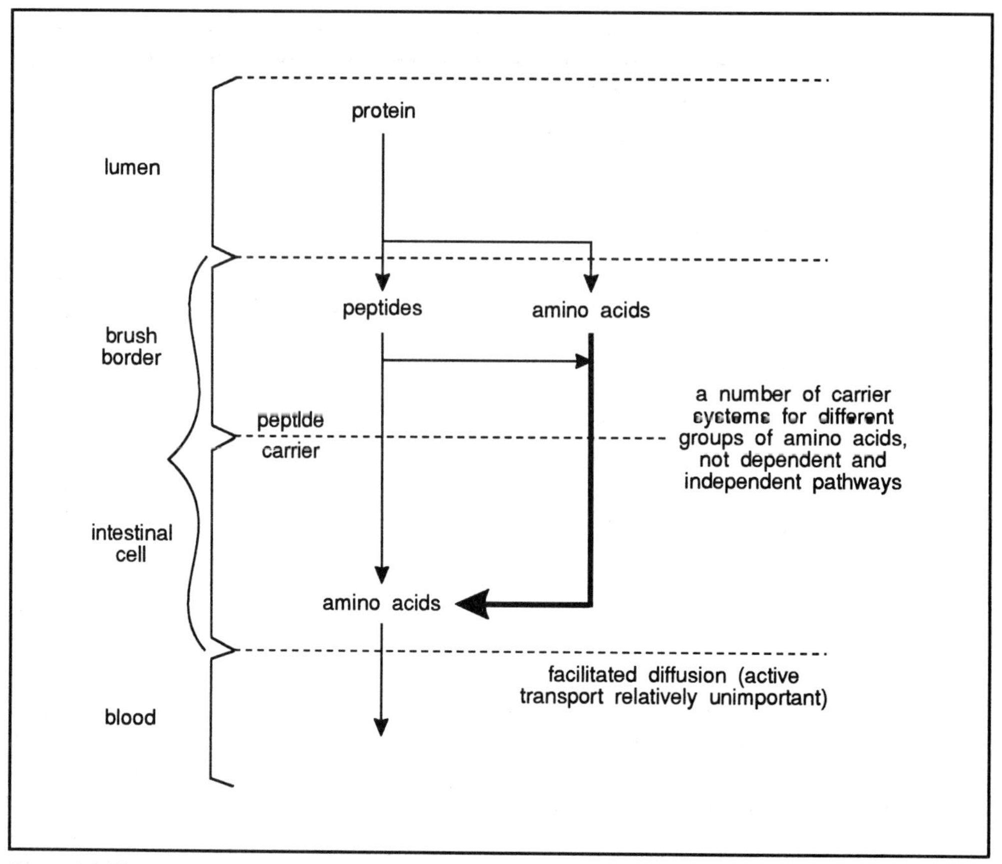

Figure 4.6 Digestion and absorption of proteins and amino acids (see text for further details).

- for neutral amino acids with short (eg alanine) or polar side chains (serine, threonine);

- for neutral amino acids with aromatic or hydrophobic side chains (phenylalanine, tyrosine, methionine, valine, leucine, isoleucine);

- for the amino acids proline, hydroxyproline;

- for the β-amino acids (β alanine and taurine);

- for basic amino acids and cystine (lysine, arginine, cystine);

- for acidic amino acids (aspartate, glutamate).

The peptides absorbed by the epithelial cells (by a peptide transporter) are almost all hydrolysed to amino acids within the cytoplasm before they leave the cell. Exceptions to this are peptides containing proline, hydroxyproline, or β-alanine which are transported into the blood.

Π It would be a good form of revision to redraw Figures 4.5 and 4.6 for yourself but this time put on the digestive enzymes that are found in the lumen that are involved in the hydrolysis of carbohydrates and proteins including their sites of production. In this way you will provide yourself with a comprehensive picture of how each is hydrolysed and adsorbed.

4.4.3 The digestion and absorption of lipids

Π Can you suggest why the digestion and absorption of lipids is likely to be different from that of carbohydrates and proteins?

emulsion formation

lipase action

Lipids are defined as having good solubility in organic solvents, and consequently are only sparingly soluble or insoluble in water. This presents a problem for digestion because the lipid molecules are not readily accessible to the digestive enzymes which are in the aqueous phase while both the lipids and their constituent molecules tend to form large complexes. This problem is overcome by emulsion formation, ie increasing the contact area between the aqueous and lipid phases and by solubilising the products of digestion with detergents/surfactants. The enzyme lipase adsorbs to the lipid/water interface and hydrolyses some triglyceride to fatty acids and monoacylglycerol. These lipid degradation products and also phosphilipids which are present in food act as surfactants and tend to break large drops of triglyceride into smaller droplets (forming an emulsion) making the triglyceride more available for further hydrolysis. The pancreatic lipase requires the presence of a small protein called colipase for full activity. Colipase binds to both the lipid/water interface and to lipase, thus holding the enzyme in its most effective environment.

bile salts

emulsion formation

apolipoproteins chylomicrons

Bile acids are also detergents which readily form mixed micelles with other lipids such as phospholipids and fatty acids. These micelles are spherical shaped aggregates with the phospholipids and fatty acids forming a bilayer and the bile acids forming the outer and inner surface of the sphere and making them hydrophilic. These micelles are much smaller than lipid droplets, and the components are in equilibrium with molecules in solution. Water insoluble lipids such as cholesterol, which occupy positions in the lipid core of the micelles, can therefore, be solubilised. During digestion of triglyceride, the fatty acids and monoglycerides released by the action of lipase at the surface of the lipid emulsion droplets equilibrate with those in aqueous solution, and the latter equilibrate with those in bile acid micelles. The micelles diffuse to the surface of the intestinal epithelial cells where the fatty acids, monoglycerides and some cholesterol are absorbed by diffusion across the cell membrane. Within the cell, long chain fatty acids (more than 12 carbon atoms) and monoglyceride are recombined to form triglyceride in the endoplasmic reticulum. Small droplets of these reformed lipids, together with cholesterol, are in turn coated with phospholipids and special proteins called apolipoproteins, and packaged into vesicles which fuse with the cell membrane. The fusion of these vesicles with the cell membrane releases the droplets into the intercellular space (exocytosis). From here, the droplets, called chylomicrons, move into the lacteal and then into the intestinal lymph vessels (see Figure 4.3) and thence along the thoracic duct to enter the blood in the left subclavian vein (at the base of the neck). In contrast, medium chain fatty acids (6-10 carbon atoms) pass through the intestinal epithelial cell directly into the blood (Figure 4.7).

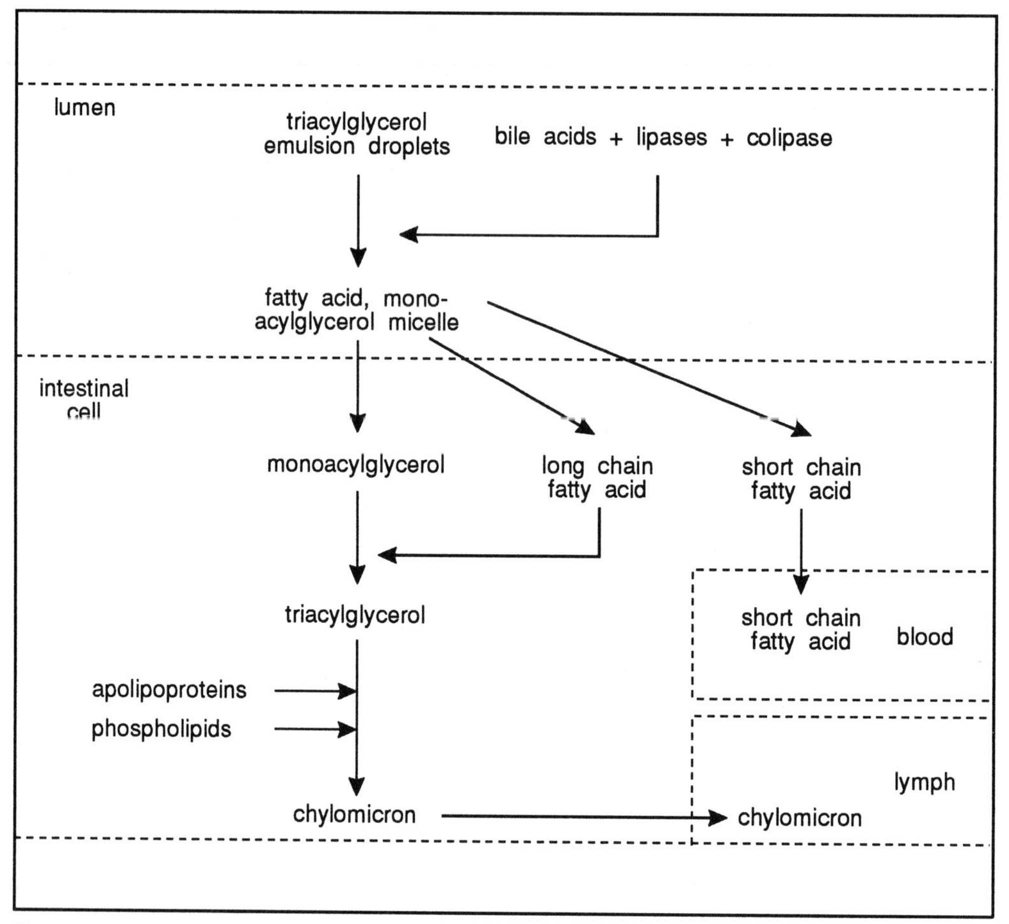

Figure 4.7 Digestion and absorption of lipids (see text for description).

4.4.4 The mechanisms of absorption of water, electrolytes and vitamins.

Π Examine Table 4.2. What general statement can you make about the way in which each of the groups as listed in the table is absorbed?

The table shows that generally, electrolytes are absorbed by active transport mechanisms as are most of the water soluble vitamins (at least when they are present at low concentrations). By contrast most of the fat soluble vitamins are absorbed by diffusion. It is thought that most fat soluble vitamins require the presence of both fat and bile salts for their efficient uptake.

Nutrient	Mechanism of absorption	Comments
Water	Diffusion	Moves in either direction according to osmotic pressure
Electrolytes		
Na^{+}	Active transport & diffusion	
Cl^{-}	Active transport & diffusion	
Ca^{2+}	Active transport	Rate regulated by parathormone and vitamin D
Fe^{2+}	Active transport	
K^{+}	Active transport	
Mg^{2+}	Active transport	
Water soluble vitamins		
B_1 (thiamin)	Active transport & diffusion	
B_2 (riboflavin)	Carrier mediated, some active	
B_3 (niacin)	Active transport & diffusion	
Pantothenic acid	Diffusion?	
Biotin	Active transport	
B_6 (pyridoxine etc)	Diffusion	
Folic acid	Active transport	
B_{12}	Involves receptor, mechanism not known	Require a glycoprotein, intrinsic factor, secreted by gastric parietal cells
C	Active transport & diffusion	
Fat soluble vitamins		
A	Facilitated diffusion	Require fat and bile salts for absorption
D	Diffusion	Require fat and bile salts for absorption
E	Diffusion	Require fat and bile salts for absorption
K	Active transport & diffusion	Require fat and bile salts for absorption

Table 4.2 Mechanisms of absorption of water, electrolytes and vitamins.

SAQ 4.3

To test your understanding of the processes of digestion and absorption, indicate which of the following statements are correct.

1) Absorption refers to the process of food molecules binding to the membranes of the cells lining the intestine.
2) Digestion is the process of breakdown of food into small molecules which can subsequently be absorbed.
3) Carbohydrate that is digested in the small intestine is hydrolysed ultimately to monosaccharides (simple sugars) which are absorbed by active transport.
4) Proteins are digested to amino acids in the lumen of the small intestine and absorbed by active transport.
5) Lipids are absorbed by simple diffusion.
6) Bile acids are necessary for the effective absorption of fats and fat soluble vitamins.
7) Most electrolytes are absorbed by facilitated diffusion.
8) Most water soluble vitamins are absorbed by active transport.
9) The enterohepatic circulation refers to the movement of bile acids from their formation in the liver, storage in the gall bladder, release into the small intestine and excretion with the faeces.

SAQ 4.4

Complete the sentences below by writing an appropriate word in the blanks using the words provided. Note not all words may be relevant and some may be used more than once. The sentences all relate to the absorption of nutrients by an intestinal epithelial cell.

The mucosal surface of the small intestine is modified to form numerous 1 [ ], each of which is covered with epithelial cells, but contain 2 [ ] and 3 [ ] vessels. Each absorptive epithelial cell has on the mucosal surface numerous 4 [ ]. These give the cell their 5 [ ] [ ], and it is at this site that many enzymes for the final stages of digestion are located. These modifications of the surface serve to increase the 6 [ ] [ ] availa ble for absorption. The absorptive cells of the small intestine are responsible for the uptake of most of the nutrients, with the exception of some of the water, electrolytes, and products of 7 [ ] fermentation in the large intestine. Absorption of nutrients usually occurs by one of three mechanisms: simple diffusion, facilitated diffusion, or active transport. Active transport of nutrients such as glucose and amino acids into the absorptive cells relies on the different distribution of pumps between the 8 [ ] and 9 [ ] membranes of the cells. Active transport of glucose is not directly linked to ATP break down, but requires the pumping of 10 [ ]. The intracellular concentration of this is kept low by a 11 [ ] pump on the 12 [ ] membrane, allowing 13 [ ] and glucose to enter the cell together via a carrier on the 14 [ ] membrane. Thus Na^+ enters the cell 15 [ ] its concentration gradient.

Word list: basal; basolateral; musocal; fibre; surface area; down; up; lymph; blood; villi; microvilli; brush border; Na^+; glucose; artery.

4.5 Regulation of digestion and absorption

Feeding, digestion and absorption are obviously complex processes that need to be regulated. For example, it would be wasteful to secrete enzymes in the absence of food.

Π Make a list of five processes involved with feeding, digestion and adsorption that require a control mechanism?

A useful way of approaching this is to consider what and when each part of the system performs its specific functions. For example, the mouth, ingests the food. When? For most people, the answer is several times each day, rather than continuously. This illustrates the fact that there must be some mechanism which determines the need for food (hunger) and stimulates the desire to find food and start eating. A mechanism must also provide a signal to stop eating when sufficient food has been ingested. In humans, the regulation of food intake is extremely complex, because people do not eat simply to relieve their hunger, however, this topic will only be briefly discussed in this chapter.

hunger

Some other processes requiring regulation are:

- secretion by the exocrine glands of the gastrointestinal tract;
- emptying of the stomach;
- motility of the stomach (rate of contractions);
- motility of the intestine.

The regulation of these processes is brought about by a combination of nervous and endocrine control. Table 4.3 lists the main hormones thought to be involved in the control of gastrointestinal secretions.

Hormone	Site of production	Action
Gastrin	Pyloric gland in the stomach	Stimulates secretion of gastric, pancreatic juice, and bile.
Secretin	Intestinal mucosa	Inhibits secretion of gastric juice, stimulates secretion of pancreatic juice, bile and intestinal juice.
Cholecystokinin (CCK) (pancreozymin)	Intestinal mucosa	Stimulates secretion of pancreatic juice and intestinal juice, increases delivery of bile from gall bladder.
Gastric inhibiting peptide (GIP)	Intestinal mucosa	Inhibits secretion of gastric juice.

Table 4.3 Hormones involved in the control of gastrointestinal secretions.

Π In the following sections we will be discussing the control mechanisms that operate various processes. As you read the sections involving the control of secretion (eg saliva, gastric), you may find it helpful to construct yourself summary diagrams like this one we have produced to summarise the regulation of acid secretion in the stomach (see Figure 4.9).

neurocrines

In addition to these hormones, other peptides have been found to be released from the nerves of the gut, and are termed neurocrines. These include vasoactive intestinal peptide (VIP), gastric releasing peptide (GRP, or bombesin), GIP (gastric inhibiting peptide), neurotensin and substance P. The last two have also been found in the brain. The nervous system is intimately involved in the control of secretion, both directly and via the release of these neurocrines.

secretagogues

All of the substances that bring about secretion are termed secretagogues, and they interact with receptors on the cell surfaces of the exocrine cells. Different exocrine cells usually possess different sets of receptors. When the secretagogues bind to the receptors, a chain of events is started that, in the case of cells secreting enzymes, ends with the fusion of intracellular membrane bound granules with the plasma membrane and the release of the granular material (the enzymes) into the extracellular space. Proteolytic enzymes and phospholipase A are produced and stored as inactive precursors (zymogens or proenzymes) and so the storage vesicles are termed zymogen granules.

nerve:gastrointe stinal tract interactions

As mentioned earlier there exist two separate networks of nerve fibres in the gastrointestinal tract: the myenteric nerve plexus and the submucosal plexus. These plexuses are connected and contain nerve cells with processes that come from receptors in the wall of the gut or the mucosa. The musocal receptors include those sensitive to stretch (mechanoreceptors) and also ones sensitive to the composition of the chyme (chemoreceptors). The nerve cells innervate the hormone secreting cells of the gut and the muscle layers. The intestine is also innervated by the both types of autonomic nervous system. The parasympathetic nerve activity generally increases muscle activity whilst sympathetic nerve activity generally decrease it. The contraction of the various sphincters is under sympathetic nerve control.

4.5.1 Saliva secretion

reflex and learnt reactions

The secretion of saliva appears to be entirely under nervous control and is a reflex action. Impulses from the sense organs of smell (olefactory) and taste (eg taste buds) are conducted to the salivary nucleus in the brain stem from which parasympathetic nerves carry the returning impulses to the salivary glands and stimulate secretion. Sympathetic nerves probably inhibit secretion. Other sensory stimuli associated with food can also bring about secretion. These involve learned behaviour. (You may remember the famous experiment by Pavlov who conditioned dogs to salivate at the sound of a bell which they associated with the arrival of food).

4.5.2 Gastric secretion

influence of taste, sight and smell

role of vagus nerve and digestion products

The initial part of gastric secretion also occurs by a reflex, started by the taste of food (or through conditioning, the sight or smell of food). The vagus nerve stimulates both the acid and pepsinogen producing cells. As some protein digestion occurs the products together with the vagus nerve, stimulate the secretion of gastrin from the pyloric glands of the stomach. The gastrin released is carried via the blood to the oxyntic (parietal) cells and this stimulates further acid production. As the chyme enters the duodenum, the distension may also cause further stimulation of gastric juice secretion but the mechanism is not understood. When the chyme in the duodenum has a pH less than 5 or contains fatty acids, gastric secretion is inhibited by the action of secretin and possibly by a hormone called enterogastrone.

4.5.3 Pancreatic and intestinal secretion

control of biocarbonate and enzyme secretion

Secretin and cholecystokinin (CCK) both stimulate the production of pancreatic juice, the former controlling water and bicarbonate secretion, and the latter controlling enzyme secretions. Enzyme secretion is also stimulated by nervous impulses from the vagus nerves. CCK also causes the gallbladder to contract, releasing bile into the common bile duct and then into the small intestine. Both CCK and secretin stimulate secretion of intestinal juice, and their release is stimulated by the presence of chyme in the duodenum, probably by acid and partially digested protein and fat.

Overall, therefore, you can see that food is the stimulus for saliva secretion and some gastric secretion and the digestion products are in turn stimulus for further gastric secretion and secretion from both the pancreas and intestine. The pH in the lumen of the intestine is auto-regulated because acidity stimulates the secretion of more bicarbonate until neutralisation occurs.

neurocrine influences

The exact roles of the neurocrines are not known, although VIP (vasoactive intestinal peptide) is known to dilate blood vessels and increase intestinal secretion. GIP (gastric inhibiting peptide) release is stimulated by glucose and fat in the duodenum and it inhibits both gastric secretion and gastric motility and stimulates insulin secretion. Neurotensin release is stimulated by fatty acids and it inhibits gastrointestinal motility and increases ileal blood flow. Substance P increases the motility of the small intestine. The interrelationships between GRP (gastric releasing peptide), GIP, the vagus nerve, somatostatin, gastrin, insulin and acid production are shown in Figure 4.8.

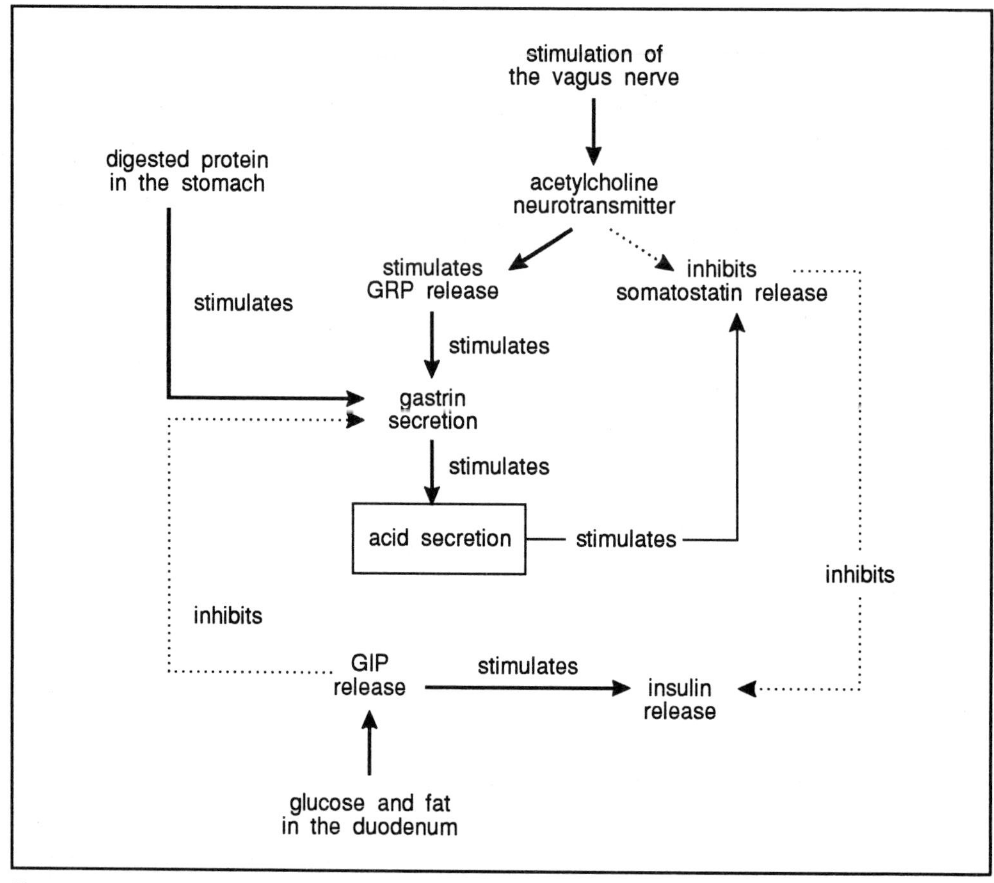

Figure 4.8 Regulation of acid secretion in the stomach.

4.5.4 Regulation of gastric emptying

In general the factors that are involved in the regulation of gastric emptying are the same factors that control gastric and pancreatic secretion. Although it might appear that the effects of the regulatory factors just cancel each other out, it is necessary to realise that chyme is moving from the stomach, through the duodenum, into the remainder of the small intestine (Figure 4.9). Thus, when some chyme is forced into the duodenum, inhibitory signals will slow further gastric emptying until the chyme has moved further along the gastrointestinal tract.

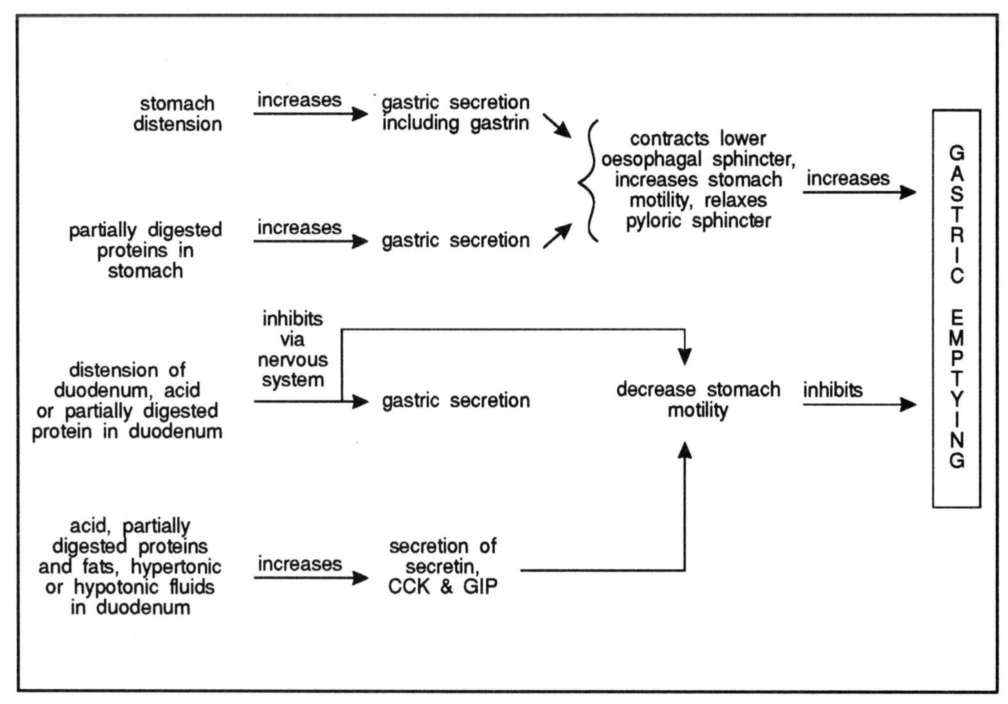

Figure 4.9 Regulation of gastric emptying.

4.5.5 Regulation of food intake

role of hypothalamus

VMN

Food intake is controlled both centrally and peripherally. The factors which influence food intake are integrated by the brain and, in particular by the hypothalamus. Within the hypothalamus is an area known as the ventromedial nucleus (VMN). This area is considered to be a satiety centre. By this we mean that stimulation of this area creates the feeling of being well fed (sated). Stimulation of the VMN leads to a sympathetic nervous system output and subsequent decrease in feeding.

LH

Also present in the hypothalamus is a centre which stimulate feeding. This area in the lateral hypothalamus (LH) is connected to the parasympathetic nervous system. Stimulation of this area leads to a parasympathetic nervous system output and increased feeding.

DMN

Signals from peripheral sensory nerves appear to be received in the hypothalamus in the so called dorsomedial nucleus (DMN). This interacts with the VMN and LH. In this way feeding is related to signals received from the peripheral sensory system. Thus we can see the chain of control as:

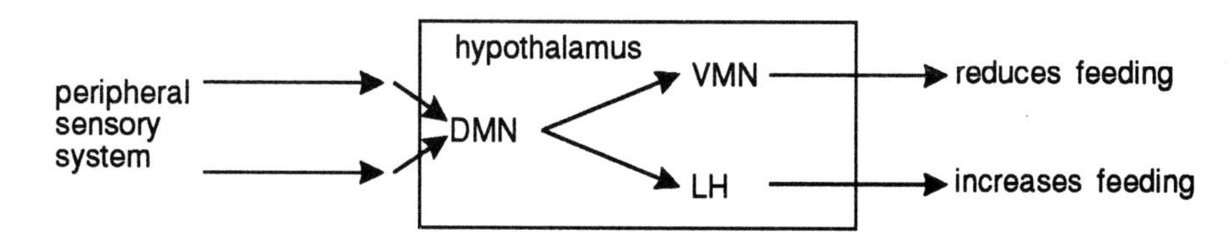

This is however only part of the story. We would like to have greater knowledge of how the signals received in the DMN are intepreted by the VMN and the LH and how stimulation of these areas actually controls feeding rates. We know for example that the paraventricular nucleus (PVN) adjacent to the VMN in the hypothalamus sends signals to the sympathetic and parasympathetic nerves and to the pituitary gland which itself secretes hormones which influence food intake.

PVN

Interestingly, some peptide hormones found in the gastrointestinal tract, including GRP, CCK and neurotensin appear to be involved in the brain centres which influence feeding behaviour. The cerebral cortex receives and integrates external influences on feeding behaviour. The vagus nerve appears to be involved in sending information from the gastrointestinal tract to the brain.

Before continuing with the final section which discusses disorders of the gastrointestinal tract, try the following SAQ to test your understanding of the regulatory mechanisms.

SAQ 4.5

Indicate which of the following statements are correct.

1) A secretagogue is a substance that is secreted by a cell in response to a stimulus.

2) A zymogen granule is a granule of enzyme precursor molecules, which can be released from a cell.

3) Saliva secretion is entirely under nervous system control and is stimulated by sympathetic nerves.

4) Gastric acid secretion is stimulated by the vagus nerve and by the hormone gastrin which is released into the stomach from the pyloric glands.

5) Gastric acid secretion is inhibited by the action of secretin.

6) A signal for the inhibition of gastric secretion is the presence of a high pH in the duodenum.

7) CCK stimulates the secretion of pancreatic enzymes, and also causes the gall bladder to contract.

8) Gastrin, secretin, CCK and GIP all increase gastric emptying in order to speed up the delivery of nutrients to the small intestine.

9) Several neuropeptides are found in the gut and are probably involved in the regulation of secretion and gut motility.

10) The control of feeding is centred in the hypothalamus, which receives signals from sense organs and other parts of the brain, but not from the gut.

4.6 Disorders of the gastrointestinal tract

Not surprisingly with such a complex system, and particularly one exposed to the environment, a number of things can go wrong. The following sections relate the symptoms of a number of diseases to their causes.

4.6.1 Peptic ulcer - a secretory disorder?

A peptic ulcer is a crater like lesion in the mucosa in a part of the gastrointestinal tract exposed to gastric juice. The pain caused by the effects of the acid on the mucosal wall is the most obvious symptom. Bleeding may occur, and sometimes the ulcer erodes a portion of the stomach or duodenal wall completely allowing bacteria and partially digested food into the peritoneal cavity, which then becomes actively inflamed. Normally the mucus barrier is sufficient to prevent irritation and autodigestion of the gastric or duodenal mucosa by the gastric secretions. In patients with duodenal ulcers the secretion of gastric acid is often greater than normal in response to a meal, as is the blood level of gastrin, though the latter is usually normal when fasting. Hypersecretion of pepsin may also be a cause, as may hyposecretion of mucus. Factors involved in the hypersecretion of acid include emotional disturbance, alcohol, caffeine and aspirin.

causes of peptic ulcer

Treatments are based on reducing the nervous stimulation of acid secretion by using neurotransmitter receptor blocking drugs, by blocking the acid secreting pump, by neutralising the acid with bases such as aluminium hydroxide or magnesium hydroxide or by increasing the effectiveness of the mucus using a drug which makes proteins adhere to the ulcer site.

4.6.2 Cholera

Cholera is a disease characterised by very severe diarrhoea with a water loss of up to 20 litres a day, abdominal cramps and severe dehydration. The rapid loss of fluid can result in death in less than 24 hours. The disease is caused by a bacterium, *Vibrio cholera*, transmitted in contaminated food or water, that multiplies in the small intestine. The organism adheres to the mucosal surface but does not actually enter the body. The bacteria secrete a toxin (choleragen) that binds to the mucosal cell surface and increases the production of cyclic AMP (cAMP) in the cell. This increases the secretion of Cl^- from intestinal glands and inhibits the carrier for Na^+, thus increasing the NaCl content of the lumen. The subsequent increase in osmotic pressure causes the diarrhoea. Fortunately, the Na^+ pump and the glucose/Na^+ carrier are not affected, so administration of solutions containing glucose (with NaCl) facilitates the reabsorption of Na^+ and hence water, reducing the diarrhoea.

effects of choleragen

4.6.3 Pernicious anaemia

The anaemias are a group of diseases characterised by tiredness and lethargy, which are caused by a reduced oxygen carrying capacity in the blood, in turn due to reduced numbers or function of the red blood cells. One type is associated with an inadequate supply of vitamin B_{12} which is necessary for the normal production of red blood cells. An inadequate dietary intake is unusual, except in some vegetarians and the normal cause is therefore inadequate absorption.

inadequate B_{12} absorption

Vitamin B_{12} requires the presence of a glycoprotein called intrinsic factor for its absorption. This is secreted by the parietal cells in the gastric mucosa. The intrinsic factor binds tightly to the vitamin B_{12} and the complex binds to specific receptors in the ileum, with the subsequent transfer of the vitamin across the epithelium. In most people with pernicious anaemia, there is a deficiency in the production of intrinsic factor. Trypsin is required for the process of vitamin B_{12} absorption to be efficient, and thus absorption is sometimes decreased in people with inadequate pancreatic secretions.

4.6.4 Lactase deficiency

Most of the carbohydrate content of milk is in the form of the disaccharide lactose. As shown previously, this requires the presence of the enzyme lactase which is located on the brush border of the absorptive cells. Lactase hydrolyses lactose to glucose and galactose. In many adults of certain ethnic groups, consumption of milk can cause abdominal pain and diarrhoea. This is because they lack the enzyme lactase, so the lactose is not absorbed, and becomes available to the bacteria in the colon for fermentation. There is a rapid fermentation and production of gas and this, together with the osmotic effect, causes the pain and diarrhoea. Lactase deficiency is very rare in infants who, of course, normally exist entirely on milk for several months. However, it can develop during childhood. Both children and adults can become temporarily deficient in enzymes of the brush border following an infection of the small intestine such as gastroenteritis.

The final SAQ tests your ability to relate the symptoms of disease states to the major defects which causes them.

SAQ 4.6

Answer each of the following questions by drawing an arrow to indicate an increase (↑), decrease (↓) or no change (→).

1) In patients with peptic ulcers, what would be the most likely effect on their pain of

   a) blocking neurotransmission in the vagus nerve?

   b) eating pickles?

   c) drinking coffee?

2) In a person with vitamin B_{12} deficiency anaemia, what would be the effect on red blood cell production of

   a) giving an oral dose of vitamin B_{12}?

   b) increasing the habitual dietary intake of vitamin B_{12}?

   c) injecting vitamin B_{12}?

3) In a person suffering from cholera, what is the likely effect on the diarrhoea of

   a) drinking water?

   b) drinking a solution of NaCl?

   c) drinking a solution of glucose and NaCl?

Summary and objectives

In this chapter we have examined the structure of the gastrointestinal tract and its associated organs, the processes of digestion and absorption, and the control, and some disorders, of these processes.

Now that you have completed this chapter you should be able to:

- describe the functions of the organs of the gastrointestinal tract;
- discuss the relationship between the structure of absorptive cells and their functions;
- draw flow diagrams and describe the processes of digestion and absorption of the major nutrients;
- describe the major mechanisms controlling these processes;
- explain the physiological basis for some disorders of these processes.

The circulation

The circulation

Cells are metabolically active and as a result require a continuous supply of oxygen and nutrients. These metabolic processes also generate waste products which must be removed from the cells. However, most cells are located at some distance from nutrient sources such as the digestive tract and sites of waste disposal such as the kidneys. This problem is solved by the circulatory system which comprises the blood, blood vessels through which the blood flows and the heart which produces the flow. The heart and blood vessels together form the cardiovascular system, which provides a connection between various tissues. The substance that bathes cells is called interstitial fluid (or tissue fluid). The interstitial fluid is serviced both by blood and lymph. The blood absorbs oxygen from the lungs and nutrients from the digestive tract and transports these and other substances to the tissues where they diffuse from the capillary blood vessels into the interstitial fluid. The interstitial fluid passes these substances in turn to the cells and exchanges them for wastes which are then removed from the tissues by blood and lymph. The blood, interstitial fluid and lymph constitute the internal environment and therefore its composition must be kept between specific values.

cardiovascular system

In this chapter we will examine the problems of obtaining an appropriate flow of blood through the tissues and of controlling the flow to meet the demands of the various tissues.

5.1 The blood

Blood is a specialised connective tissue, consisting of cells (and other structures or formed elements) suspended in a liquid matrix, the plasma. Together with the interstitial fluid and lymph it has a major role in maintaining the cellular environment and performs a number of important functions. It transports oxygen from the lungs to the cells of the body: carbon dioxide from the cells of the body to the lungs; nutrients from the digestive organs to the cells of the body and waste products from the cells of the body to the kidneys, lungs, liver and sweat glands.

Π Make a list of as many other functions performed by the blood as you can.

The sort of function you could have attributed to blood include:

- transport of hormones from endocrine glands to the cells;
- regulation of body pH through buffers;
- regulation of normal body temperature through the heat absorbing and cooling properties of its water content;
- protection of the body against blood loss through the haemostatic mechanisms;

- defense of the body against foreign microbes using circulating antibodies and specialised defence cells.

Blood is composed of two portions: cell or cell-like structures and plasma (a liquid containing dissolved substances). The cellular elements make up about 45 percent of the blood volume and plasma comprises the remainder.

The general physical characteristics of blood are give in Table 5.1.

Π From the data provided in the table and from the information give in the text, calculate how much plasma is contained in a body (a % of body weight). You will have an opportunity to check your answer in the next section.

Temperature	38°C
pH	7.35 - 7.45
Salinity	0.9%
As % of total body weight	8%
Volume	5-6 l for males:
	4-5 l for females.

Table 5.1 Physical characteristics of blood.

5.1.1 Plasma

Plasma is part of the total extracellular fluid (ECF) and is the source of the interstitial fluid. It is the medium through which all materials are exchanged between cells and the circulating blood. It is a straw coloured fluid consisting principally of water (90 percent) together with various dissolved or suspended substances. The volume of the plasma is approximately 2.8-3.0 litres in men and 2.4 litres in women, and it represents approximately 4.5 percent of the total body weight.

temperature regulation

The importance of plasma to the maintenance of the internal environment depends on the high proportion of water it contains. Many substances are simply dissolved in the plasma water for transport to the tissues. The high specific heat of water is important for temperature regulation, the plasma acting as both a heat distribution and heat dissipation system.

The plasma contains a variety of dissolved substances which are principally electrolytes and food materials. These are listed below:

	Plasma g (l^{-1})	Interstitial fluid g (l^{-1})
bicarbonate	1.5	1.9
chloride	3.6	4.1
phosphorus	0.1	0.1
sodium	3.2	3.5
calcium	0.1	0.1
potassium	0.16	0.16
magnesium	0.02	0.02
urea	0.15	0.15
protein	76.00	4.14
glucose	0.8-1.5	0.9

SAQ 5.1

1) List four categories of substances that are transported by the blood.

2) List four other functions of the blood other than transport.

plasma proteins

Major constituent of plasma are the plasma proteins. They fulfil a variety of very important functions. For example, the plasma proteins exert an osmotic pressure which affects the exchange of fluid between plasma and the tissue. The plasma proteins also act as carriers for a variety of substances such as hormones and drugs.

Some of the plasma proteins comprise the soluble proteins which in conjunction with the platelets provide the factors required for clotting during haemostasis. An additional group of the plasma proteins include the gamma globulins which are essential components involved in the body resistance to infection. The total concentration of the plasma proteins is normally between 6.4 and 8.3g per 100 ml, and can be divided into 3 major groups; albumin, globulins and fibrinogen.

5.1.2 Albumin

maintenance of plasma volume

Albumin is the most abundant plasma protein constituting 55 percent of plasma proteins and because of its molecular weight which is approximately 66 000 D, is too large to pass easily across capillary walls. It consequently accumulates in the plasma and is the principal protein responsible for the osmotic pressure which promotes the movement of water from the interstitial fluid to the plasma thus helping to maintain plasma volume. Patients with a low plasma protein level, possible suffering from nephrotic syndrome (kidney disease), often develop severe oedema (the loss of water from the plasma into the interstitial fluid within the tissues). Albumin is also important as a transport protein responsible for the carriage of fatty acids, bilirubin, thyroxine, cortisol and various drugs.

transport proteins

5.1.3 Globulins

immunoglobulins

Globulins subdivided into α, α_2, β and γ globulins, comprise 38 percent of plasma proteins. This fraction also contains several antibody proteins released by plasma cells. The gamma globulin, immune globulin G (Ig G), is especially well known because it is able to form an antigen-antibody complex with the proteins of many viruses and bacteria. Other immune globulins (eg Ig, M, IgD, IgA, and IgE) are also important in body defenses. Note that an in depth discussion of this aspect of physiology is given in the BIOTOL text 'Cellular Interactions and Immunology'.

5.1.4 Fibrinogen

blood clotting

Fibrinogen comprises about 4 percent of plasma proteins and is associated with the blood clotting mechanism in conjunction with the platelets. It is also involved in inflammatory processes.

For blood to function as the body's transport medium its volume must be maintained within defined levels and it must remain in the liquid state. However, blood has an inherent ability to take on a semi-solid state by a process known as haemostasis. This rapid process is triggered in response to a break in the wall of a blood vessel to prevent the loss of circulating fluid.

We will now examine the haemostatic mechanism and some pathological conditions that can occur if this system is faulty. You should realise that blood clotting is not exactly the same as haemostasis. Haemostasis is the stasis (= stopping or stagnation) of blood and is due to the *combined* action of the clotting system and platelets. It is incorrect to use haemostasis and clotting as being synonymous.

5.2 Haemostasis

Haemostasis or cessation of bleeding is a complex process involving at least four interrelated mechanisms which operate to prevent blood loss:

- contraction of the injured vessel;
- adhesion and aggregation of platelets at the site of the lesion;
- clot formation;
- clot dissolution.

5.2.1 Contraction of the injured vessel

When a blood vessel is damaged, the smooth muscle in its wall contracts immediately. Such a vascular spasm reduces blood loss for up to 30 minutes, during which time the other haemostatic mechanisms go into operation. The spasm is caused by damage to the vessel wall and by reflexes initiated by pain receptors. Potent vasoconstrictor substances such as adrenaline or 5-hydroxytryptamine (serotonin) are also involved in this vasoconstriction. You should note that contraction of the injured vessel can only take place in vessels with a smooth muscle layer. The fine capillaries have no such smooth muscle layer. In these cases, the first reaction is the contraction of the appropriate precapillary sphincter(s). These sphinters consist of small circles of muscle which can control the opening into the capillary much as the pyloric sphincter controls the movement of food from the stomach to the intestines (see Chapter 4).

5.2.2 Adhesion and aggregation of platelets

platelet activation

An initial plug is formed by platelets. When platelets come into contact with the damaged parts of a blood vessel, their characteristics change. They enlarge and their shapes become more irregular. They also become sticky and adhere to exposed collagen at the site of damage. As the platelets adhere to the collagen protein, they become activated and secrete a variety of chemicals including adenosine diphosphate (ADP), 5-hydroxytryptamine and prostaglandins. These act locally to attract more and more platelets which aggregate or clump together to produce a viscous mass. The platelets stick closely together and provide an effective plug to the damaged blood vessel. The plug is reinforced by fibrin threads formed during coagulation.

5.2.3 Clot formation

fibrin and fibrinogen

The platelet plug on its own is inadequate to stop bleeding from any sizeable damage to the blood vessels, particularly where there has been damage to an artery. The simple plug is not able to withstand the pressure inside the vessel. There has to be some form of reinforcement. This is accomplished by the formation of insoluble fibrin from the soluble plasma protein fibrinogen. The fibrin forms a network of sticky fibres over the injured portion of the vessel and over the platelet plug. It traps more platelets plus some red cells and it gradually takes on the appearance of a recognisable clot.

clotting factors

The conversion of fibrinogen to fibrin is a very complicated procedure involving many clotting factors. These are recognised by a Roman numeral according to the recommendations of an International Committee. Haemophilia refers to several different hereditary deficiencies of coagulation in which bleeding may occur spontaneously or after only minor trauma. Persons with the most common type of haemophilia, haemophilia A, lack factor VIII, the anti-haemophilic factor. Persons with haemophilia B and C lack factors IX and XI respectively.

5.2.4 Clot dissolution

Assuming that all the factors are present and a suitably strengthened clot has been formed, various processes quickly follow on to generate a complete and permanent repair. Soon after its formation the clot shrinks or retracts to about 40 percent of its original volume, squeezing out serum in the process. Serum is simply plasma but without most of the clotting factors which are largely trapped in the clot. The final clot is tougher but more elastic than the original one and as a result is a much more effective plug to the wound.

serum

plasminogen and plasmin

As the clot forms and retracts, a plasma protein, plasminogen, is activated by factors released from the endothelium of the blood vessel. The active enzyme produced is called plasmin and it breaks down the fibrin fibres of the blood clot so there is no longer a strong fibrous network and the clot disintegrates.

In the interval between clot formation and clot dissolution, special cells called fibroblasts have migrated to the site of damage from the surrounding tissue. These secrete protein fibres of connective tissue and establish the framework for the final scar tissue formation.

Π So far we have only considered the idea that clotting is a useful activity preventing blood loss from damaged vessels, occasionally it can occur within the intact vessel. What would be the consequences of this?

thrombosis

Clotting can occasionally occur in unbreached vessels giving rise to the pathological condition of thrombosis or intravascular clotting which may result in coronary thrombosis or cerebral thrombosis. The clot may completely occlude the vessel and as a consequence of this there is an interruption of the blood supply to the circulated organ and possible ischaemic damage is so severe that it results in death. This is the case when a thrombus forms in a major blood vessel to the heart or the brain stem. Fortunately a variety of physiological inhibitors of clotting are present in the plasma and these normally prevent intravascular clotting and so maintain the liquid state of the blood. The term ischaemic damage simply means damage arising from the staunching or stopping of the blood supply (Greek, iskhaimos from iskhein, to keep back + haima, blood).

Π Do you know the names of any anticoagulants?

Heparin is a powerful natural anticoagulant which is produced and stored in the mast cells of the connective tissue and is therefore present throughout the body. Coumarin derivatives of which Warfarin is an example, are orally administered anticoagulants; their adminstration results is less active clotting.

SAQ 5.2

Select the appropriate term.

1) The loss of water from the plasma into the interstitial fluid is called (the nephrotic syndrome, oedema, haemophilia).

2) The group of plasma proteins present in greatest quantities are (albumin, globulus, fibrinogen).

3) (Adenosine diphosphate, 5 hydroxytryptamine, adrenalin, prostaglandins) is not secreted by platelets activated by adhesion to collagen at the site of a damaged blood vessel.

4) Plasmin is (a hormone, a scar protein, a proteolytic enzyme, a clotting factor) involved in the later stages of haemostrasis.

5) Thrombosis is a pathological condition which arises from production of (heparin, warfarin, a blood clot, fibroblasts).

5.3 Blood cells

The most common classification of the formed elements of the blood is as follows:

Erythrocytes (red blood cells).

Leucocytes (white blood cells), which may be subdivided in granular and agranular leucocytes.

5.3.1 Formation of blood cells

haemopoiesis

The process by which blood cells are formed is called haemopoiesis. In the adult red blood cells, granular leucocytes and platelets are produced in red bone marrow (myeloid tissue). Agranular leucocytes arise from both myeloid (bone marrow) tissue and from lymphoid tissue (lymph nodes, spleen and tonsils).

5.3.2 Erythrocytes

These account for about 99 percent of blood cells. A healthy male has about 5.4 million erythrocytes per cubic millimetre (μl) and a healthy female has about 4.8 million μl^{-1}. The higher value in the male is because of a higher rate of metabolism in males. The body produces about 2 million new mature blood cells every second. This production takes place in the red bone marrow in the spongy bone of the skull, sternum, vertebrae and ribs as well as the epiphyses of the long bones.

erythropoiesis

The process of erythrocyte formation is called erythropoiesis. Normally erythropoiesis and red cell destruction proceed at the same rate, but if the body needs more erythrocytes then a homeostatic mechanism increases red cell production. In one haemostatic mechanism the kidneys release an enzyme called renal erythropoietic factor that converts a plasma protein into the hormone erythropoietin. This stimulates the red bone marrow to produce more erythrocytes.

Microscopically, erythrocytes (red blood cells) appear as biconcave discs about 8 μm in diameter. They lack a nucleus and mitochondria and can therefore neither reproduce nor carry out extensive metabolism. The red blood cell membrane encloses cytoplasm containing a large amount of red protein called haemoglobin which is responsible for the red colour of blood.

5.3.3 Functions of erythrocytes

oxygen and carbon dioxide transport

The haemoglobin molecule consists of a protein called globin made up of four subunits and each subunit contains a pigment called haem, which contains an iron atom. As the erythrocytes pass through the lungs each of the iron atoms in the haemoglobin molecule combines with one molecule of oxygen to form oxyhaemoglobin. The oxygen is transported in this state to the tissues of the body. In the tissues the oxygen is released to diffuse into the tissue fluid and the globin portion combines with a molecule of carbon dioxide from the tissue fluid to form carbaminohaemoglobin. This complex is transported to the lungs where the carbon dioxide is released. Only 25 percent of the carbon dioxide is transported by haemoglobin in this form, the rest is transported in blood plasma as the bicarbonate ion HCO_3^-.

Erythrocytes are highly specialised for their gas transport function. Each erythrocyte contains 280 million molecules of haemoglobin to increase the oxygen carrying capacity. Typical arterial blood contains about 20 ml of dissolved oxygen in each 100ml (note that plasma alone can dissolve maximally about 0.5 ml of oxygen per 100 ml that is about 1/40th the capacity). The biconcave shape of the erythrocytes maximises the surface area available for the diffusion of oxygen and carbon dioxide into the cell where combination with haemoglobin occurs. This shape also permits temporary distortion when the erythrocytes pass through the capillaries.

Π What is the condition called where there is a deficiency of red blood cell? Can you think of any possible causes and the consequences of such a deficiency?

A general condition in which there is either a deficiency of red blood cells or a reduction in haemoglobin is called anaemia. This results in poor oxygen transport. Anaemia can be brought about by either iron or vitamin deficiencies. Anaemia can also be due to blood loss, or the destruction of the bone marrow. Bone marrow malfunctions in which formation of abnormal cells can also occur as a consequence of some hereditary disorder. Also kidney disease may result in anaemia. This is because the kidneys may fail to produce the renal erythropoietic factor and this in turn leads to a failure to stimulate the red bone marrow to produce erythrocytes.

5.3.4 Leucocytes

Unlike erythrocytes, leucocytes (white blood cells) have nuclei and do not contain haemoglobin. Leucocytes are far less numerous than red blood cells, averaging from 5000 to 9000 cells per µl (=mm^3).

There are two major groups of leucocytes:

- granular leucocytes which develop from red bone marrow, have lobed nuclei and granules in the cytoplasm. The three kinds of granular leucocytes are neutrophils, eosinophils and basophils. These three kinds are distinguished by their interaction with dyes. Thus for example basic dyes stain bascophils;
- agranular leucocytes which develop from myeloid and lymphoid tissue and have regular nuclei and do not posses cytoplasmic granules. The two kinds of agranular leucocytes are monocytes and lymphocytes.

5.3.5 Functions of leucocytes

The main function of leucocytes is to combat any microbial infection by phagocytosis, antibody production and by cytotoxic effects. The roles and interactions involved in these processes are quite complex and are dealt with in detail in the BIOTOL text 'Cellular Interactions and Immunobiology'. Here we will provide a general over-view.

neutrophils

Neutrophils are about 10-20µm in diameter and contain lysosomes in which are found a large number of hydrolytic enzymes. These cells represents over 90% of the granular leucocytes and are the most active leucocytes responding quickly to tissue destruction caused by bacteria and other pathogens. They are capable of phagocytosis and engulf particulate matter, especially bacteria. They produce an enzyme lysozyme (muraminidase) which hydrolyses the cell walls of bacteria invaginated organisms are retained and hydrolyted in vacuoles.

lysozyme

eosinophils

Eosinophils make up about 25% of the leucocytes. These cells seem to be able to kill large targets which cannot be phagocytosed such as helminths. When they are stimulated they degranulate and the contents of the granules are released to the outside of the cell. Amongst these products is a toxic protein (the so called major basic protein) and also components which inactivate some of the agents which are involved in the inflammatory response (for example histamine).

basophils

Basophils make up a very small portion of the leucocytes (less than 0.2%). We should also include in this group the so called mast cells. These cells are involved in allergic reactions. They leave the capillaries, enter the tissue and liberate heparin, histamine, seratorin (5-hydroxytryptamine) and a variety of other pharmacologically active agents. They are stimulated to do this by the presence of particular antigens.

monocytes

macrophages

Monocytes migrate into infected tissue and clean up debris following an infection (note that not all monocytes migrate freely some become resident in particular tissues eg Kupffer cells in the endothelial layer of the liver, intraglomerular mesangial cells in kidney and the alveolar macrophages in the lungs). These cells which are called macrophages when they enter peripheral tissue are involved in presenting antigens to the antibody producing systems of the body. Note these cells will more readily engulf cells that are coated with antibodies (opsonisation).

Lymphocytes can be broadly divided into two groups B-cells and T-cells. B-cells are produced in the myeloid and lymphoid tissues and in man are matured before they leave the bone marrow. These cells are the antibody producing cells. (In birds these cells are produced in the bone marrow but have to be processed by a gland called the Bursa of Fabricius before they become competent to produce antibodies).

sub population of T-cells

The T-cells are also produced in the myeloid tissue of the bone marrow but have to be processed by the thymus before they will become competent to fulfil their function. These cells are sometimes called thymic - derived lymphocytes - hence T-cells for short. T-cells can be further subdivided into T_{helper} cells, $T_{cytotoxic}$ cells and $T_{suppressor}$ cells.

roles of T-cells

The T_{helper} cells help to process antigens and to stimulate B-cells to make antibodies. $T_{suppressor}$ cells restrict B-cells production of antibodies. Thus these two sub-sets of T-cells control B-cells.

$T_{cytotoxic}$ cells recognise and kill cells which contain intracellular infection such as viruses. They also appear capable of recognising and killing host cells which carry particular mutations. Essentially what the T-cells do is to act as a surveillance mechanism. Thus these cells are able to

- identify infected host cells and to destroy them;
- identify erranous (mutant) host cells and destroy them;
- process antigens (foreign chemicals) and present them to antibody producing cells, B-cells.

functions of B-cells

The B-cells consists of a large number of different cells each capable of producing antibody molecules with a particular specifity. On infection, the antigens, produced by the infection, are processed by both macrophages (antigen presenting cells) and T_{helper} cells and stimulate the appropriate B-cells to proliferate. These B-cells colones each produce antibodies which will combine with the antigens. The combination of antibodies with their antigens has several important consequences. For example the antigens may become inactivated or they are more readily phagocytosed by the macrophages or if the antigens form part of a cell the cell may become lysed. This latter reaction involves a group of blood proteins called complement.

Π It would be helpful to you to draw yourself a summary chart of these various leucocytes. We will start the chart for you.

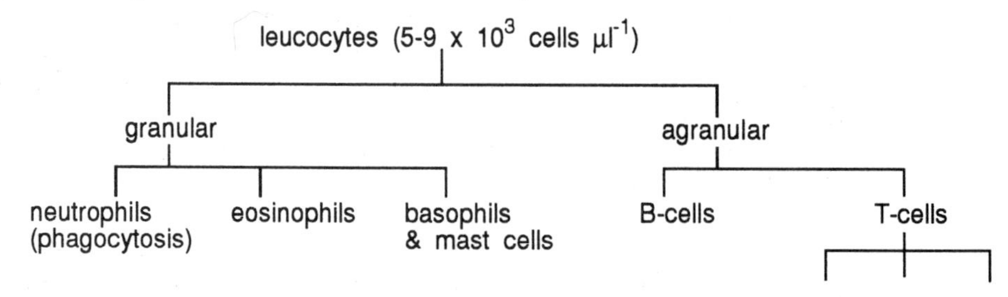

We will not extend the discussion of immunity here as this has been adequately covered in the BIOTOL texts 'Defence Mechanisms' and 'Cellular Interaction and Immunobiology'. Nevertheless you should realise that the antigen-antibody response helps the body combat infection and provides, because of its inherent memory function, long term immunity to some diseases, preventing or reducing the magnitude of a second infection. An increase in the number of white cells present in the blood typically indicates a state of inflammation or infection. A high neutrophil count (high numbers of cells) indicates damage by invading bacteria. An increase in the number of monocytes generally indicates a chronic infection. High lymphocyte counts indicate antigen-antibody reactions and eosinophils and monocytes are elevated during allergic reactions.

SAQ 5.3

Match the white blood cell listed below with the appropriate descriptive statement which follows.

a) neutrophils

b) eosinophils

c) basophils

d) monocytes

e) leucocytes

1) [ ] on entering the peripheral tissues are known as free macrophages.

2) [ ] representing about 2.5 percent of the total white cells, these cells may supress inflammation and may counteract large invading organisms.

3) [ ] release histamine at sites of inflammation.

4) [ ] usually the first white cells to arrive at the site of injury.

5) [ ] cells produce antibodies.

5.4 Interstitial fluid and lymph

The substance that baths cells is called interstitial or tissue fluid. This fluid provides cells with nutrients and removes their waste products. It also protects the cells from extreme temperatures and changes in pH. The interstitial fluid is in turn serviced by blood and lymph. The blood provides oxygen from the lungs, nutrients from the digestive tract, hormones from endocrine glands and various enzymes. It transports these substances to all the tissues where they diffuse from the capillaries into the interstitial fluid. In the interstitial fluid, the substances are passed into the cells and exchanged for wastes.

The blood can also function as the transport system for disease causing organisms. To protect the organism from foreign invasion a system of defence cells (see 5.3) is present in the blood. To further protect itself from the spread of infection the body has a lymphatic system, which is a collection of vessels containing a fluid called lymph. The lymph transports various materials, including waste, viruses and bacteria from the interstitial fluid. The lymphatic system removes the viruses and the bacteria and transports the remaining materials to the blood. The blood then carries the waste to the liver, lungs, kidneys and sweat glands, where they are removed from the body.

5.4.1 Formation of interstitial fluid and lymph

interstitial fluid become lymph in lymphatic vessels

Whole blood does not flow into the tissue space but remains in closed blood vessels. However, certain constituents of the plasma do move through the capillary walls, and once they move out of the blood they form part of the interstitial fluid. Interstitial fluid and lymph are basically the same. The fluid which bathes the cells is interstitial fluid and that which flows through the lymphatic vessels is called lymph. Both fluids are similar in composition to plasma although they contain less protein. Interstitial fluid and lymph will be discussed in greater detail later in this chapter.

5.5 Heart: structure and function

The heart is the centre if the cardiovascular system and is responsible for the circulation of the blood. The heart is actually two pumps in one, one pump propels blood through the pulmonary circulation (to the lungs where the blood receives oxygen and releases carbon dioxide), and the other pumps blood through the systemic circulation (to all tissues of the body). The heart is a hollow, muscular organ that beats over 100,000 times a day to pump 7,200 l of blood per day through over 60,000 miles of blood vessels. The blood vessels form a network of tubes that carry blood from the heart to the tissues of the body and then return it to the heart.

Π Examine Figure 5.1 carefully and try to remember as many of the labelled feature as you can. One way of helping you to remember these names is to trace Figure 5.1 and put all the names down on a sheet of paper. Them sometime later see if you can put the correct label on the correct item. Use Figure 5.1 to check your answer.

The heart is about the size of a clenched fist and is situated obliquely between the lungs. It is surrounded by a tough, fibrous pericardium which limits its size and anchors it to the diaphragm. The innermost membrane of the pericardium is moist, enabling the heart to beat with a minimum of friction. Inside, the heart consists of four chambers, two atria and two ventricles (Figure 5.1). There are one-way valves between the atria and the ventricles, the atrio-ventricular valves. These valves allow blood to flow from the atria into the ventricles but prevent blood from flowing back into the atrium when the ventricle contracts. The atrio-ventricular valve between the right atrium and the right ventricle has three cusps and is called the tricuspid valve. The atrio-ventricular valve between the left atrium and the left ventricle has two cusps and is called the bicuspid or mitral valve.

atria and ventricles

tricuspid and dicuspid valves

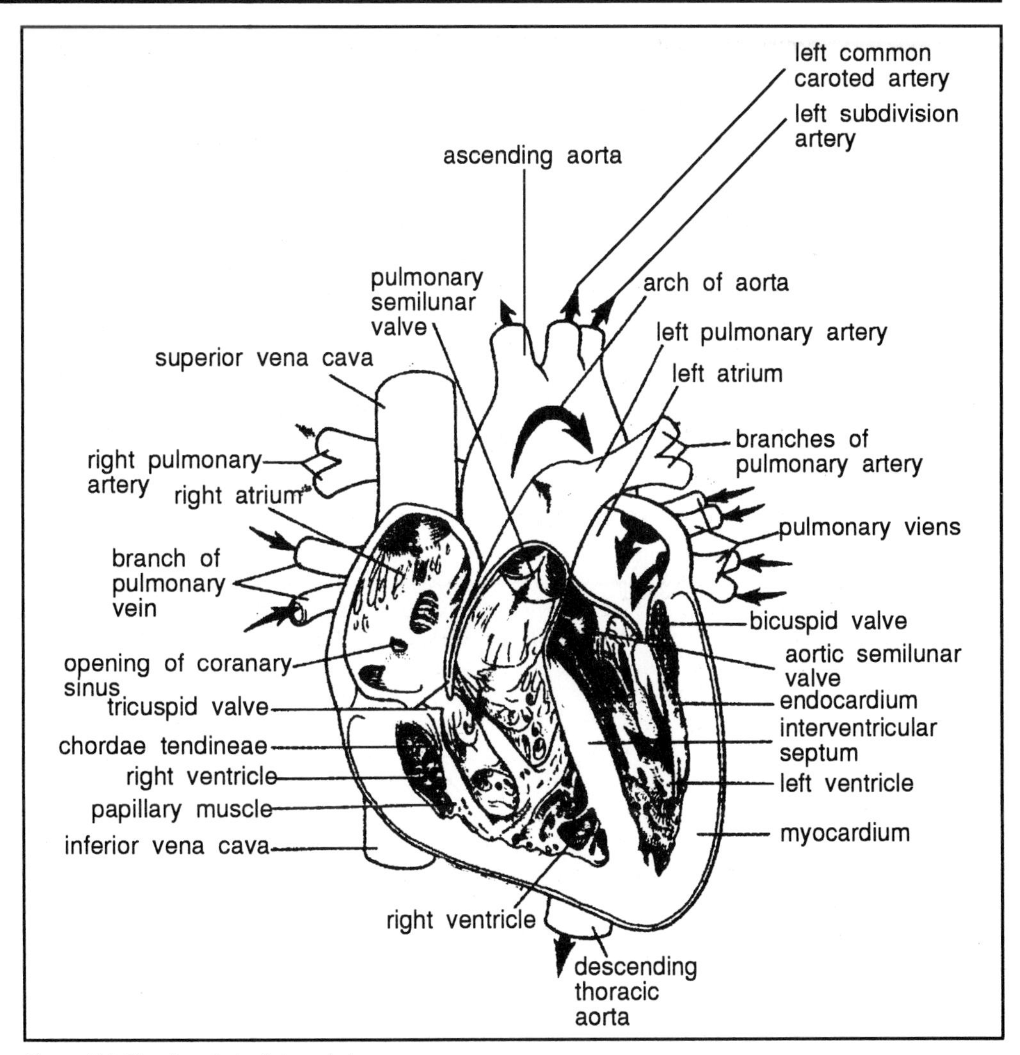

Figure 5.1 Heart: anterior internal view.

Each ventricle contains papillary muscles. These muscles are attached by thin, strong connective tissue strings called chordae tendineae to the cusps of the atrio-ventricular valves. The papillary muscles contract when the ventricles contract and prevent the valves from opening back into the atria by pulling on the chordae tendineae attached to the valve cusps. Blood flowing from the atrium into the ventricle forces the valve open into the ventricle, but when the ventricle contracts, blood pushes the valve back towards the atrium. The atrio-ventricular valve closes off as the valve cusps meet.

The semilunar valves separate the ventricles from the aorta and the pulmonary artery. Each valve consists of three pocket like semilunar cusps. Blood flowing out of the ventricles pushes against each valve, forcing it open; but when the blood flows from the aorta or pulmonary artery towards the ventricles, it enters the pockets of the cusps, causing them to meet in the centre of the aorta or pulmonary artery, so closing them and stopping blood from flowing back into the ventricles.

There are no valves at the entrances of the venae cavae into the right atrium and of the pulmonary veins into the left atrium.

SAQ 5.4

Insert the correct number from Figure 5.2 to match the major structures given below. (Do this without referring to Figure 5.1).

aorta	aortic valve
superior vena cava	right atrium
left atrium	left ventricle
pulmonary artery	pulmonary valve
bi-cuspid (mitral) valve	tri-cuspid valve
inferior vena cava	right vontriolc
pulmonary veins	

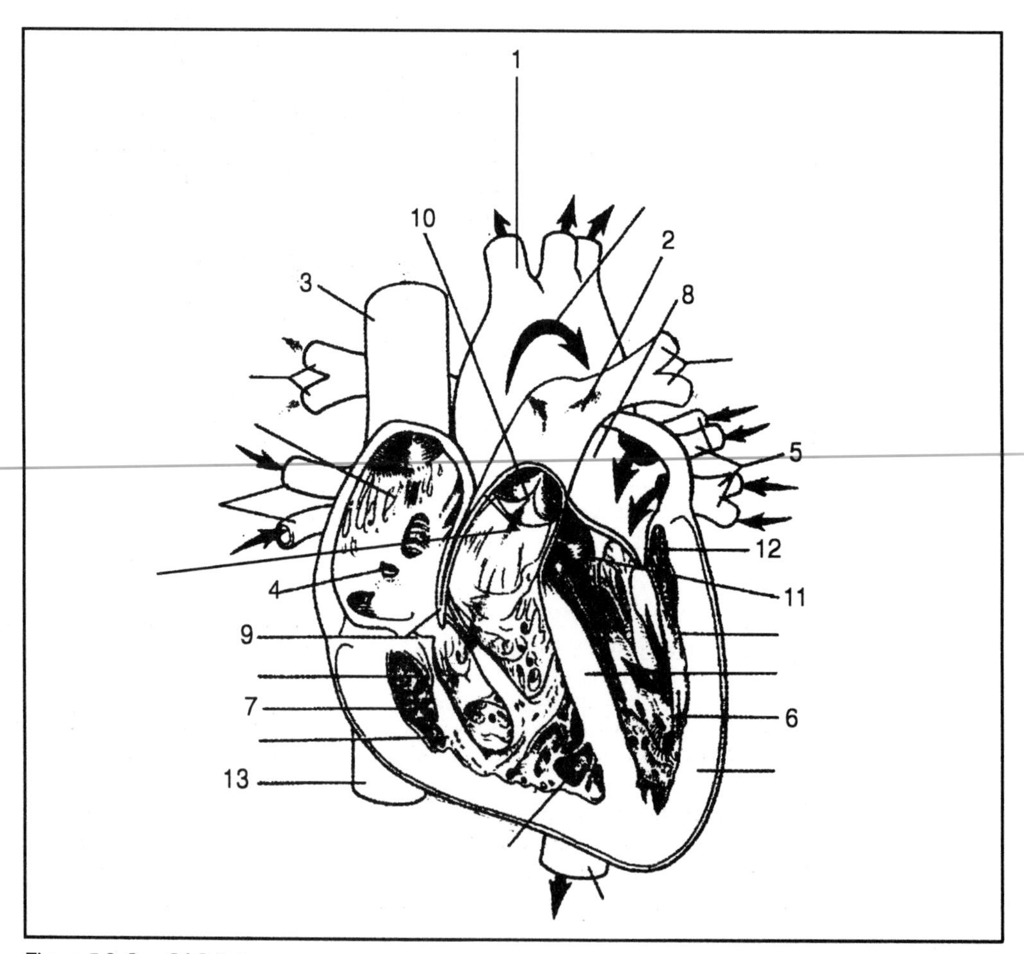

Figure 5.2 See SAQ 5.4.

5.5.1 Blood flow through the heart

Blood flow through the heart is shown in Figure 5.3. Remember that both atria contract at the same time and both ventricles contract together following a slight delay.

Blood enters the right atrium from the systemic circulation which supplies oxygenated blood to all the tissues of the body, through the superior and inferior venae cavae. Most of the blood in the right atrium passes straight into the right ventricle as the ventricle relaxes following the previous contraction. When the right atrium contracts, the blood still remaining in the atrium is pumped into the ventricle.

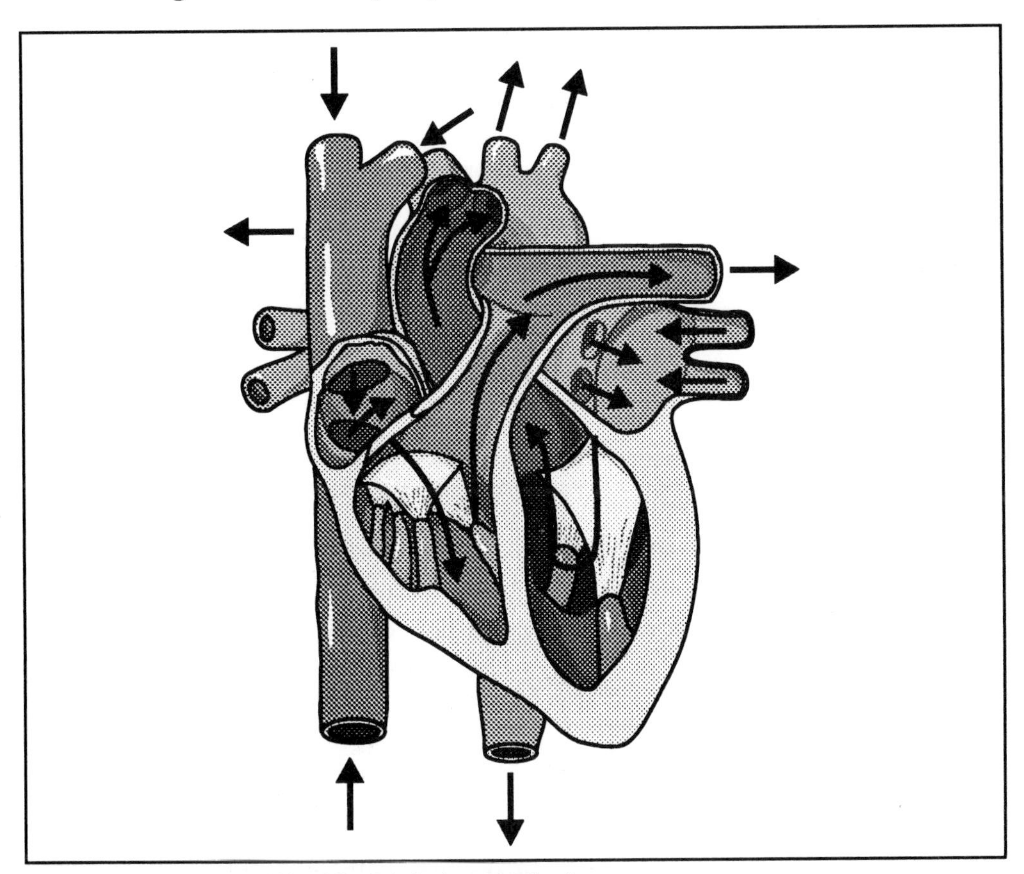

Figure 5.3 The passage of blood through the heart (stylised).

Contraction of the right ventricle pushes blood into the pulmonary artery. The atrio-ventricular valve ensures that there is no backflow of blood into the right atrium.

The pulmonary artery carries blood to the lungs where oxygen is picked up and carbon dioxide is released. Blood returning from the lungs enters the left atrium through the four pulmonary veins. The blood passes through the bicuspid valve and contraction of the left atrium completes filling of the left ventricle.

Contraction of the left ventricle closes the atrio-ventricular valve and pumps blood into the aorta. Blood flowing through the aorta is distributed throughout the whole body except for the respiratory vessels in the lungs.

5.5.2 Coronary circulation

coronary arteries

myocardium

The heart is always beating and therefore must receive a steady supply of blood to enable the cardiac muscle to perform this work. The heart muscle cannot make direct use of the blood that is flowing through its chambers to receive the oxygen and nutrients which provide the energy for contraction. Instead the coronary arteries supply blood to the heart muscle. They have their origin at the start of the aorta, behind the cusps of the aortic valve. The main coronary arteries are visible on the surface of the heart. These divide into smaller vessels which then penetrate into the heart muscle (myocardium). There are two main arteries, the left coronary artery supplies the anterior part of the left ventricle while the right coronary artery supplies the posterior part of the left ventricle and most of the right ventricle. The venous blood from the coronary capillaries then drains directly into the chambers of the heart.

angina

If narrowing develops in one of the coronary arteries then the blood supply to the corresponding part of the heart muscle may be inadequate. At rest the blood supply carried through a narrowed artery may be sufficient to nourish the heart's muscular wall. However, during exercise or strong emotional stress the heart has to do more work and requires a larger blood supply. When an artery is narrowed the blood flow is impeded and the required increase in blood supply cannot be met. A point is reached when the heart muscle becomes ischaemic and pain is produced. This chest pain is called angina.

Π Do you know of any causes of narrowing of the coronary arteries and how may angina be treated?

The commonest causes of narrowing of the coronary arteries are atherosclerosis or hardening of the arteries. Another important but uncommon cause of narrowing of the coronary arteries is coronary artery spasm. Angina may be treated by using vasodilators and calcium antagonist drugs, which in effect open up the coronary vessels and reduce the force of cardiac muscle contraction.

atherosclerious

arteriosclerosis

Here we should distinguish between atherosclerosis and arteriosclerosis, two conditions which are often confused with each other. Arteriosclerosis is the term given to the thickening and loss of elasticity of the walls of vascular tissue. It is a common condition of aging. Atherosclerosis on the other hand begins with a fatty streak consisting of lipid loaded macrophages (= foam cells) under the endothelial monolayer. As atherosclerosis develops, this early leasion develops into a fully fledged atherosclerotic leasion in which we find cellular debris (platelets, phagocytes, lymphocytes) cholesterol droplets and Ca^{2+}. The leasions thus clearly show that inflammatory processes are involved in their genesis.

5.5.3 Electrical activity of the heart

The heart is composed of special muscle called cardiac muscle. When the heart beats a wave of excitation spreads from muscle cell to muscle cell and this is closely followed by a contraction wave of the muscle.

The wave of excitation originates at the sinoatrial node (SA) which is located at the top of the right atrium (Figure 5.4).

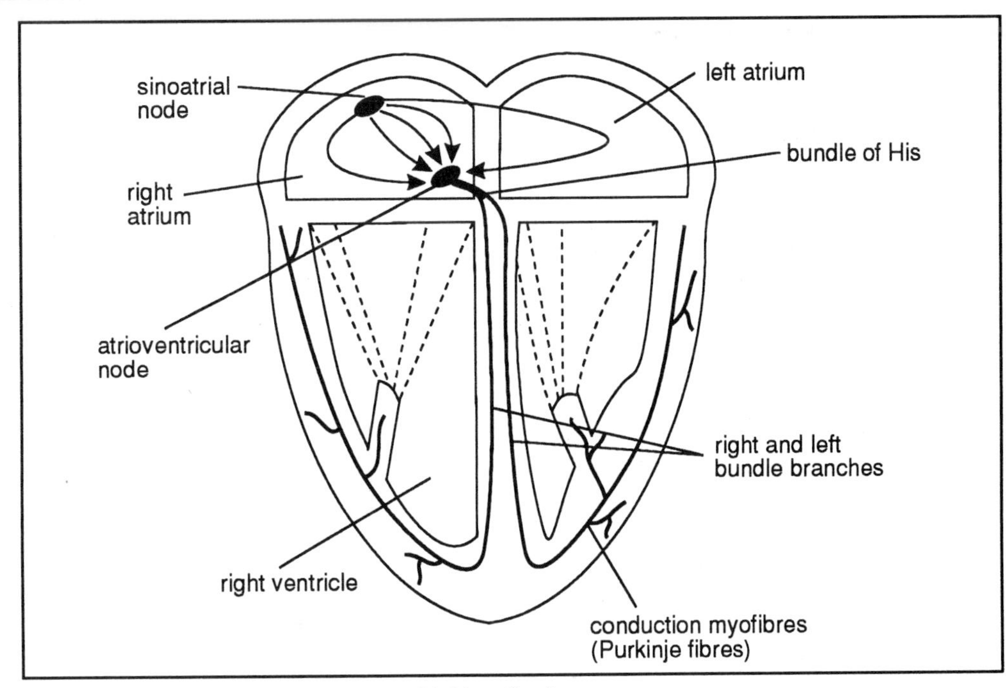

Figure 5.4 Conduction system of the heart (highly stylised).

pacemaker

role of sinoatrial node

The sinoatrial node is known as the pacemaker because it sets the pace of the heart which the rest of the heart muscle follows. The resting potential of the sinoatrial node cells is not steady but undergoes a slow spontaneous depolarisation. This gradual depolarisation is known as the pacemaker potential and it brings the membrane potential to threshold and so a cardiac action potential is initiated. After membrane repolarisation, the gradual depolarisation begins again. The rate of change of the pacemaker potential determines how quickly threshold is reached and the next action potential generated. The inherent rate of the sinoatrial node is approximately 100 depolarisations per minute, in the absence of any neural or hormonal influence. The normal resting heart rate is, however, less than this (72 beats per minute) due to the presence of parasympathetic tome which will be discussed later.

flow of nervous pulses in heart

The wave of depolarisation generated at the sinoatrial node spreads in an ordered and consistent manner throughout the whole of the heart muscle. At first it spreads from the sinoatrial node radially around the two atria to give rise to atrial contraction or systole. The fibrous disc which separates the atria and the ventricle does not allow the spread of this excitation to the ventricles because it is non-conducting. Therefore the only route to the ventricles is via the atrioventricular node. The special group of cells in the node detects the depolarisation and conducts the depolarisation through the fibrous disc which is in the right atrium directly above the internal wall (septum) between the right and left ventricles. From the atrioventricular node there is a tract of conducting fibres known as the Bundle of His. These fibres are modified cardiac muscle cells arranged in a bundle. The Bundle of His runs through the atrioventricular septum and into the interventricular septum. It then continues down both sides of the septum as the right and left bundles branches. Purkinje fibres then conduct the wave of depolarisation throughout the ventricles. These fibres conduct the depolarisation wave to both ventricles which contract (ventricular systole). The ventricles relax during diastole and await for the next wave of depolarisation to arrive from the atria.

Bundle of His

Purkinje fibres

ventricular systole

Π The transmission of nerve impulses in the heart are quite complex. It might be helpful to look at Figure 5.4 again and retrace the flow of the nerve impulse from the sinoatrial node through the atrioventricular node and the Bundle of His to the right and left bundle branches and subsequently to the Purkinje fibres.

5.5.4 The electrocardiogram (ECG)

The wave of depolarisation which spreads throughout the heart during systole involves depolarisation of the cardiac muscle cells. Immediately following this depolarisation wave is a wave of repolarisation which restores the membrane potential back to its resting level so that it can respond to the next wave of depolarisation sent out by the pacemaker.

This process of depolarisation and repolarisation is similar to the way that neurons conduct action potentials along their axons. These action potentials involve electrical changes at the membranes of the cardiac muscle cells.The electrical changes occurring in the heart can be detected by electrodes placed on the body surface and the recordings produced are known as electrocardiograms or ECGs.

A typical trace is seen in Figure 5.5 and such recordings can be obtained if the electrodes are placed on the chest or on the limbs, typically four electrodes are used one on each wrist and each ankle of the patient.

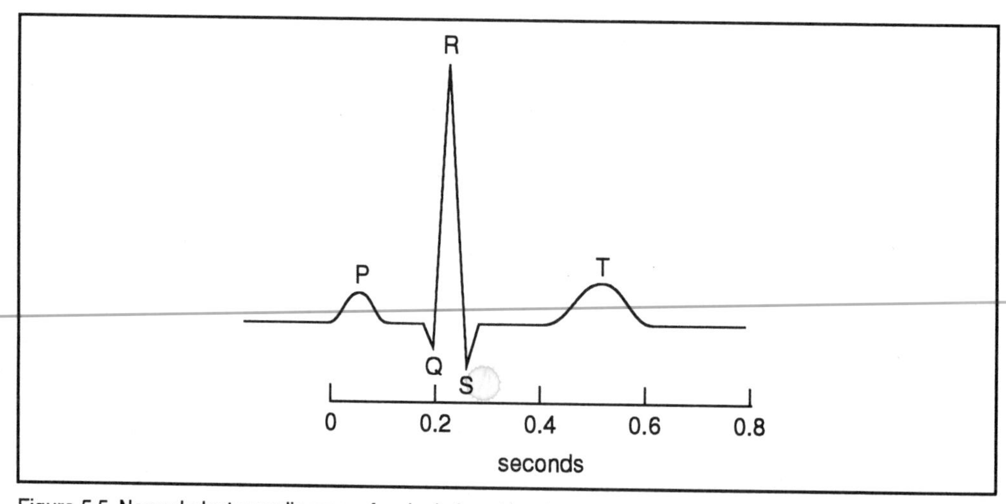

Figure 5.5 Normal electrocardiogram of a single heart beat.

The following table lists the relationship between the events in the heart and the components of the ECG.

P	spread of depolarisation over the atria
P-Q	depolarisation wave passing down the Bundle of His, atria repolarised
QRS	spread of depolarisation over the ventricles
T	repolarisation of the ventricles
T-P	diastole

Π For what diagnostic purposes may ECG recordings be used?

The ECG is useful in diagnosing abnormal cardiac rhythms and conduction patterns and following the course of recovery from a heart attack. It can also determine the presence of foetal life. It is also used to correlate rhythm disorders and symptoms and to follow the effectiveness of drugs.

5.5.5 Cardiac output

The heart has properties that allow it to beat independently but its operation is regulated by events happening in the rest of the body. All body cells must receive a controlled amount of blood to function normally. When cells are very active, as during exercise, then they will need more blood. During periods of inactivity, the cellular needs are reduced and the heart will reduce its output.

CO

The volume of blood pumped by each ventricle per minute is called the cardiac output (CO), expressed as l min^{-1}. It is also the volume of blood flowing through either the systemic *or* the pulmonary circulation per minute.

The cardiac output is determined by multiplying the heart rate (HR) and the stroke volume (SV), the volume of blood ejected by each ventricle with each beat.

$$CO = HR \times SV$$

For example, if the heart rate is 72 beats per minute and each ventricle eject 70 ml of blood with each beat, the cardiac output is

$$CO = 72 \text{ beats/min} \times 0.07 \text{ l/beat} = 5.0 \text{ l min}^{-1}$$

These values are approximate for a normal resting adult. Total blood volume is approximately 5 l and this means that essentially all the blood is pumped around the body once every minute. During periods of exercise cardiac output may reach 35 l min^{-1}.

Factors that increase stroke volume or heart rate tend to increase cardiac output. Factors that decrease stroke volume or heart rate tend to decrease cardiac output.

Heart rate and stroke volume do not always change in the same direction. For example, stroke volume decreases following blood loss while the same condition causes the heart rate to increase. These changes produce opposing compensatory effects in cardiac output.

5.5.6 Control of stroke volume

Stroke volume is the volume of blood ejected by each ventricle during each contraction. The ventricles, however, never completely empty during contraction and so a more forceful contraction causes the ventricles to empty more completely and therefore increases the stroke volume.

Changes in force of contraction can be produced by two important factors:

end diastolic volume

1) Changes in the volume of blood within the ventricles just prior to contraction, that is changes in the filling (end diastolic) volume. One factor that determines the force of ventricular contraction is the length of cardiac muscle fibres. Within

limits, the greater the length of stretched fibres, the stronger the contraction. This relationship is known as Starling's law of the heart and represents an intrinsic mechanism for regulating cardiac output. The significance of this mechanism is that an increased flow of blood returning from the veins into the heart (venous return) forces an increase in cardiac output by distending the ventricle and increasing cardiac output. An important function of this relationship is to maintain the equality of right and left cardiac outputs. If the output of the right heart was suddenly greater than that of the left, then the increased blood flow to the left ventricle would automatically produce an equivalent increase in left ventricular output and so ensuring that blood does not accumulate in the lungs.

2) Changes in the sympathetic nervous system input to the ventricles and also in the levels of circulating adrenaline. The effect of the sympathetic nervous system and of an increased concentration of adrenaline in the plasma is to increase ventricular contractility at any given muscle fibre length. Therefore, the increased force of contraction and stroke volume resulting from sympathetic nerve stimulation or increased adrenaline is independent of a change in end diastolic ventricular volume. This represents an intrinsic mechanism regulating stroke volume.

5.5.7 Control of heart rate

Cardiac output depends on heart rate as well as stroke volume. Changing heart rate is the body's principal means of short term control over cardiac output and blood pressure.

In the absence of any nervous or hormonal influences on the sinoatrial node, the heart will beat rhythmically at a rate of approximately 100 beats min^{-1}. This is the inherent, automatic discharge rate of the sinoatrial node. However, the heart rate may be much higher or lower than this due to the influence of the autonomic nervous system and hormones.

influence of the brain on heart role

The cardioacceleratory centre lies within the medulla of the brain. Arising from this centre are sympathetic nerve fibres which pass down the spinal cord and then pass outwards in the cardiac nerves and innervate the SA node, the AV node and the heart muscle. Stimulation of the cardioacceleratory centre will increase the heart rate and the strength of contraction.

The medulla also contains the cardioinhibitory centre. Arising from this centre are parasympathetic fibres that innervate the heart via the vagus nerve. These fibres also innervate the SA node and the AV node. Stimulation of the parasympathetic fibres decreases the heart rate.

The autonomic control of the heart is therefore the result of opposing sympathetic (stimulatory) and parasympathetic (inhibitory) influences. Sensory impulses from blood pressure receptors (baroreceptors) also affect the rate of heart beat. These reflexes will be discussed later under the control of blood pressure.

Figure 5.6 illustrates how sympathetic and parasympathetic (autonomic) activity influence sinoatrial node function. Parasympathetic stimulation decreases the slope of the pacemaker potential so that threshold is reached more slowly and heart rate decreases. Parasympathetic stimulation also hyperpolarises the membrane so that the pacemaker potential starts from a lower value. Sympathetic stimulation increases the slope of the pacemaker potential so that threshold is reached more rapidly and the heart rate increases.

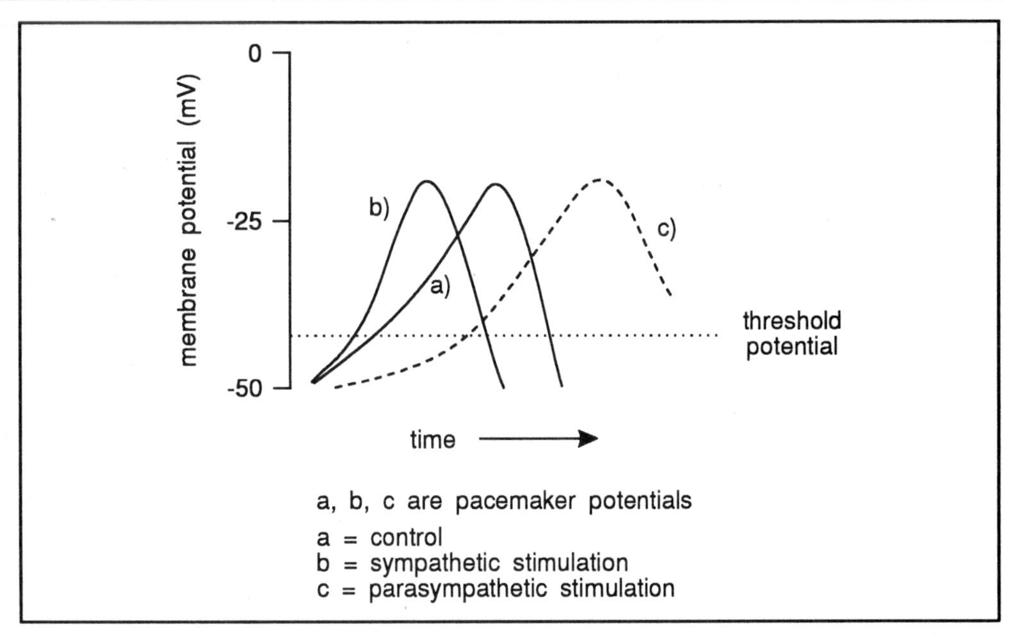

Figure 5.6 Effects of sympathetic and parasympathetic nerve stimulation on the slope of the pacemaker potential of an SA nodal cell.

Figure 5.7 summarises the major determinants of stroke volume and heart rate.

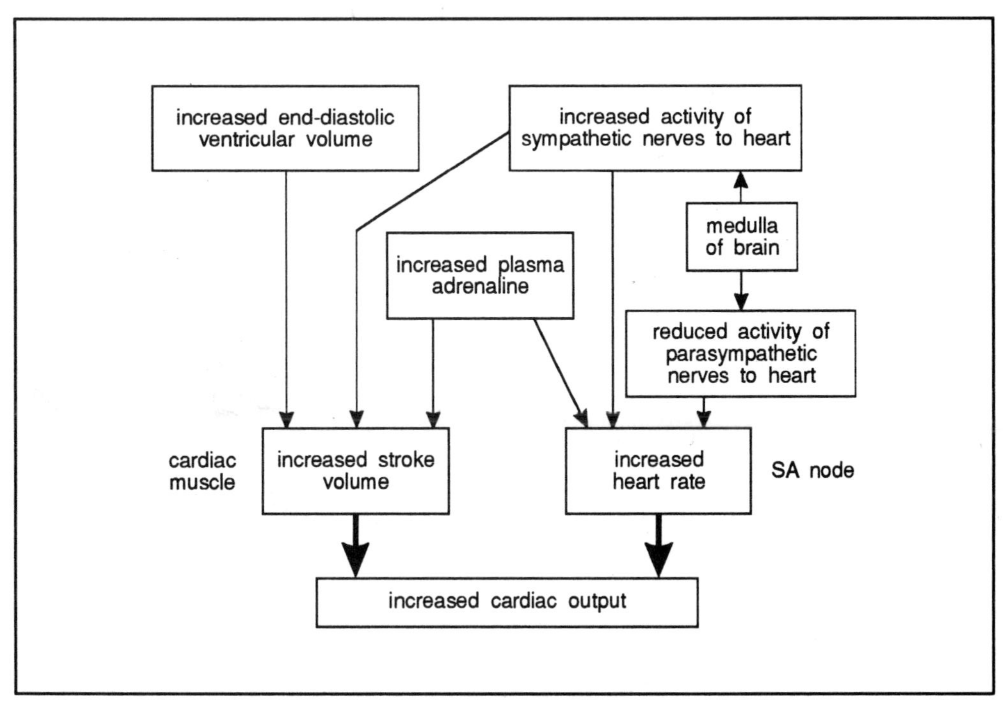

Figure 5.7 Major factors determining stroke volume and heart rate.

∏ Make a list of any other factors which might influence heart rate?

You may consider that some of the following factors will influence heart rate. Temperature: increased body temperature occurring during fever or strenuous exercise increases heart rate. Decreased body temperature decreases heart rate and force of contraction.

Emotions: strong emotions such as fear, rage, anxiety and various stressors increase heart rate. Mental depression and grief decrease heart rate.

Sex and age: the heart beat is a little faster in females than in males and faster in young children than old people.

Chemical factors: adrenaline, produced by the adrenal medulla in response to sympathetic stimulation, increases the rate and strength of the heart beat. Elevated levels of potassium and sodium decrease the rate and strength of the heart beat. An excess of calcium increases the rate and strength of contraction.

SAQ 5.5

Indicate if the following statements are true or false.

1) The Bundle of His conducts nerve impulses from the atrioventricular node to Purkinje fibres.

2) If the heart rate (HR) is 80 beats per minute and each of the two ventricles ejects 80 ml of blood for each beat, then the cardiac output (CO) is 80 x 80 x 2 = 12800 ml min^{-1}.

3) Starling's law of the heart indicates that the strength of the contraction of cardiac muscle fibres is inversely proportional to the length of the fibres.

4) Parasympathetic stimulation of the sinoatrial node lowers the threshold potential needed to generate an action potential in the heart.

5) The volume of blood in the ventricles influences the force of contraction of the cardiac muscles.

6) Cations such as potassium, sodium and calcium all increase the rate and strength of contraction.

5.6 Blood vessels: structure and function

The heart pumps blood under high pressure into the vascular system. The vascular system is the distribution network of blood vessels which brings blood close to each cell in every organ and tissue of the body. This is a vital function because without an adequate blood supply the cells would soon die.

5.6.1 Types of blood vessels

systemic circulation

Blood is pumped out of the left ventricle into the aorta at the start of the systemic circulation. The aorta has to accept this sudden injection of blood at high pressure, and pass it along to the large arteries. A number of large arteries then supply the major tissues and organs of the trunk (abdomen) and thorax, head and limbs. The arteries divide into smaller arteries and arterioles which in turn supply blood to a large number of capillaries. The aorta, arteries and arterioles make up the arterial tree or system. We shall look at the circulation in the capillary bed, known as the microcirculation, in more detail later. The capillaries are the smallest blood vessels and are present in the greatest number. Another important feature of the capillaries is that they are the only blood vessels where exchange of materials can occur. Fluid, containing oxygen and nutrients moves from the capillaries into the tissue spaces and supplies the tissue cells. Some of this tissue or interstitial fluid is reabsorbed back into the capillary and in this way removes waste products such as carbon dioxide from the cells and their environment. A number of capillaries then join up into venules and small veins which then go on to amalgamate into larger veins and eventually all the blood returning from the lower body feeds into the inferior vena cava while that from the upper body flows into the superior vena cava. Both the superior and inferior vena cava empty into the right atrium. The venules, veins and vena cavae constitute the venous system.

arteries

arterioles

capillaries

venules

veins

vena cavae

Π Use the above description to draw yourself a flow diagram summarising this information. We have begun one for you.

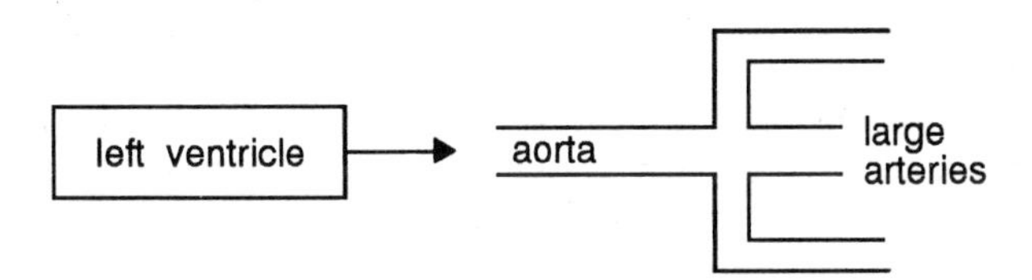

Blood leaving the right side of the heart enters the pulmonary system through the pulmonary arteries and after circulating through the pulmonary capillaries the blood is returned to the left atrium via the pulmonary veins.

5.6.2 General features of blood vessels

Before looking in detail at the structure and function of the different blood vessels it is worth considering some general structural features of blood vessels. The different materials which are used to make up the walls of the vessels determine the structure and ultimately its function (Figure 5.8).

3 layered vessel walls

Arteries and veins are similar in general structure, possessing the same three layers: tunica interna of endothelial cells on the inside lining the lumen, tunica media which contains smooth muscle and connective tissue and the tunica externa on the outside composed mainly of connective tissue.

All the blood vessels have a smooth inner lining (tunica interna) made up of a single layer of endothelial cells. The 'smoothness' of the endothelium prevents blood cells and other substances which could form a blood clot and block the blood vessel from sticking to the surface.

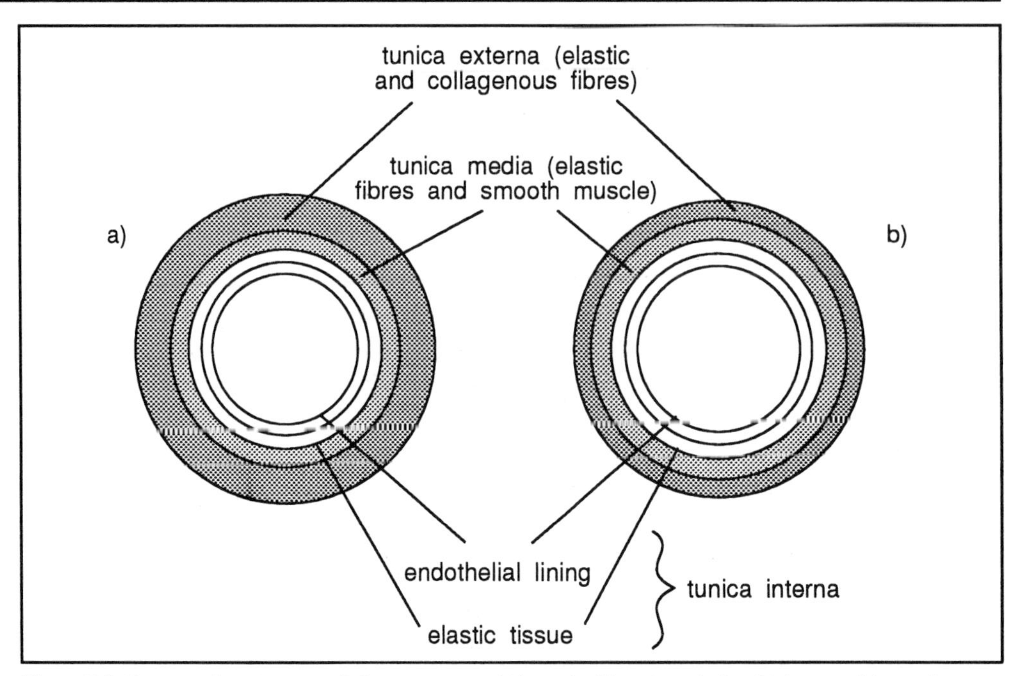

Figure 5.8 Comparative structure of a) an artery, and b) a vein. Note the relative thickness of the tunica externa in each case. The figure is a highly stylised representation of these blood vessels.

smooth muscles of blood vessels

The other major cellular component of blood vessels is smooth muscle. Blood vessels are not just passive pipe-like structures, but have to respond to the varying needs of the organs and tissues. This results in the requirement to increase or decrease the blood supply depending on the level of activity in the tissue. For example, skeletal muscle does not require a large blood supply at rest because the rate of metabolism is very low, but during exercise, metabolism in the muscle is very high and to supply sufficient energy required for muscle contraction the blood flow must increase considerably. The greater blood flow provides the large amounts of oxygen and glucose which are required to sustain the increased level of muscle activity.

This increased supply can be controlled by the relaxation of the smooth muscle in the walls of the blood vessels (arterioles) which supply the skeletal muscle with arterial blood. This has the effect of increasing the internal diameter (and therefore the cross sectional area) of the arteriole. An increase in cross sectional area provides less resistance against the blood flowing into the arteries and therefore more blood will be diverted along this easier path. The functional significance of this will be discussed later.

Besides the endothelial and smooth muscle cells there are varying amounts of connective tissue containing elastic fibres and collagen. The elastic fibres allow the walls to be stretched and then recoil to their original shape without being damaged. The collagen provides support and strength to the blood vessels.

5.6.3 Aorta and large arteries

The internal diameter of the aorta and the large arteries is large and therefore offers little resistance to the flow of blood. The walls contain large amounts of elastic fibres, especially in the tunica media, so they can become distended without damage. This has particular relevance to the aorta because blood is pumped out of the left ventricle during systole, under high pressure. When the heart relaxes during diastole the elastic recoil of the walls of the aorta will help to maintain the forward flow of blood away from the heart. A similar situation occurs in the large arteries.

The aorta and large arteries are referred to as elastic (conducting) arteries because they convert a pulsatile flow into a more continuous flow and because they conduct blood from the heart to the distribution system of the arteries.

5.6.4 Small arteries and arterioles

As the arteries become smaller so the relative proportion of smooth muscle present in the walls increases. This is particularly noticeable in the case with arterioles, the smallest arteries. These arteries have very small internal diameters and, therefore, present a large resistance to the flow of blood. The arterioles, therefore, can determine the distribution and local flow of blood to a particular tissue or organ by changing the degree of contraction/relaxation of the smooth muscle in their walls. Relaxing the smooth muscle will increase the internal diameter of the arteriole and thus will allow the blood to flow more easily (decreased resistance to flow). Conversely the contraction of the smooth muscle decreases the diameter of the arteries and this makes blood flow more difficult (increased resistance to flow).

The blood supply to the systemic circulation is arranged as a parallel distribution system so that if the supply to a particular organ or tissue is increased, for example that to skeletal muscle during exercise, then less blood is available to flow to other organs.

We shall return to some of these points when we consider the regulation of arterial blood pressure later in the chapter.

5.6.5 Capillaries

capillaries consist of endothelial cells

The capillaries are the vessels where exchange of materials takes place between the blood and the tissues. This is brought about by the filtration of the plasma into the tissues to form tissue fluid (interstitial fluid). The wall of the capillary presents a minimum barrier to the tissue fluid formation. This is because the wall of the capillary is a single layer of flattened endothelial cells, and the plasma is filtered through small gaps between adjacent endothelial cells to form tissue fluid.

The capillaries are the smallest blood vessels and the internal diameter is just large enough to allow red blood cells to pass through in single file. They are by far the most numerous vessels and each organ and tissue bed in the body is well supplied with capillaries.

5.6.6 Microcirculation

The microcirculation refers to the detailed organisation of blood flow in the tissue bed (Figure 5.9).

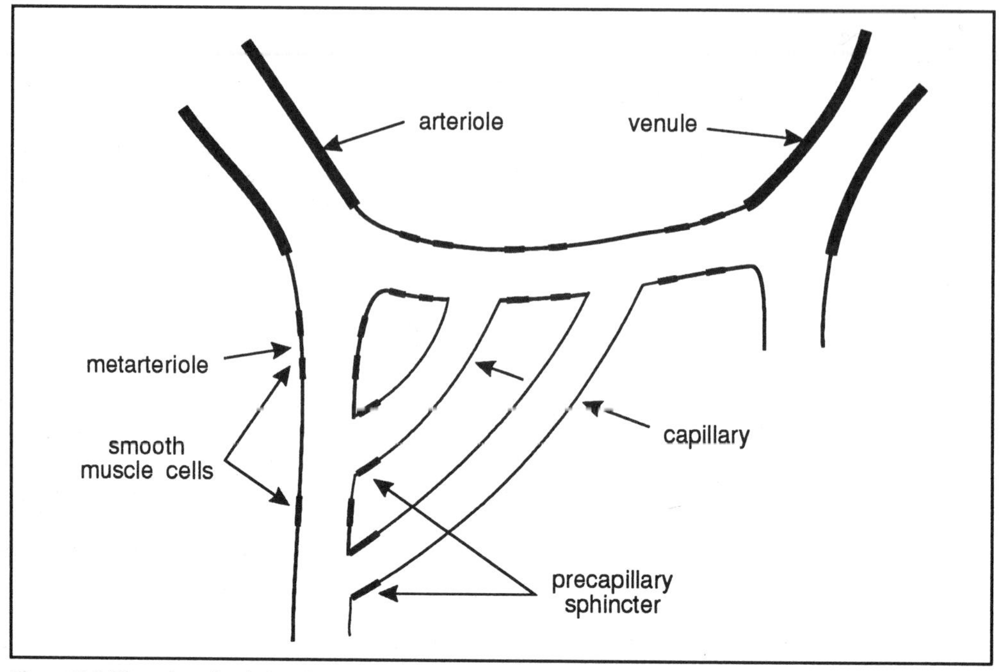

Figure 5.9 The microcirculation. Note the absence of smooth muscle in the capillaries.

metarterioles

precapillary sphincters

Most tissues have connecting vessels called metarterioles which join the arterioles to the venules directly and are therefore preferential channels for the flow of blood. The capillary bed is where the majority of exchange processing occur. Capillaries do not have smooth muscle cells along their length, whilst metarterioles have some smooth muscle in their walls. The amount of blood that flows along a capillary is governed by a ring of smooth muscle known as the precapillary sphincter around the entrance to the capillary bed. The more active the tissue becomes the more the precapillary sphincters open, and the more blood flow into the capillaries increases. Therefore, as a result of this control the microcirculation can respond to local conditions, such that, if there is a need for a raised local blood flow then the smooth muscle in the arteriolar and metarteriolar walls together with the precapillary sphincters will relax leading to a dilation and greater blood flow to that local area.

large veins have valves

Blood is collected up from the capillaries, first into venules and then into small veins. These small veins converge and form progressively larger veins until finally the blood reaches the vena cavae. In general the veins offer little resistance to blood flow as the diameter of the vessels progressively increases as the veins converge on the route back to the heart. The walls of the veins are thinner than those of the corresponding arteries and can more easily be distended or collapsed. The larger veins have valves in their walls to maintain the flow of blood back to the heart. They have relatively little elastic fibre while the larger veins have a high proportion of non-elastic fibrous tissue containing collagen. Therefore the walls are tough and can be distended with blood without recoil. At rest as much as 60 percent of the blood is in the veins. This blood can be pushed towards the heart when required by the contraction of the smooth muscle found in the walls of the veins and by the contractions of skeletal muscle acting as a pump.

5.7 Exchange at capillaries

The blood vessels allow for the blood to be transported around the body to all the various organs and tissues but it is at the capillaries that exchange of substances occurs. The tissue or interstitial fluid, which is a filtrate of the blood, acts as an intermediary between the blood and the tissues.

Π What substances either required by the tissues or produced by the tissues are exchanged at the capillary/tissue interface?

You could have made quite an extensive list of substances. Essentially what we anticipated you would include is the transport of nutrients into the tissues and the removal of metabolic end products. Thus oxygen, glucose lipids and amino acids are made available to the tissue cells by diffusion from the blood in the capillaries. While substances (eg CO_2, waste produced, hormones) produced by the cells diffuse out of or are secreted into the blood for removal or use elsewhere. If waste products were not removed this could prove harmful and eventually prevent normal cellular function.

Tissue fluid is formed at the capillaries and circulates around the tissue cells before being reabsorbed back into the blood by the capillaries. Any tissue fluid that fails to be reabsorbed back into the capillaries is drained away by the lymphatic capillaries.

Tissue fluid circulation is a most important mechanism involved in homeostasis as it provides the internal environment of the tissue cells. The total volume of tissue fluid is about 12 l and it has been estimated that about 20 l of tissue fluid is formed and reabsorbed each day, therefore a large proportion of the total body fluid is involved in this circulation around the tissues. There is a delicate balance between the rates of formation and reabsorption, which when uncontrolled quickly leads to oedema (accumulation of interstitial fluid).

5.7.1 Formation and reabsorption of tissue fluid

The main force for the formation and removal of tissue fluid is the pressure of the blood in the capillaries, the hydrostatic pressure. This pressure at the arterial end of the capillary is about 30-35 mmHg. The capillary wall is composed of a single layer of endothelial cells with pores present at the junction of these cells. These pores are large enough to allow all the constituents of plasma except the proteins to move out into the tissue spaces along with water by filtration. This process is sometimes described as ultrafiltration and the solutes move as a result of bulk flow. Besides this bulk flow through the pores, substances such as glucose can move into the tissue spaces by passage across the thin endothelial cells. The absence of plasma proteins in the tissue fluid sets up an osmotic gradient or suction force or oncotic pressure, equivalent to about 25 mmHg. In the arteriole hydrostatic (outward) pressure is greater than the osmotic suction pressure and so fluid moves out of the capillary to form tissue fluid (Figure 5.10).

ultrafiltration through capillary walls

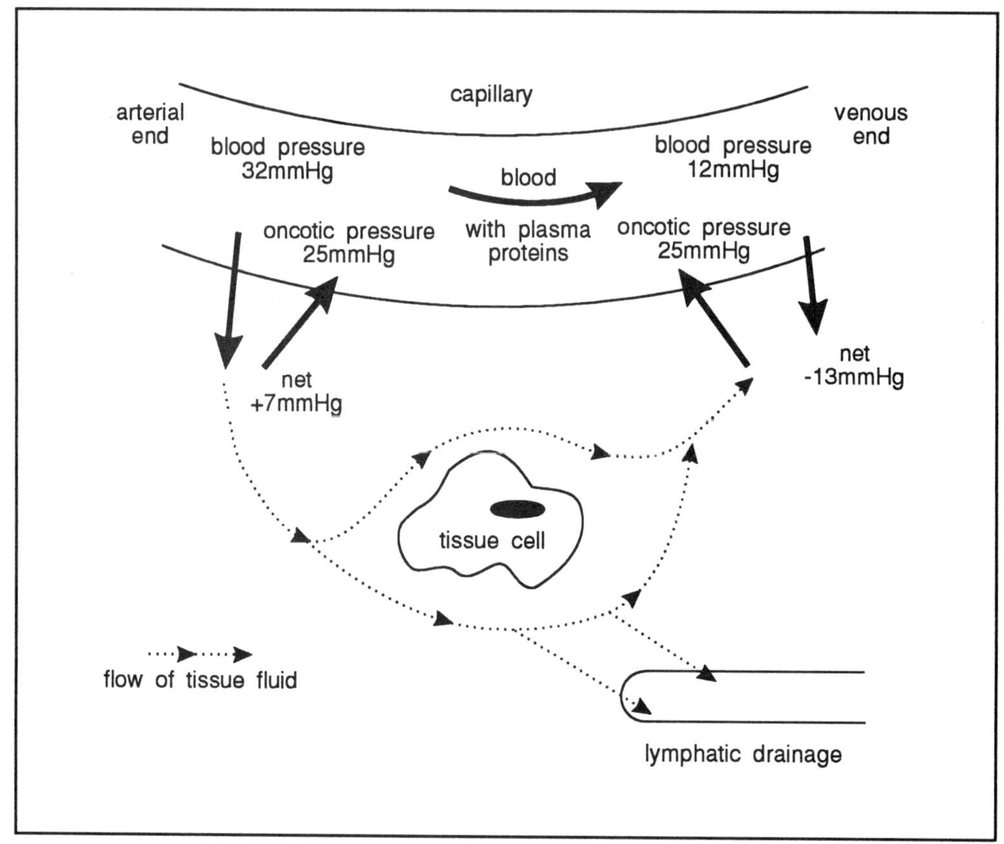

Figure 5.10 Tissue fluid formation and reabsorption.

As a result of the resistance offered by the capillary to the flow of blood, the blood pressure drops along the capillary. At the venous end the blood pressure has dropped to about 10-15 mmHg. This outward pressure from the capillary is now less than the suction pressure (25 mmHg) back into the capillary. Therefore, fluid moves back into the capillary at the venous end. This explanation of the reabsorption of tissue fluid was first suggested by Starling an eminent physiologist.

5.7.2 Lymphatic system

Sometimes the tissue fluid is not reabsorbed at the venous end of the capillary and the lymphatic capillaries are required to remove the excess tissue fluid (Figure 5.10). The lymphatic capillaries are not connected anatomically to the blood capillaries in the tissues and have a different structure from them. The small lymphatic ducts are blind at the end and larger. Their walls are composed of single layer of endothelial cells which overlap to form valve like openings at their junction. If the fluid volume and pressure increases within the tissues then the lymphatic valves are forced open. Any excess fluid enters the lymphatic capillary and is now described as lymph. It has been estimated that about 3 l of lymph are formed per 24 hours.

The lymph in the lymphatic capillaries passes in one direction into larger vessels which join up to form a network. These lymphatic vessels run from the limbs and all organs into larger trunks which eventually return the lymph to the great veins of by means of ducts situated on the neck and thorax. During this circulation, the lymph passes through lymph nodes which are accumulations of lymphoid tissue. At these lymph nodes micro-organisms and other harmful materials are removed from the lymph. In the lymph nodes lymphocytes are added to the lymph. Lymphatic vessels contain valves along their length similar to those of the veins and the movement of lymph and its eventual return to the veins is brought about by the contraction of skeletal muscle acting as a pump.

5.8 Blood pressure and flow in the vascular system

Blood flows through the vascular system because of the differential pressures in various parts of the system. Blood always flows from regions of higher pressure to regions of lower pressure. The mean pressure in the aorta is about 100 mm Hg. This pressure decreases rapidly through the arterial system and more slowly through the venous system to be only 2-3 mm Hg in the vena cava (Figure 5.11 and 5.12).

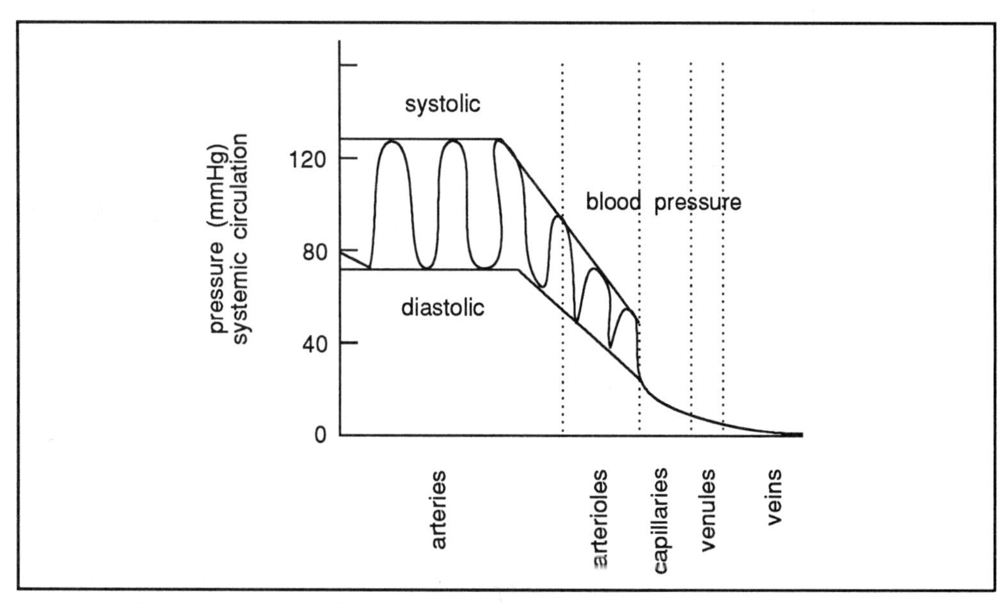

Figure 5.11 Pressures in the vascular system.

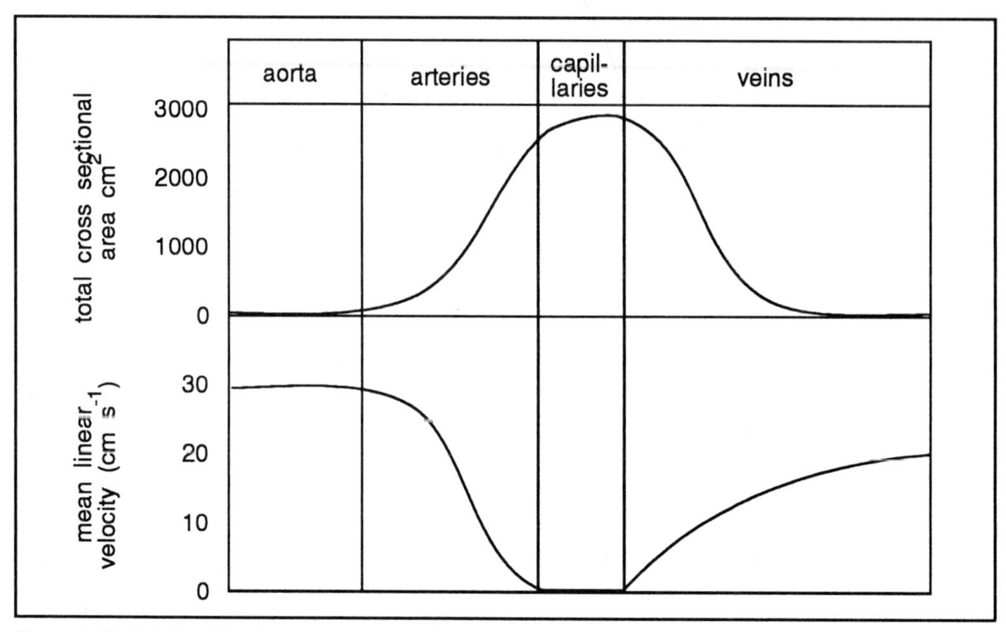

Figure 5.12 Relationships between cross sectional area and velocity of blood flow in vessels at the systemic circulation.

Pressure drops in the various parts of the cardiovascular system:

Vessel	Pressure (mm Hg)
aorta	100
arteries	100 - 40
arterioles	40 - 25
capillaries	25 - 12
venules	12 - 8
veins	10 - 5
vena cavae	2
right atrium	0

Other mechanisms also aid the unidirectional flow of blood. Blood leaving the capillaries enters the venules and veins, which are larger in diameter and therefore offer less resistance to flow. Contraction of skeletal muscles around the veins also helps drive blood towards the heart.

5.9 Factors affecting blood pressure

Blood pressure (BP) is defined as the outward pressure exerted by the blood on the wall of any blood vessel. There are several factors that influence arterial blood pressure.

5.9.1 Cardiac output (CO)

Cardiac output (CO) is the principal determinant of blood pressure. It is the amount of blood ejected by the left ventricle into the aorta each minute and is calculated by multiplying heart rate by stroke volume. Blood pressure varies directly with CO. If CO is increased by increasing stroke volume or heart rate, then blood pressure increases. A decrease in CO causes a decrease in blood pressure.

5.9.2 Peripheral resistance

vasomotor centre of medulla oblongata

Peripheral resistance is the resistance to blood flow offered by the size of the blood vessels. The smaller the diameter of a blood vessel, the more resistance it offers to the flow of blood. A major function of arterioles is to control the level of peripheral resistance by changing their diameters and, therefore, blood pressure and flow. The vasomotor centre in the medulla oblongata of the brain regulates the degree of constriction or dilation of the arterioles.

5.9.3 Blood volume and viscosity

Blood pressure is directly proportional to the volume of blood in the cardiovascular system. A decrease in blood volume (normally 5 l), for example as a result of a haemorrhage, results in a drop in blood pressure. Conversely, anything that increases blood volume, such as a high salt intake which will cause water retention, will increase blood pressure.

A decrease in the viscosity of the blood, caused by a depletion of plasma proteins or red blood cells, will also decrease blood pressure.

5.9.4 The control of blood pressure

Arterial blood pressure (BP) must be maintained between fixed levels to ensure that there is continuous perfusion of the organs and tissues. If this perfusion should cease even for a short time then irreparable damage can result. This is particularly crucial in the brain where an inadequate blood flow quickly results in the loss of consciousness. The two major factors responsible for the control of arterial blood pressure are cardiac output and peripheral resistance. The relationship between factors is:

Arterial BP = cardiac output x peripheral resistance

response to fall in BP

If blood pressure (BP) were to fall then either an increase of the cardiac output or the peripheral resistance would cause the arterial BP to rise to its normal level. The cardiac output tends to remain relatively constant at rest (about 5 l min^{-1}), therefore, the control of arterial BP is brought about mainly by adjustments of the peripheral resistance. The control of blood pressure occurs without our conscious awareness of the events involved and is carried out by the autonomic nervous system.

baroreceptors

As with any control system there are sensors within the system, these sensors continually monitor blood pressure. The sensors which monitor arterial BP are the baroreceptors. They are found in the walls of the aortic arch and the carotid sinus. They are specialised stretch receptors which respond to the degree of stretching in the arterial walls. The degree of stretching depends on the pressure of the blood in the lumen of these arteries. The baroreceptors send their information about the blood pressure along afferent nerve fibres to the vasomotor centre (VMC) in the medulla of the brain which ultimately controls the blood pressure. Information is also sent to the cardioinhibitory and cardioacceleratory centres in the medulla.

The VMC controls arterial BP by sending signals via the sympathetic efferent nerve fibres to the smooth muscle in the arteriolar walls. The state of contraction or relaxation of the smooth muscle depends on the level of nervous activity sent out by the VMC. The peripheral resistance, which has a major influence on the arterial BP, is therefore, determined by the state of contraction or relaxation of the smooth muscle in the arteriolar walls.

The cardiac centre sends signals via the efferent nerve fibres to the heart which either increase or decrease cardiac output to help regulate arterial BP.

SAQ 5.6

Arrange the following into the sequence of events that occurs if the arterial BP falls. Begin with 1) and finish with 8).

1) A fall in arterial BP will lead to less stretching of the arterial walls.

2) The VMC receives less inhibitory input from the baroreceptors.

3) The VMC increases sympathetic activity in the efferent fibres to the smooth muscle in the arterioles.

4) Vasoconstriction leads to an increased peripheral resistance.

5) Stretching of arterial walls detected by the baroreceptors.

6) The increased activity in the efferent sympathetic fibres causes more contraction of the smooth muscle of the arterioles and so vasoconstriction.

7) Baroreceptors send fewer nerve impulses along the afferent nerves to the VMC and the cardiac centre.

8) Increased peripheral resistance leads to an increase in arterial BP.

Note, read our response carefully it has some important additional information.

The opposite series of responses are set into operation if the arterial BP rises too high. However, in this situation the end result will be relaxation of the smooth muscle in the arterioles (vasodilation) and a reduction in peripheral resistance, as well as a reduced cardiac output. These changes will correct the initial rise in blood pressure.

carotid bodies, chemoreceptors and arterial BP

Various chemoreceptors which are sensitive to chemicals in the blood, are also involved in the regulation of arterial BP. Chemoreceptors near or on the carotid sinus and aorta are called carotid and aortic bodies, respectively. An oxygen deficiency (hypoxia), an excess of carbon dioxide (hypercapnia) or an increase in hydrogen ion concentration will stimulate the chemoreceptors and send impulses to the VMC. In response the VMC increases sympathetic stimulation to the arterioles to bring about vasoconstriction and an increase in blood pressure.

Higher brain centres such as the cerebral cortex can also influence blood pressure. During anger the cerebral cortex stimulates the VMC to increase sympathetic discharge to the arterioles causing vasoconstriction and an increase in blood pressure. During emotional upset impulses from the higher centres may inhibit the VMC to cause vasodilation and a decrease in blood pressure.

5.9.5 Local control of blood flow

The VMC is the central means of controlling arterial BP and thus it determines the overall basal level of arteriolar vasoconstrictor tone. However, there is also a means of local control in any individual organ or tissue which can override the VMC basal tone locally. The reason for this arrangement is that the more active tissues require a greater supply of blood to enable their higher rate of metabolism to be maintained. Therefore, the tissues must be able to control local vasodilation which in turn will cause more blood to flow into the tissue. This vasodilation is often due to the increase in the products of the tissue's increased metabolism. An example of this control is the production of lactic acid which acts directly on the smooth muscle in the walls of the blood vessels during exercise. In this case there is a vast increase in the flow of blood through the skeletal muscle beds due to local dilation caused by lactic and whilst a generalised vasoconstriction is controlled by the VMC which restricts blood flow to other organs and tissues and therefore maintains arterial BP at approximately normal levels.

5.10 Circulatory changes during exercise

response to low oxygen tensions

vasodilation

Blood pressure is regulated within a range of normal values, and blood flow through tissues is closely matched to the metabolic needs of the tissue. During exercise blood flow through skeletal muscle is dramatically changed. The rate of blood flow through exercising skeletal muscle may increase 20 fold compared to that of resting muscles. The increased blood flow is brought about by both local and systemic regulatory mechanisms. At rest only 20-25 percent of skeletal muscle capillaries are open. During exercise this proportion increases to 100 percent. Low oxygen tensions resulting from increased muscular activity and the release of vasodilators such as lactic acid and carbon dioxide increase dilation of the capillaries. Increased sympathetic nerve stimulation causes vasoconstriction in the blood vessels of the abdominal organs and viscera, whereas the blood vessels of skeletal muscles dilate. Consequently blood is shunted from the viscera and abdominal organs to the skeletal muscles. Sympathetic stimulation of the heart results in an increased cardiac output and the systolic blood pressure increase by between 20 and 60 mmHg. Also, the movement of skeletal muscles and the constriction of veins greatly increases the venous return to the heart. The distribution of the systemic cardiac output at rest and during strenuous exercise is presented below.

	rest (ml min^{-1})	strenuous exercise (ml min^{-1})
Brain	750	750
Heart	250	750
Muscle	1200	12 500
Skin	500	1900
Kidney	1100	600
Abdominal		
Organs	1400	600
Other	600	400
Total cardiac		
Output	5800	17 500

SAQ 5.7

Explain why there is an increased blood flow to the skin and a reduced blood flow to the abdominal organs during exercise?

5.11 Aging and the cardiovascular system

arteriosclerosis and atherosclerosis

Changes associated with aging include loss of elasticity of the aorta, reduction in cardiac muscle cell size, loss of cardiac muscle strength, a reduced cardiac output and an increase in blood pressure. There is an increased incidence of coronary artery disease, the major cause of heart disease and death in older people. Congestive heart failure, where the pumping performance of the heart is impaired, also occurs. Hardening of the arteries and deposition in arteries that supply the brain (arteriosclerosis and atherosclerosis see 5.5.2) result in malfunction or death of brain tissue.

5.12 Hypertension

Hypertension, or high blood pressure, affects approximately 20 per cent of the population at some time in their lives. A person is generally considered hypertensive if the systolic blood pressure is greater than 150 mmHg and the diastolic blood pressure is greater than 90 mmHg. However, as normal blood pressure is age dependent, classification of an individual as hypertensive depends on his age.

Chronic hypertension has an adverse effect on both the function of the blood vessels, the heart, and some other organs. The heart has to perform an increased amount of work which leads to hypertrophy of the left ventricle and possible heart failure. Hypertension also increases the rate of atherosclerosis development which increases the probability of blood clot formation or blood vessel rupture.

Some conditions leading to hypertension include production of excess aldosterone or angiotensin, reduced renal blood flow and some adrenal medullary tumours. Any of these conditions increases the cardiac output and peripheral resistance, both of which result in a greater blood pressure. Although these conditions result in hypertension, about 90 per cent of hypertensive patients suffer from idiopathic or essential hypertension, in which the cause of the condition is not known.

Π Can you think of any factors which may be controlled by drugs which could be used in the treatment of hypertension?

Drugs that dilate blood vessels or those which increase the rate of urine production (diuretics), or drugs which decrease cardiac output are used to treat essential hypertension. The vasodilator drugs reduce the peripheral resistance and improve renal blood flow. Diuretics increase the rate of urine production and reduce the extracellular fluid volume thereby reducing the load on the heart. Beta blocking drugs decrease the cardiac output by reducing the heart rate and force of contraction.

Summary and objectives

In this chapter we have examined the circulatory system. We first gave a brief description of blood including both its composition and its functions. This description included an outline of the types of cells that are present and the relationship between blood and interstitial fluid and lymph. We then went on to examine the circulatory aspects of blood including the structure and function of the heart. We included a description of the enervation of the heart and the control of cardiac output. We then described the peripheral blood vessels with particular emphasis on the microcirculatory system through the capillaries. In the final part of the chapter we discussed blood pressure and the flow of blood in the vascular system with some comment on the consequences of exercise, aging and hypertension.

Now that you have completed this chapter you should be able to:

- list the major functions of blood;
- list the major components of blood;
- describe in outline the processes of clot formation and dissolution;
- ascribe functions to various blood cell types;
- identify and correctly label a large number of structures associated with the circulatory system;
- calculate cardiac output from supplied data;
- describe the process involved in heart contraction;
- describe the sequence of events in which arterial blood pressure is self regulating;
- use a wide variety of terms associated with the description of the circulatory system.

Respiration

Respiration

6.1 Introduction

oxygen

carbon dioxide

Respiration is the process by which gases are exchanged between the body and its environment. The two principal gases involved are oxygen (O_2) and carbon dioxide (CO_2). Oxygen is required for the production of energy, via oxidation of ingested foodstuffs, while carbon dioxide is produced as a waste product of this process and must be eliminated from the body. The organs responsible for gas exchange in the mammalian body are the lungs. In order for the body to work efficiently, there is a close relationship between the lungs and the circulatory system since it is this latter system which transports the gases around the body. Furthermore, the respiratory system must also be flexible in that it must adapt to the differing demands placed upon it by a functional body. For example, when we exercise the body requires increased amounts of oxygen to oxidise foodstuffs and produces increased amounts of carbon dioxide which must be removed.

In this chapter we will begin by looking at the structure of the respiratory system and relating its structure to its function. Secondly, we will consider how gases are exchanged, what volumes are exchanged, and how they are measured. Next we will look at how these gases are transported from the lungs to the cells of the body and vice-versa. Finally, we shall consider how respiration is regulated and how it can be altered to meet the changing needs of the body.

6.2 Structure of the respiratory system

6.2.1 Gross structure

It can be seen from Figure 6.1 that the respiratory system consists of two lungs contained within the thoracic cavity. Air enters the lungs (via the nose, mouth) and through the trachea. Surrounding the majority of the circumference of the trachea are rings of cartilage.

Π Can you think of an explanation for the presence of rings of cartilage in the trachea (an otherwise soft tissue)?

bronchi

bronchiole

alveoli

The function of the tracheal cartilage is to provide support, keeping the trachea open at the same time allowing flexibility to accomodate head and neck movements. The trachea divides into two bronchi supplying either the right or the left lung. Each bronchus sub divides into smaller tubes called bronchioles or (singular = bronchiolus). Eventually this sub division into increasingly finer tubes terminates in a series of small sac-like structures called alveoli (singular = alveolus). It is in the alveoli that gas exchange between the atmosphere and blood takes place. Most of the lung tissue consists of alveoli. These small structures represent a very large surface area for gas exchange. In fact, the total surface area of the alveoli is about 80 m^2, the size of a tennis court.

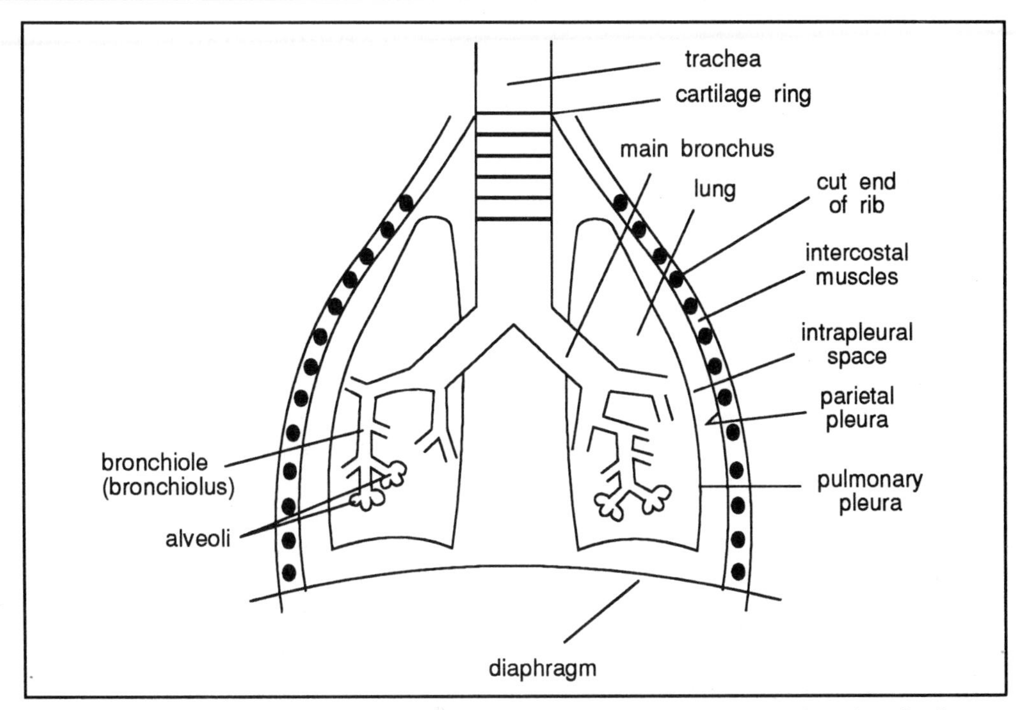

Figure 6.1 Stylised diagram illustrating the structure of the human respiratory system (greatly stylised).

pulmonary and parietal pleura

The external surface of the lungs is covered by a thin layer of tissue called the pulmonary pleura. Lining the inside of the thoracic cavity is the parietal pleura. The two pleura are in contact with each other but are not joined together. They are separated and lubricated by a very small volume of fluid. We shall see the significance of this later. The base of the chest cavity is closed by the diaphragm. This is a large sheet of striated muscle which separates the thorax from the abdomen. The external framework of the chest cavity consists of the ribs which form a hollow cage. Lying between the ribs are the intercostal muscles which are important for movement of the rib cage during respiration. The top of the chest cavity is closed by the muscles at the neck. Thus we can think of the lungs as lying in a fully enclosed cavity.

diaphragm

intercostal muscles

6.2.2 Histology (microstructure) of the respiratory system

Figure 6.2 shows a cross section through a bronchiolus. This is one of the many small tubes produced by the divisions of the bronchi, before they terminate in the alveoli.

ciliated epithelial

Surrounding the bronchiole is layer of smooth muscle. This muscle is not under voluntary control. The degree of contraction of this muscle layer alters the effective diameter of the bronchiole. What effect does this have on the respiratory system? It limits the amount of air which can enter (and leave) the lungs. People who suffer from asthma, have particularly sensitive smooth muscle, which contracts very rapidly in response to a variety of stimuli (for example allergens). It is this so called bronchoconstriction which leads to their breathing difficulties. Also shown are goblet cells and ciliated epithelial cells. These two cell types constitute part of the defence system of the lungs. The goblet cells secrete mucus which will trap any inhaled foreign objects eg dust particles. The cilia waft any such particles back to the nose and throat, from where they are disposed. The terminals of the bronchiolar divisions are the alveoli,

goblet cells

the sites where gas exchange occurs. The structure of an alveolar terminus is shown in Figure 6.3.

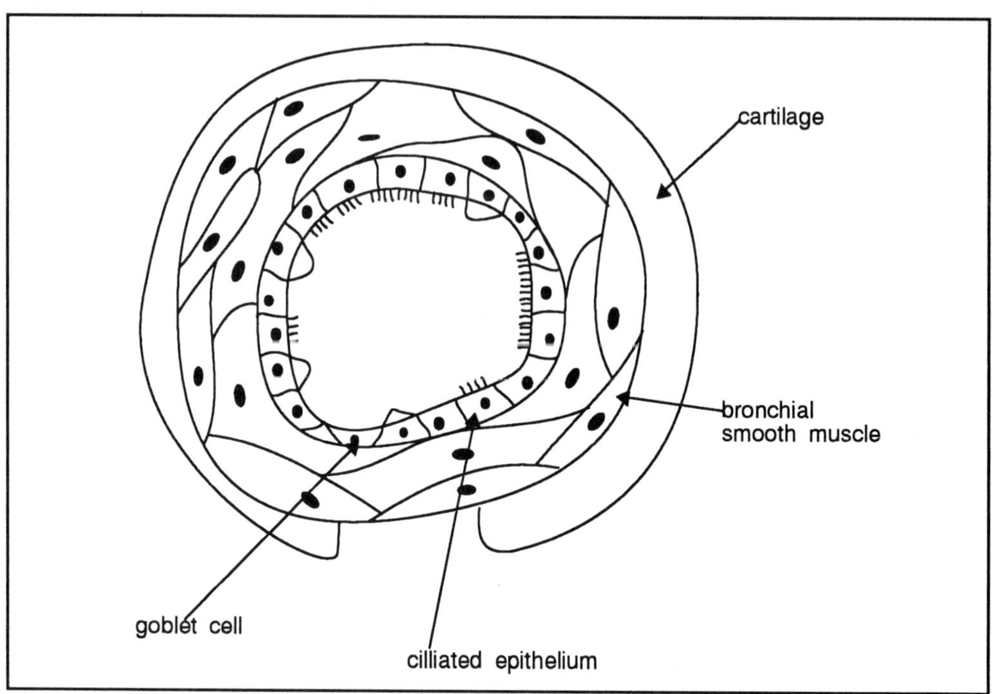

Figure 6.2 Stylised diagram showing a cross section through a bronchiolus.

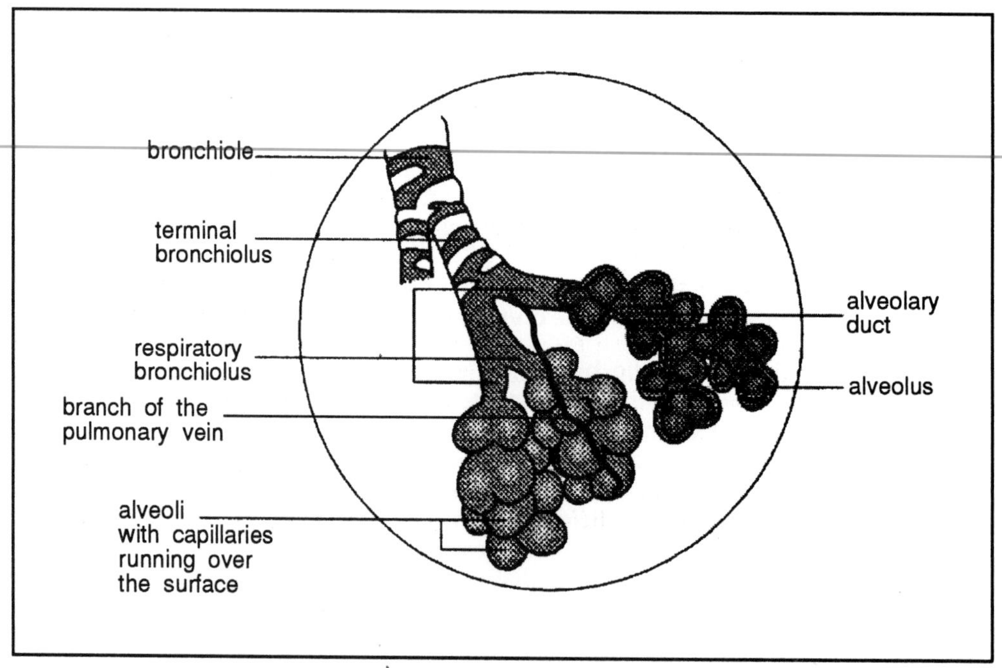

Figure 6.3 The structure of alveolar termini showing the relationship to the capillaries of the pulmonary circulation.

Type I cells

The alveoli are hollow sacs which are continuous with the bronchioles. They are lined with epithelial cells, called Type I cells. These cells are in close contact with the blood capillaries of the pulmonary circulation. The pulmonary circulation contains the blood which is pumped from the right ventricle of the heart through the lungs and back to the left atrium of the heart. The distance separating the air in the alveoli and the blood in the capillaries is about 0.2 mm, an extremely small distance.

Π Why do you think the distance between blood and air is so small?

The distance between blood and air is so small because it permits rapid diffusion of gases between the two compartments. The gases in the alveoli are also moist, another factor which enhances their diffusion. Within the alveoli there are also a few Type II epithelial cells. These cells produce a detergent like substance called surfactant. The role of surfactant in inspiration will be discussed later.

SAQ 6.1

List three factors which ensure that the process of gas exchange (and transport of gases) is efficiently achieved.

6.3 Ventilation of the lungs

inspiration and expiration

Ventilation of the lungs describes the physical process whereby air enters and leaves the lungs. Air entering the lung is called inspiration, air leaving the lung is called expiration. Air will move from regions of high pressure to regions of low pressure, rather analogous to a substance moving down its concentration gradient. Since air in the environment is at atmospheric pressure (about 760 mm Hg) for air to enter the lungs, the pressure within the lungs must be at sub-atmospheric pressure. In this section we will consider how this sub-atmospheric pressure is generated and how the process of inspiration and expiration is achieved.

6.3.1 Pulmonary pressures

When considering movements of the lungs (and of the thoracic cavity as a whole) during inspiration and expiration, it is important to realise that the structures involved are elastic in nature. This means that these structures have an optimum position and shape to which they will always return if they are moved from that optimum position. In the case of the lungs they have a natural tendency to collapse, whilst the chest wall has a tendency to expand outwards. This is most readily observed when the chest wall and pleural membranes are punctured, for example during surgery of the chest wall. This situation does not occur in normal life. Therefore, these two opposing forces must be balanced out. What is the basis of the forces involved?

Let us consider the situation occurring at the end of a normal expiration. This phase of respiration can be considered to be the system at rest. In particular let us look at the pressures within different parts of the respiratory system. This is shown in Figure 6.4.

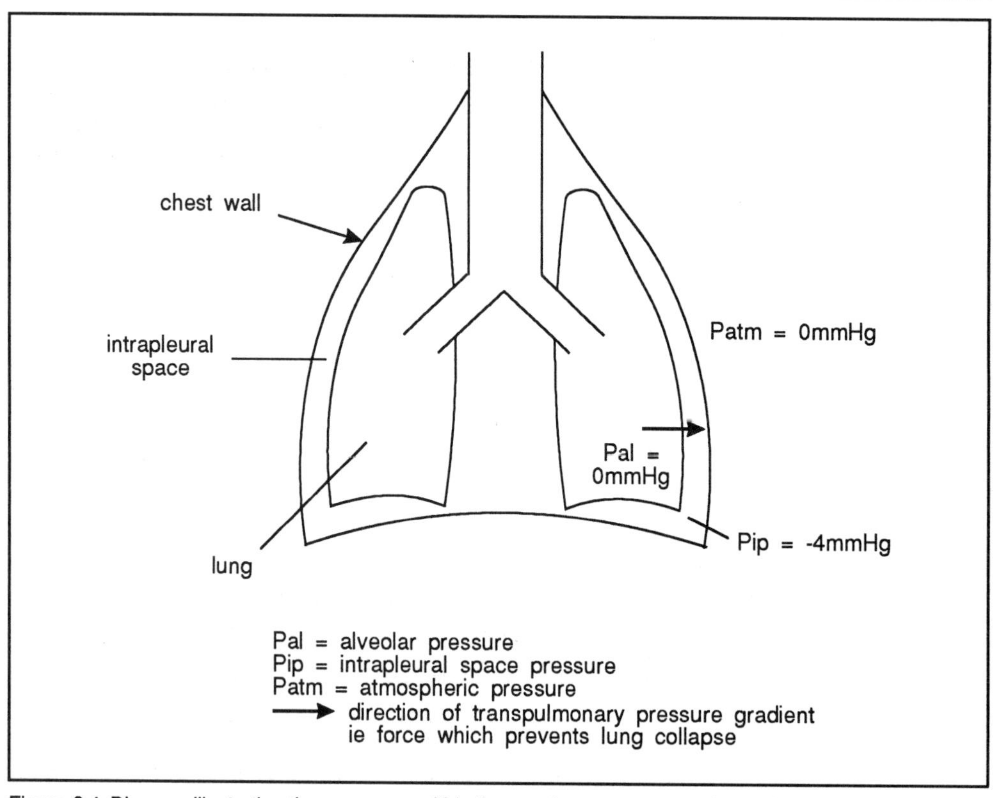

Figure 6.4 Diagram illustrating the pressures within the respiratory system at the end of a normal expiration (see text for details). Adapted from Vander, AJ, Sherman JH and Luciano DS (1990) Human Physiology: The Mechanism of Body Function, McGraw Hill, New York.

All the pressure quoted in the Figure are relative to atmospheric pressure (760 mm Hg). (Note 760mmHg = latmos. In SI units this is 101 325 N sm^{-2}. In this text we will confine our discussion to using mmHg as the units of pressure since this is the unit commonly used in respiratory studies). Thus a pressure of +10 mm Hg is really 770 mm Hg, whilst a pressure of -10 mm Hg is really 750 mm Hg. At the end of an expiration the pressure within the alveoli of the lungs is 0 mm Hg (ie Pal = 0). The pressure within the pleural space (ie between parietal and pulmonary pleura) is 4 mm Hg below that of atmospheric pressure, ie -4 mm Hg. Thus there is a pressure gradient across the wall of the lung.

Π Can you calculate the pressure difference and the direction in which it is operating?

transpulmonary pressure

intrapleural pressure

The pressure difference is 0 - (-4) mm Hg = +4 mm Hg. This force is acting outwards. It is this force which prevents the lungs from collapsing (as is their natural tendency). It is called the transpulmonary pressure. Equally, there is also a force which prevents the chest wall from expanding outwards. This force is generated by the pressure difference between atmospheric pressure on the outside of the chest wall and the intrapleural pressure on the inside of the chest wall. This force is acting inwards and exactly balances the outward acting forces. Thus, the lungs and chest wall are at equilibrium. The question that needs to be answered is why should the intrapleural pressure be sub-atmospheric? We have stated above that the natural tendency of the lungs is to collapse whilst that of the chest wall is to expand outwards. Furthermore, we have said this does not happen because of equal and opposite forces balancing out the tendencies for movement to occur. However, in reality there is a very small movement of the lungs and chest wall in their relative directions. The result of this is that there is a very small increase in the volume of the pleural space. You should remember that this space is fluid filled. Because it is fluid filled, the small increase in volume causes a small drop in its pressure. Hence the pressure in the pleural space becomes sub-atmospheric.

SAQ 6.2

Given the above information explain why damage and opening of the pleural space causes the lungs to collapse.

6.3.2 The process of ventilation

Ventilation is as we have already said the mass movement of air into and out of the lungs on a rhythmic basis, ie breathing. For air to enter the lungs, the pressure within the alveoli must become sub-atmospheric. Only when this happens can air enter the lungs since air moves from regions of high pressure to low pressure, in this situation from atmospheric pressure to sub-atmospheric pressure. How does alveolar pressure become sub-atmospheric? Changes in alveolar pressure are caused by changes in the volume of the lungs. Pressure and volume are related to each other by Boyles Law. At its simplest this tells us that for an increase in the volume of a closed container the pressure within that container decreases. Conversely a decrease in volume causes an increase in pressure. As an analogy, think of a given amount of gas in a given volume as the concentration of that gas. Think of concentration as being equivalent to pressure. Therefore, without changing the amount of gas, increasing the volume of the container in which the gas is held decreases its concentration and therefore decreases its pressure. Let us now consider how air enters the lungs. Remember the lungs are air filled sacs contained within a closed container, the thoracic cavity. At the end of an expiration/beginning of an inspiration alveolar pressure is 0 mm Hg ie the same as atmospheric pressure and intrapleural pressure is -4 mm Hg. Inspiration is initiated by contraction of the diaphragm and the inspiratory intercostal muscles. What effect does this produce? The diaphragm flattens and the inspiratory intercostals moves the chest wall upwards and outwards causing an increase in the volume of the thoracic cavity. As a consequence of this the inside of the chest wall moves away from the lungs. Thus the pressure in the intrapleural space becomes more negative as its volume increases. Therefore, the transpulmonary pressure also increases, which causes the lungs to increase their volume. Thus, the lungs move to the same extent as the chest wall, by virtue of the two pleural membranes being in close proximity to each other. Because the volumes of the lungs increase so the pressure within them decreases (Boyles Law). Thus, alveolar pressure becomes sub-atmospheric and therefore air is drawn into the lungs. At the end of inspiration, alveolar pressure is again at atmospheric pressure (air has moved in to equalise the pressure with that on the outside).

Boyles Law

How is the air expelled from the lungs? The first stage in this process is that the activity of the nerves which cause contraction of the diaphragm and inspiratory intercostal muscles are inhibited. Therefore, both the diaphragm and the intercostal muscles relax. This causes the lungs to collapse inwards and as they collapse the pressure of the air within the lungs increases. The internal pressure then exceeds the atmospheric pressure and air moves out of the lungs to the environment. Expiration is a passive process compared to inspiration which is an active process. In certain circumstances expiration can be an active process for example during coughing. In these cases, rapid contraction of the expiratory intercostal muscles causes the chest wall to return to its original position. Simultaneous contraction of muscles in the abdominal wall forces the diaphragm upwards. Both these processes actively deflate the lungs and move air out

expiration

of them with considerable force. Exercise is another example of a situation in which expiration is an active process.

SAQ 6.3

Draw a flow chart to show the steps involved in inspiration.

6.4 The role of surfactant

Surfactant is a detergent like compound produced in the alveoli.

Π Where is surfactant produced in the lung?

surfactant and Type II cells

Surfactant is produced by the Type II cells of the alveolar epithelium. We described above how the degree of lung expansion, that is the increase in volume, is dependent upon the degree of change of the transpulmonary pressure. One factor affecting the expansion of lungs for a given change in transpulmonary pressure is the compliance of the lung. By compliance we mean the stretchability of the lungs. The higher the value of this compliance then the easier it is for the lungs to expand and thus take in air. When the compliance is reduced it is more difficult to breath in. In this case there must be stronger contraction of the diaphragm and inspiratory intercostals to cause changes in the pressure of the pleural space and therefore in the transpulmonary pressure and lung expansion. What factors determine lung compliance? The first, and least important, factor is the elastic tissue content of the respiration system. An example is the thickening of the lung tissue which is associated with decreased elasticity and therefore decreased lung compliance. The second and most important determinant of lung compliance is the surface tension within the alveoli. Surface tension arises at any air/water interface. Because of surface tension water dripped onto a dry surface remains as drops rather than spreading out in a thin film. The inner lining of the alveolar epithelium is moist and is in direct contact with air. The effect of surface tension is due to the attractive forces between individual water molecules. In the alveoli, the effect of surface tension tends to collapse the alveoli and this would make them unable to carry out their function of gas exchange.

elastic tissue

Any expansion of the lungs has to overcome this potentially collapsing force. The role of surfactant is to reduce the surface tension within the alveoli. We can think of the surface tension within the alveoli as tending to collapse these sacs, rather like the tension in a blown up balloon. Thus:

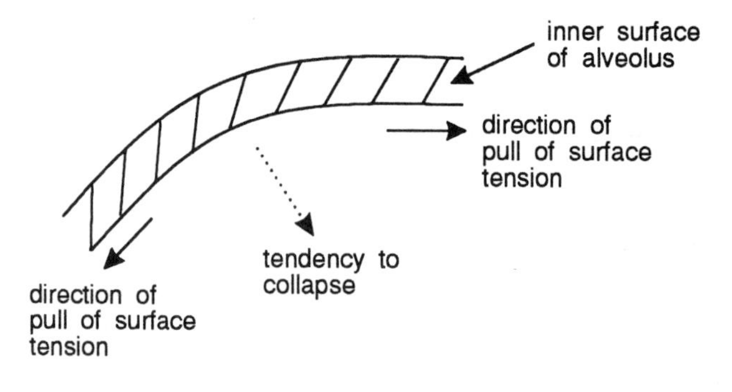

The surfactant produced in the alveoli reduces the surface tension. Therefore there is a reduction in the force trying to collapse the alveoli. Thus it reduces the potential of the alveoli to collapse.

SAQ 6.4

Some people may not produce sufficient quantities of surfactant. In order to maintain normal respiration, would the change in their intrapleural pressure have to be greater or smaller than that of a normal person.

6.5 Lung volumes

spirometer

So far we have described the process of ventilation but have ignored the volume of air which is inspired and expired during respiration or the actual volume of the lungs themselves. In this section we will consider these topics. Such measurements are carried out using a spirometer which produce traces like the one shown in Figure 6.5.

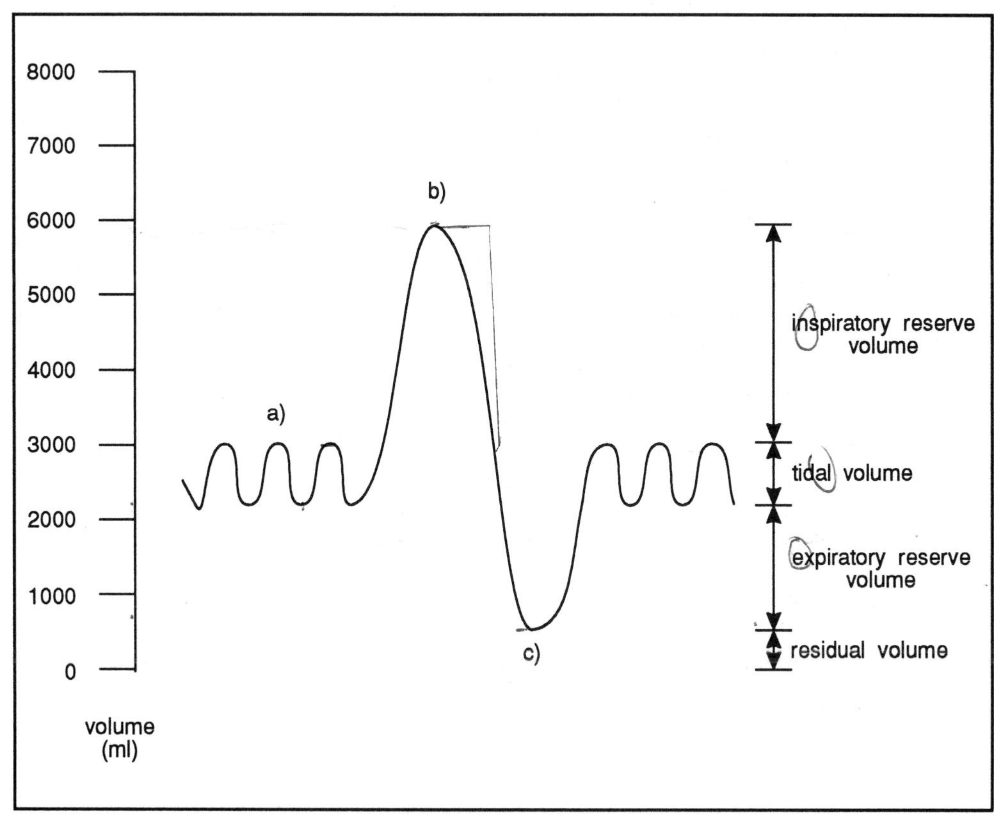

Figure 6.5 A typical spirometer trace showing the various lung volumes (see text for discussion).

tidal, inspiratory reserve and expiratory reserve volumes

The initial oscillations (A) of the spirometer trace represent the volume of air that is taken in or given out with each normal breath. This is called the tidal volume (TV) and in an average male is about 500 ml. If, instead of taking a normal breath, a subject takes the biggest inspiration that is possible for the him/her to take the volume over and above tidal volume is called the inspiratory reserve volume (IRV), (see B in Figure 6.5). Similarly, there is also an expiratory reserve volume (ERV) which is the maximum amount of air that can be forcibly exhaled (C) over and above a normal expiration. The sum of tidal volume, inspiratory reserve volume and expiratory reserve volume is called the vital capacity (VC). The vital capacity is one lung parameter which is routinely measured during clinical investigation of the lung. Also measured during a clinical investigation of lung function is the forced vital capacity (FVC), forced because a subject goes from maximum inspiration to maximum expiration (B to C) as fast as possible. Even after a maximal expiration there is still air left in the lungs. This is called the residual volume. This cannot be measured using a spirometer. Actual lung volume is influenced by such factors as age, build, sex, degree of fitness and presence of disease.

vital capacity

SAQ 6.5

From Figure 6.5 calculate the approximate volumes of:

1) Tidal volume.

2) Inspiratory reserve volume.

3) Expiratory reserve volume.

4) Vital capacity.

A proportion of the tidal volume of air which is taken in with each breath, will not reach the alveoli to be used in gas exchanges. This is because some air remains in the trachea and bronchioles themselves. The volume of this portion of the respiratory system, ie the part which does not permit gas exchange, is called the dead space. A typical value for the volume of the dead space is 150 ml. The total ventilation of the lungs is called the minute ventilation. This is calculated from the tidal volume multiplied by the number of breaths taken per minute.

SAQ 6.6

A person takes 15 breaths per minute with a tidal volume of 600 ml. Calculate the minute volume. If the dead space of this person is 150 ml what volume of air is then actually used for gas exchange?

6.6 The transport of gases

Having transferred air into the alveoli within the lungs we must now examine how gases are transported around the body.

Oxygen must be supplied to the tissues for the process of energy production (catabolism). Carbon dioxide is a waste product of these processes and must be removed. Thus oxygen must be transferred from alveoli to blood and finally to cells. Carbon dioxide passage occurs in the opposite direction. The actual physical process whereby gases are exchanged between these different sites is by simple diffusion. Diffusion is defined as the passive movement of substances from regions of high concentration to regions of low concentrations. When we talk about the concentration

diffusion

of a gas we usually refer to its partial pressure. This in the case of oxygen is written in shorthand form as pO_2. What do we mean by partial pressure? The law of partial pressures states that each gas in a mixture of gases exerts the same pressure it would if it were the only gas present. Thus atmospheric air has a pressure of 760 mm Hg and has the following composition: 78.9% nitrogen; 21% oxygen; 0.03% carbon dioxide, the remainder being other inert gases. Thus the partial pressure of nitrogen (pN_2) in atmospheric air is $760 \times \frac{78.9}{100} = 599.6$ mm Hg.

Π Calculate the partial pressure of O_2 (pO_2) and the partial pressure of carbon dioxide (pCO_2) in atmospheric air from the figures given above.

Your calculations should have shown $pO_2 = 159.6$ mm Hg (from $\frac{21}{100} \times 760$) and $pCO_2 =$ 0.228 mm Hg (from $\frac{0.03}{100} \times 760$).

Gases will move from regions of high partial pressure to regions of low partial pressure. Let us consider the pO_2 in different regions of the body.

Alveolar $pO_2 = 106$ mm Hg; Pulmonary capillary blood $pO_2 = 40$ mm Hg; Cell $pO_2 = <$ 40 mm Hg.

Thus the direction of oxygen transfer is from alveoli into the capillaries of the pulmonary circulation which then passes back to the heart where it is pumped around the body where the oxygen is given up to the cells.

6.6.1 The transport of oxygen

haemoglobin

Oxygen is carried in the bloodstream from the lungs to the cells of the body. It is carried bound to the oxygen transport protein haemoglobin. Haemoglobin contains iron and it is this which gives blood is characteristic red colour. The advantage of using haemoglobin is that is allows increased amounts of oxygen to be carried by the blood compared to the amount that can be carried in simple solution. The structure of haem, the iron containing moiety within the globin molecule, is shown in Figure 6.6.

oxyhemoglobin and deoxyhemoglobin

Haemoglobin is a conjugated protein. It consists of an iron containing porphyrin (haem) joined to a protein (globin). Four of these units join together to form a single molecule of haemoglobin. At the centre of each haem group is an atom of iron. Each atom of iron can bind a single molecule of oxygen. Thus a molecule of haemoglobin can carry four molecules of oxygen. It is important to note that the carriage of oxygen is not a chemical oxidation process. When haemoglobin has oxygen bound to it, it is referred to as oxyhemoglobin. Without any oxygen bound, it is referred to as deoxyhaemoglobin. In a blood sample which may contain millions of haemoglobin molecules, the proportion that is present as oxyhemoglobin is referred to as the percentage saturation of haemoglobin. This is calculated as follows:

$$\text{Percentage saturation} = \frac{\text{Amount oxygen bound}}{\text{Maximal oxygen bound}} \times 100\%$$

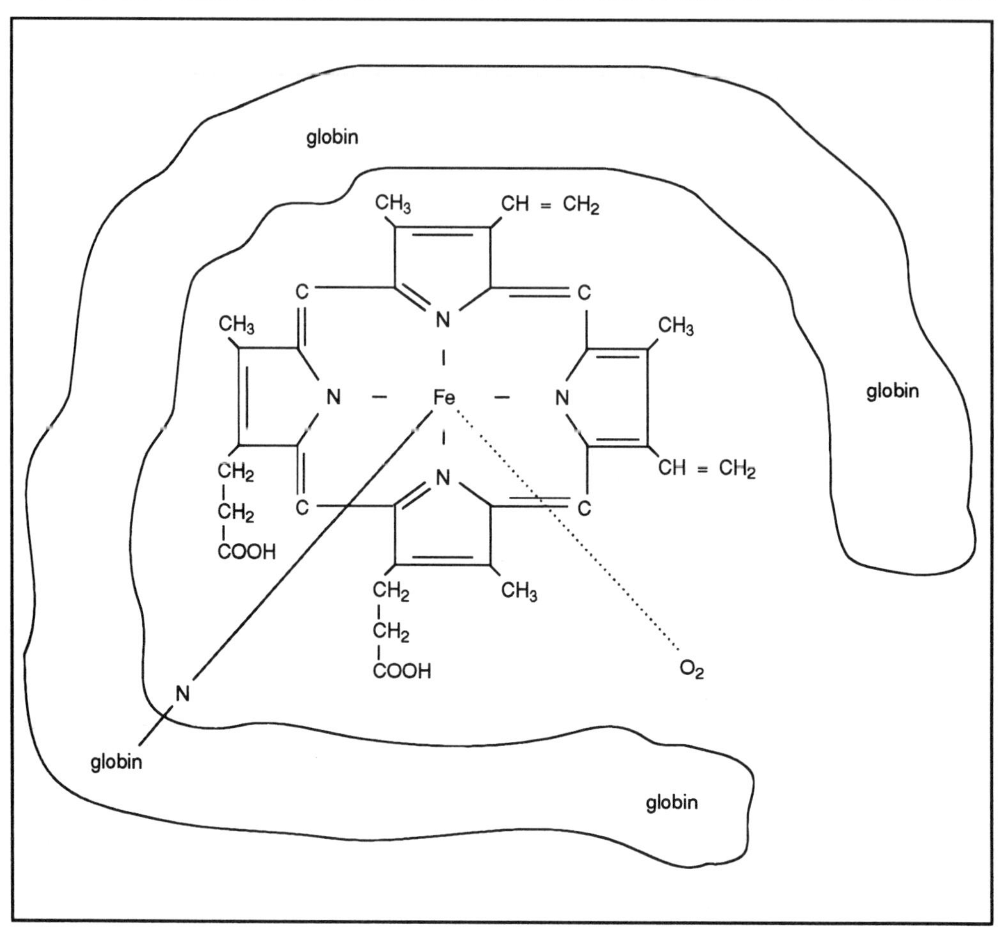

Figure 6.6 Diagram illustrating the structure of haemoglobin. The globin portion is represented only diagrammatically. It is a large protein with a molecular weight of 16000 D.

The most important factor which determines the percentage saturation of haemoglobin is the partial pressure of oxygen to which it is exposed. The relationship between the partial pressure of oxygen and the percentage saturation of haemoglobin is shown by the oxygen haemoglobin dissociation curve (Figure 6.7).

As can be seen from Figure 6.7 the relationship between partial pressure of oxygen and percentage saturation is not a simple one. The curve is sigmoid (S-shaped). This indicates a degree of co-operative binding. What do we mean by this? Simply it means that the binding of the first molecule of oxygen to a sub unit of haemoglobin, aids the binding of a second molecule to another sub unit, which in turn aids the third, which then aids the binding of the forth molecule of oxygen. The reverse is also true when the oxygen is being unloaded into the cells. When a blood sample is 100% saturated it contains about 200 ml of oxygen per litre of blood. (Note that a fuller description of co-operative binding of molecules by proteins is given in the BIOTOL texts 'Principle of Cell Energetics' and in 'Principles of Enzymology for Technological Application'. Plasma alone will only dissolve about 5 ml of oxygen per litre.

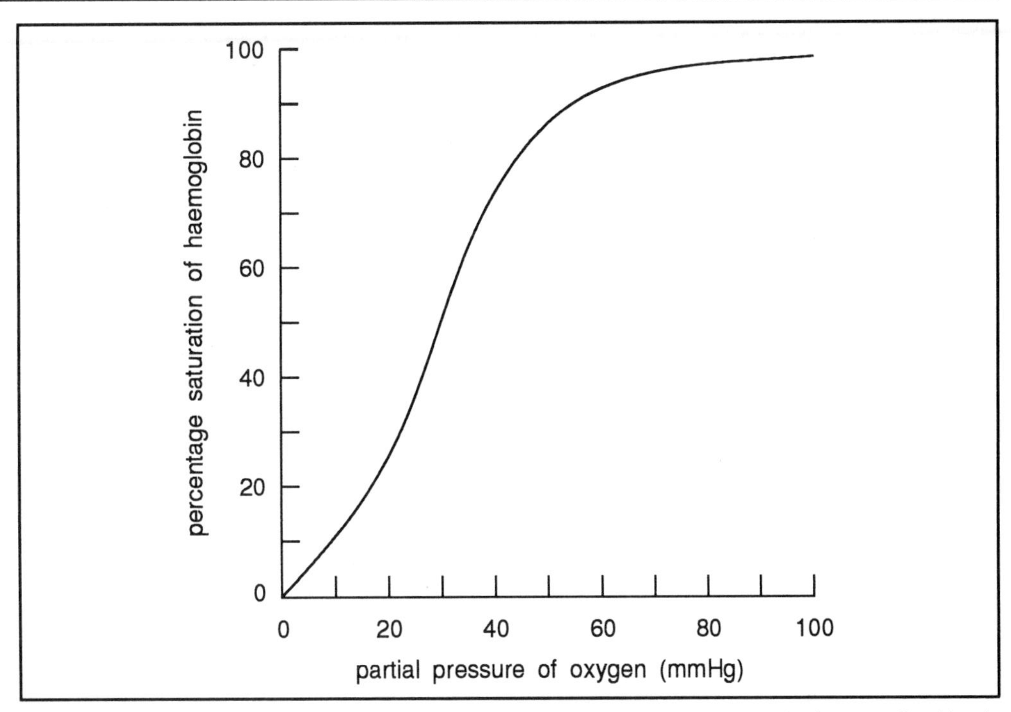

Figure 6.7 Diagram showing the oxygen haemoglobin dissociation curve for a sample of mammalian blood. Note that the actual position of this curve depends not only on oxygen partial pressure but also on the physical and chemical environment of the haemoglobin (see text).

For a given partial pressure of oxygen, the percentage saturation of a blood sample is affected by a variety of other factors. These include temperature and the presence of carbon dioxide.

Π Given below is the percentage saturation of a sample of blood at given partial pressures in the presence and absence of carbon dioxide. Plot both on linear graph paper using the same axes for both sets of data.

Partial pressure of oxygen (mm Hg)	% Saturation without CO_2 present	% Saturation with CO_2 present
0	0	0
20	35	25
40	75	55
60	90	80
80	95	85
120	100	90

Bohr effect

You should see that the presence of carbon dioxide has shifted the oxygen-haemoglobin dissociation curve to the right (the so called Bohr effect) . In reality it means that for a given partial pressure of oxygen the sample of blood is less saturated with oxygen. To put it another way, the affinity of haemoglobin for oxygen is reduced. This has important physiological implications. It means that conditions which cause carbon dioxide build up, for example, in the tissues during exercise, reduce the affinity of haemoglobin for oxygen. Therefore, it will release its oxygen to the exercising tissue more readily. This is good because exercising tissue has a high demand for oxygen. Temperature has a similar effect on oxygen binding as that of carbon dioxide. Again this is of significance since, for example, exercising tissue tends to increase local temperature.

6.6.2 The transport of carbon dioxide

Carbon dioxide is produced as a result of oxidative (energy producing) pathways. Its build up in the body would have disastrous consequences therefore it must be removed. In comparison to oxygen, carbon dioxide is transported in a variety of ways. One way in which carbon dioxide is transported is in combination with the amine groups of proteins within the red blood cells, eg haemoglobin.

The general equation for this is:

$$\text{Protein - }NH_2 + CO_2 \rightleftarrows \text{protein - }NHCOOH$$

Some transport of carbon dioxicde is achieved as a simple solution of carbon dioxide in the blood. By far the most important means by which carbon dioxide is transported is as the bicarbonate ion (HCO_3^-) in red blood cells (Figure 6.8).

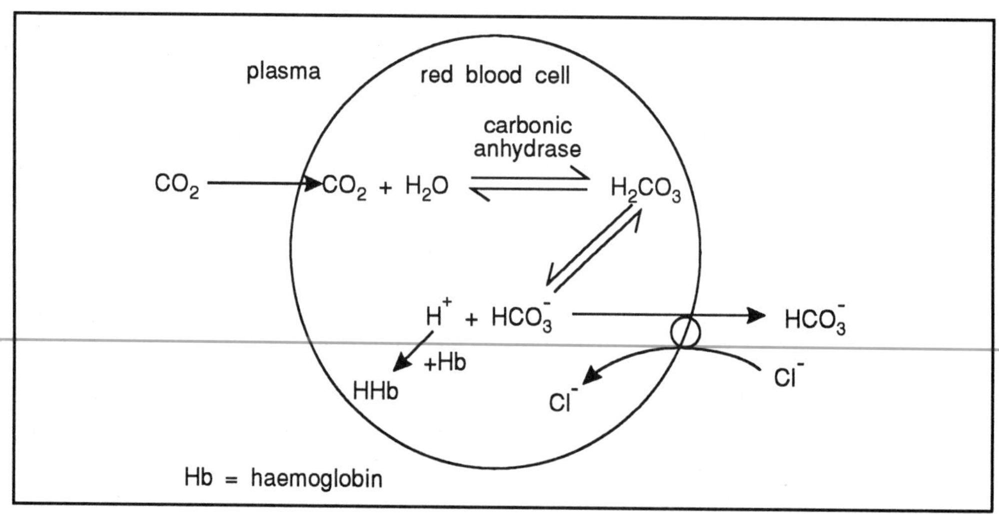

Figure 6.8 Diagram showing the carriage of carbon dioxide as the bicarbonate ion inside red blood cells. In the lungs all the reactions are reversed and CO_2 enters the alveoli and is exhaled.

carbonic anhydrase

The initial step in the formation of the bicarbonate ion is carbon dioxide dissolves to water to form carbonic acid (H_2CO_3). The formation of carbonic acid is aided by the presence of the enzyme carbonic anhydrase, which is found inside red blood cells. The carbonic acid dissociates to form both hydrogen ions (H^+) and bicarbonate ions (HCO_3^-). The bicarbonate ions leave the red blood cell and enter the blood plasma but in order to maintain electrical neutrality of the red cell membrane, chloride ions (Cl^-) simultaneously enter the red blood cell from the plasma. What happens to the H^+ ions that are produced (Figure 6.8)? Hydrogen ions are mopped up by the deoxyhaemoglobin inside the red blood cells. Remember we are now dealing with venous rather than arterial blood so the haemoglobin has given up its oxygen to the tissues. However, the haemoglobin cannot remove all the H^+ ions that are generated. Since H^+ ions determine the acidity of a solution, venous blood tends to be more acidic (have a lower pH) than arterial blood.

SAQ 6.7

A person is said to be hypoventilating ie breathing slower than normal. Explain what would happen to the level of carbon dioxide in the blood and how would this effect the pH of the blood.

6.7 Generation of respiratory rhythm

We know that respiration is achieved by alternate inspiration and expiration of a volume of air. This rhythmic inspiration and expiration is caused by contraction of the inspiratory muscles followed by their inhibition and relaxation. The action potentials which evoke the contraction of these muscles originate in a region of the brain called the brainstem, this is composed of the pons, the medulla and midbrain (Chapter 2).

site of control is the medulla

At one time it was thought that there were discrete areas in this region of the brain from which the regular pattern of breathing was originated. These were called the inspiratory and expiratory areas, which were mutually antagonistic. However, this hypothesis has been replaced. A more recent hypothesis is that there is a group of neurons in the medulla which generate action potentials causing inspiration and whose inhibition, through a number of possible mechanisms, results in expiration. There are probably more than one group of such neurons and collectively they are termed the medullary inspiratory neurons. They are thought to be pacemaker neurons which spontaneously generate action potentials. You should be aware that there are currently a number of alternative hypothesis to account for the generation of respiratory rhythm.

medullatry inspiratory neurons

6.8 Control of respiration

The physiological basis of respiratory control, in contrast to that of rhythm generation, is well understood. The overriding requirement of the respiratory system is that tissues should be adequately supplied with oxygen and that the waste product carbon dioxide should be removed. Therefore, it would seem sensible that these factors should directly control respiration. This is precisely what happens. There are specialised areas which can sense the concentration of oxygen, carbon dioxide and therefore hydrogen ions (pH) in the blood. These areas are called chemoreceptors. They can be divided into central and peripheral chemoreceptors.

6.8.1 Central chemoreceptors

The central chemoreceptors are located on the surface of the medulla of the brain. They provide a neural input onto the respiratory neurons which generate respiratory activity. The central chemoreceptors are primarily carbon dioxide/hydrogen ion sensors.

They detect changes in the concentrations of these substances within the cerebrospinal fluid (CSF). This fluid bathes the brain and spinal cord. The response of the chemoreceptors to increased levels of carbon dioxide and hydrogen ions is to increase ventilation. Central chemoreceptors are by far the most important sensors of carbon dioxide and hydrogen ions. The reason that hydrogen ions are so effective in producing ventilatory changes is that the cerebrospinal fluid contains no protein. Thus, there are few substances to buffer the effect of the hydrogen ions, (ie neutralise their action). The central chemoreceptors are not sensitive to changes in the partial pressure of oxygen in blood.

6.8.2 Peripheral chemoreceptors

carotid and aortic bodies

The peripheral chemoreceptors are situated in the carotid bodies, which are located in the neck at the point where the common carotid artery splits into the internal and external carotid arteries. Some peripheral chemoreceptors are found in the aortic bodies located in the arch of the aorta.

In both cases, the nerves which leave the carotid and aortic bodies terminate on the inspiratory neurons of the medulla. The main stimuli to these bodies are changes in the partial pressure of oxygen in the blood (pO_2). The response of these bodies to a drop in blood oxygen levels is to generate an increase in ventilation. They are also sensitive to changes in blood carbon dioxide levels (pCO_2), but are much less sensitive to this than the central chemoreceptors. The activity of the peripheral chemoreceptors can be modulated by the autonomic nervous system, either increasing or decreasing their sensitivity.

6.9 Non-respiratory function of the lungs

In addition to its primary role of gas exchange, the lungs and pulmonary circulation serve a number of other important roles. One of the most important is that the pulmonary circulation acts as a filter. It will remove small particles such as small blood clots and other particles during the passage of venous blood through the lungs.

Π See if you can write down a reason why this is such an important role of the respiratory system.

The task of filtering the blood is very important since it ensures that no particles enter the arterial circulation. Such particles could obstruct blood flow to vital organs, such as the heart itself, with disastrous consequences.

The respiratory system also functions in a metabolic capacity. For example, it activates the blood pressure regulating hormone Angiotensin II as it passes through the pulmonary circulation.

Summary and objectives

In this chapter we have examined the way in which the respiratory system functions. We briefly reviewed the structure of the main organ (the lungs) of respiration before explaining how inspiration and expiration is achieved. We also described how tidal volume, inspiratory reserve volume and expiratory reserve volume can be determined and explained how oxygen and carbon dioxide are transported between the lungs and body tissues. We briefly explored how respiration is controlled with particular emphasis on the role of the central and peripheral chemoreceptors. We also indicated that the lungs perform additional, non-respiratory, function such as microfiltration.

Now that you have completed this chapter you should be able to:

- describe in outline the gross and microscopic structure of the lungs;
- relate lung structure to its function;
- describe the generation of the intrapleural and transpulmonary pressures;
- describe how these pressures change during ventilation of the lungs;
- describe how the process of inspiration is achieved;
- contrast inspiration with expiration;
- describe where surfactant is produced and what function is performs;
- use data to calculate the different lung volumes;
- describe how oxygen is transported in the body;
- describe the way carbon dioxide is transported as the bicarbonate ion;
- describe how respiratory rhythm may be generated;
- describe the location of chemoreceptors and how they control respiration.

The kidney and its role in water and electrolyte balance

The kidney and its role in water and electrolyte balance

function of the urinary system

The urinary system is one of the major organ systems of the body that contribute to homeostasis. Through regulating the volume and composition of the body fluids the urinary system participates in the control of blood pressure, pH, water and electrolyte balance and the excretion of metabolic waste. Because of the kidney's central role in controlling so many aspects of homeostasis, any pathological change in the kidney function can disrupt the functions of many other organ systems as well. In this chapter we will examine the normal physiological role of the kidney. Its role in maintaining water and electrolyte balance and in the excretion of waste together with factors which adversely affect these functions will be explained. The consequences of abnormal renal function will also be discussed, as will be the use of diuretics.

7.1 Introduction

The many and complex activities occurring in the mammalian body require a stable environment in which to take place. Even slight chemical or physical changes may upset the smooth functioning of the cells. Many of the physiological activities of the body regulate the composition of its own internal environment and do so within relatively narrow limits. The maintenance of a stable internal environment is called homeostasis.

homeostasis

However, the very functions and activities which need a stable environment change it continuously. Chemical processes occurring in the cytoplasm of all the body's cells both use and produce a variety of substances (Figure 7.1).

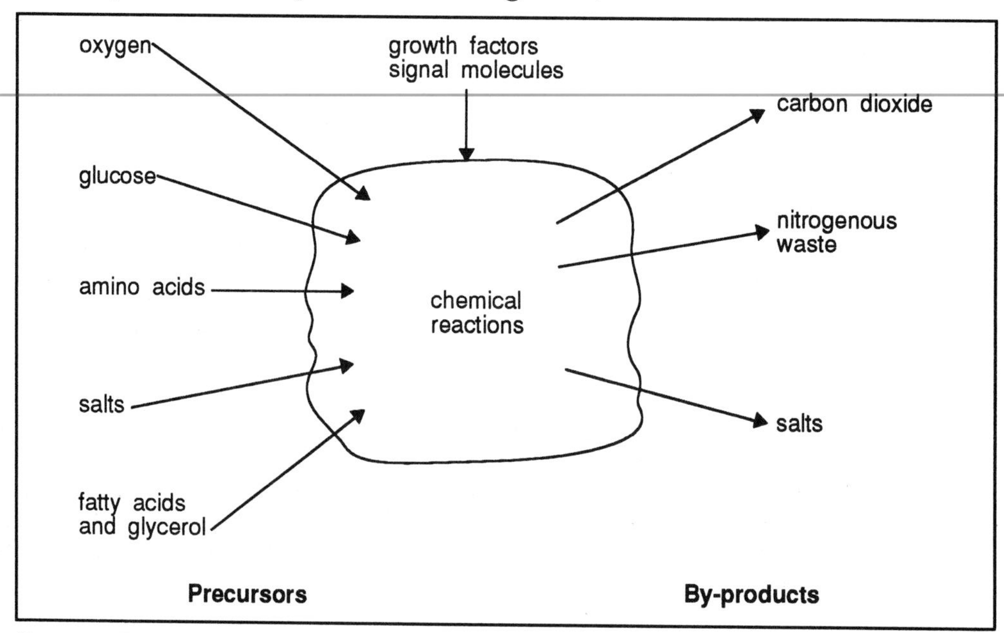

Figure 7.1 The exchange of materials by the cell.

The chemical processes which occur in all cells are collectively called metabolism.

An inevitable result of metabolism is the production of waste products which, if allowed to accumulate can become toxic. However, what is produced as metabolic waste may well be substances that are useful, but are simply present in excess of the body's requirements at that particular time. For example, the removal of calcium ions by the large intestine is homeostatically controlled; when calcium ions are plentiful in the body, more are removed; when they are in low supply, few, if any, are removed. Perspiration too, removes some of the products of metabolism and also includes salts. In times of dehydration, however, salt retention may be important to prevent further water loss; the waste salt is retained to maintain the normal water balance in the body fluids. Thus what is a waste substance at one time may be far from being so at another. Its importance to the body is dependent on the state at that moment of the internal and external environment.

So if we regard a waste as any substance that is either toxic if allowed to accumulate or is present in excess (cannot be stored or put to immediate use) it is obvious that these wastes need to be eliminated or excreted from the body.

excretion

Excretion is the elimination of any substances which are present in the body in concentrations which exceed the natural levels, whether they are metabolic or not.

7.2 Excretory products

Π Make a list of the most important wastes produced in metabolism? (Do this before reading on)

You may have included quite a few items in your list but the ones we hope you included were carbon dioxide, ammonia (or some other form of waste nitrogn such as urea) and water. Let us briefly examine each of these in turn.

7.2.1 Carbon dioxide

Carbon dioxide is produced by cells as a waste product of cellular respiration, the process in which glucose is oxidised by oxygen to produce energy, (ie glucose + oxygen = energy + carbon dioxide + water).

7.2.2 Ammonia

deamination

Unlike carbohydrates, excess protein cannot be stored in the body and so undergoes a process in the liver called deamination. During deamination, ammonia, which is toxic, is formed. The liver converts this nitrogenous (nitrogen containing) waste into the less harmful compound urea which is then eliminated from the body.

7.2.3 Water

metabolic water

During the production of energy by the cells (respiration), water is released as a waste product. This water, produced as a result of metabolism, is known as metabolic water and, depending on the state of water balance in the body, may be retained for use by the body or excreted. Water may also be gained by the body through eating and drinking and is lost from the body in urine, faeces, sweat and as water vapour in exhaled air.

7.3 Excretory organs

Humans have four excretory organs:

- lungs;
- liver;
- skin;
- kidneys.

SAQ 7.1

As a test of your previous knowledge of excretion, match these excretory organs with the following.

1) Excretion of salts, water and urea via glands.

2) Excretion of relatively large quantities of dissolved urea and other substances.

3) Excretion of carbon dioxide and water vapour from respiration.

4) Excretion of bile pigments from the breakdown of haemoglobin from worn out red blood cells.

In mammals the main organs of solute (substances dissolved in a liquid) excretion and water control are the kidneys. Within the kidneys the vast majority of the nitrogenous waste produced from protein breakdown are also excreted. Without the kidneys, accumulation of these highly toxic waste products would rapidly lead to death.

7.4 The urinary system

The kidneys, ureters, bladder and urethra are collectively known as the urinary system.

site of kidneys

Human kidneys are about 12 cm long by 7 cm wide and are dark red organs situated in the upper part of the abdominal cavity, one on either side of the vertebral column. Each kidney is covered by a tough translucent membrane which attaches it to the posterior wall of the abdomen. The kidney has a characteristic bean shape, concave on its inner and convex on its outer surface. In the tissues which surround the kidneys fat may be stored, so that the kidneys are often partially embedded in fat. This fat helps to protect the kidneys by cushioning them from impact. The kidneys are otherwise protected only by the abdominal muscle layers.

ureter
bladder
urethra

A thin muscular tube, the ureter, leaves the concave side of each kidney and extends downwards to a muscular sac, the bladder. The bladder has only one exit, a tube called the urethra which leads to the body surface. The bladder end of the urethra is normally held closed by means of a ring of muscle (a sphincter) which controls the release of urine from the bladder.

Urine is made continuously and drains continuously out of the kidneys into the ureters where it is forced downwards into the bladder by wavelike contractions (peristalsis) of the muscular ureter walls. As the urine drips in, the muscles of the bladder relax, so that it expands in volume as it fills with urine. As the volume increases, so does the pressure until a point is reached when the stretching stimulates sensory nerve endings in the walls of the bladder and thus nerve impulses are sent to the brain. The sphincter muscle around the urethra is then voluntarily relaxed to let urine drain from the bladder, through the urethra and out of the body. This is called urination (micturition).

urination

7.5 Urine composition

The composition of normal urine produced by the kidneys is shown in the table below.

Substance	Amount g/24 hours
urea	35.0
uric acid	0.8
ammonia	0.6
creatinine	0.9
sodium chloride	15.0
phosphoric acid	3.5
sodium	2.5
potassium	2.0
total solid	.60
water	1440 ml
total quantity of urine	1500 ml

Table 7.1 Composition of normal urine.

Π See if you can write down any constituents which are not normally present in the urine and whose presence may give clues to an underlying disorder?

You might think of abnormal constituents such as:

- glucose which would indicate that a person could be suffering from diabetes mellitus;
- albumin and red cells which are an indication of kidney disease;
- phenylketones which are found in phenylketonuria;
- human chronic gonadotrophin (HGC). The urine of pregnant women contains this hormone and its presence in urine forms the basis of pregnancy testing;
- the urine also contains the metabolic breakdown products of most drugs and a study of these may be important in determining how a patient's body is dealing with a drug;

- bilirubin. Some neonates have a yellowish colour in the skin after the third day. This is called physiological jaundice. It occurs because the newborn's liver is unable to break down the red blood corpuscles and bilirubin quickly enough. Excess bile salts are deposited in the skin and the whites of the eyes and they also discolour the urine. The problem normally disappears between the seventh and fifteenth day. This condition needs medical treatment since the condition can give rise to brain damage.

To be able to understand fully how urine is formed, let us look first at the structure of the kidneys.

7.6 Structure of the kidney

A vertical section through the kidney shows that there are two main regions, the dark coloured outer zone called the renal cortex and the paler inner zone called the renal medulla. The medulla is made up of several cone shaped areas called renal pyramids. Urine drains continuously from the tips of the pyramids into funnel shaped spaces (the pelvis) formed by the top of the ureter.

Π The diagram below (Figure 7.2) shows a vertical section through a kidney. Using the description just given, can you label the structures (a-e) indicated? (The answers are in the legend of the figure. Try to do this before checking with the legend).

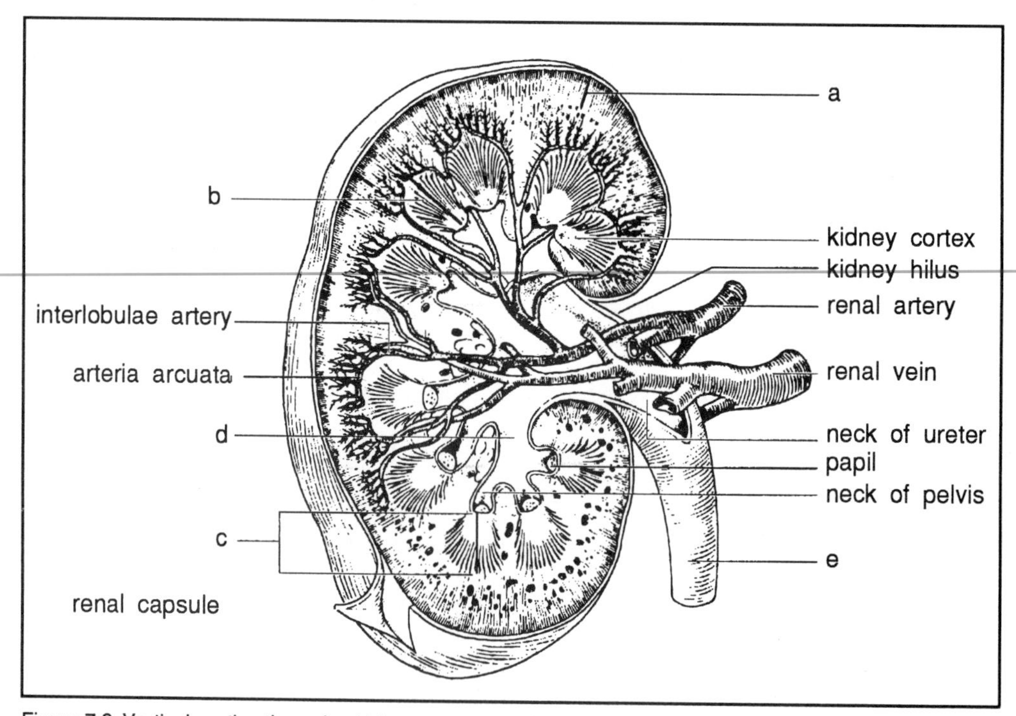

Figure 7.2 Vertical section through a kidney. a = cortex; b = medulla; c = pyramid; d = pelvis; e = ureter.

Each kidney contains:

- about 160 km (100 miles) of blood vessels and;
- more than 1 million microscopic kidney tubules (nephrons) which together have a combined length of 60 km (38 miles).

You may wish to buy a lambs or pigs kidney from a butcher's and examine its macroscopic structure. It will help you to remember the overall structure we have been describing.

7.6.1 Blood supply to the kidney

extensive blood supply to kidneys

Each kidney receives oxygenated blood at high pressure through a renal artery directly from the aorta. Inside the kidney the renal artery divides into many smaller branches, arterioles which carry the blood to the capillaries with only a small drop in pressure. The capillaries eventually join up to form venules which drain into the renal vein. The renal vein returns the deoxygenated blood to the heart via the inferior vena cava.

In relation to the size of the organ they supply, the renal blood vessels are the largest in the body. More than one-fifth of all the blood in the body passes through the kidneys in only a few minutes. This ensures that the concentration of urea does not reach dangerous levels. This high blood flow is the reason severe haemorrhage can result from trauma to the kidneys.

7.6.2 Kidneys nephrons

Malpighian corpuscle

Bowman's capsule

Each kidney contains about one million nephrons each of which is about 3 cm in length. Each nephron consists of a Bowman's capsule and a renal tubule and begins in the cortex of the kidney as an expanded, cup shaped structure (the Bowman's capsule) which is about 0.2 mm in diameter. The capsule is the blind, terminal swelling of the renal tubule and almost surrounds a small knot of about 200 capillary loops, the glomerulus. The Bowman's capsule and the glomerulus together form the Malpighian or renal corpuscle (Figure 7.3).

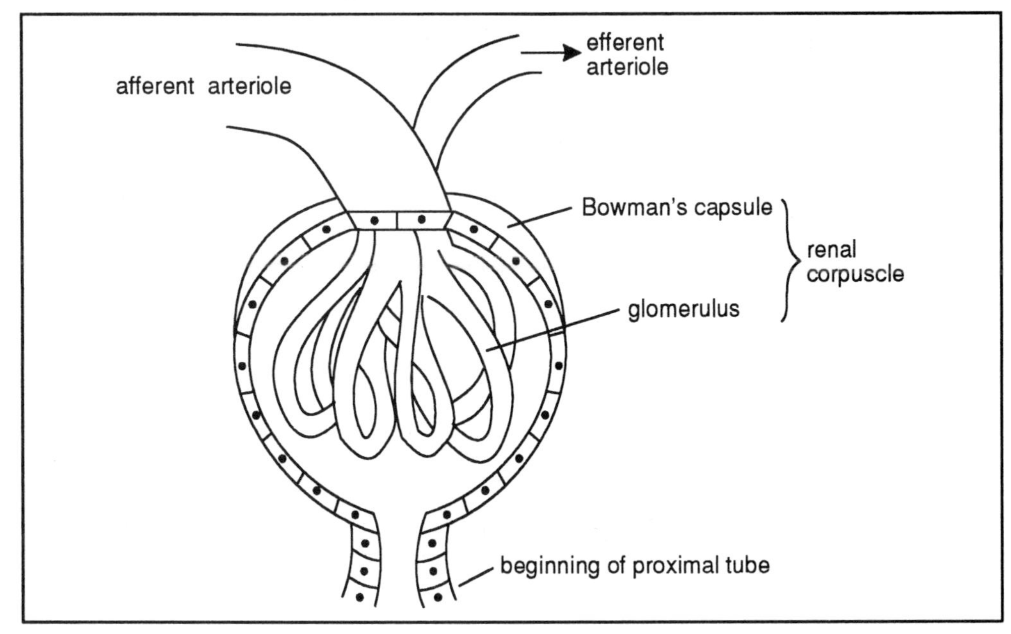

Figure 7.3 Bowman's capsule and glomerulus (the renal corpuscle).

blood supply to the glomerulus

The capillaries of the glomerulus originate from the renal artery and receive blood from a branch of this vessel called an afferent arteriole. The capillaries are drained by an efferent arteriole (Figure 7.3) which eventually joins up with the efferent arterioles from other glomeruli to form the renal vein. However, before uniting with other arterioles, each efferent arteriole first divides into capillaries which form a dense network (the peritubular network) surrounding the nephron (Figure 7.4).

peritubular

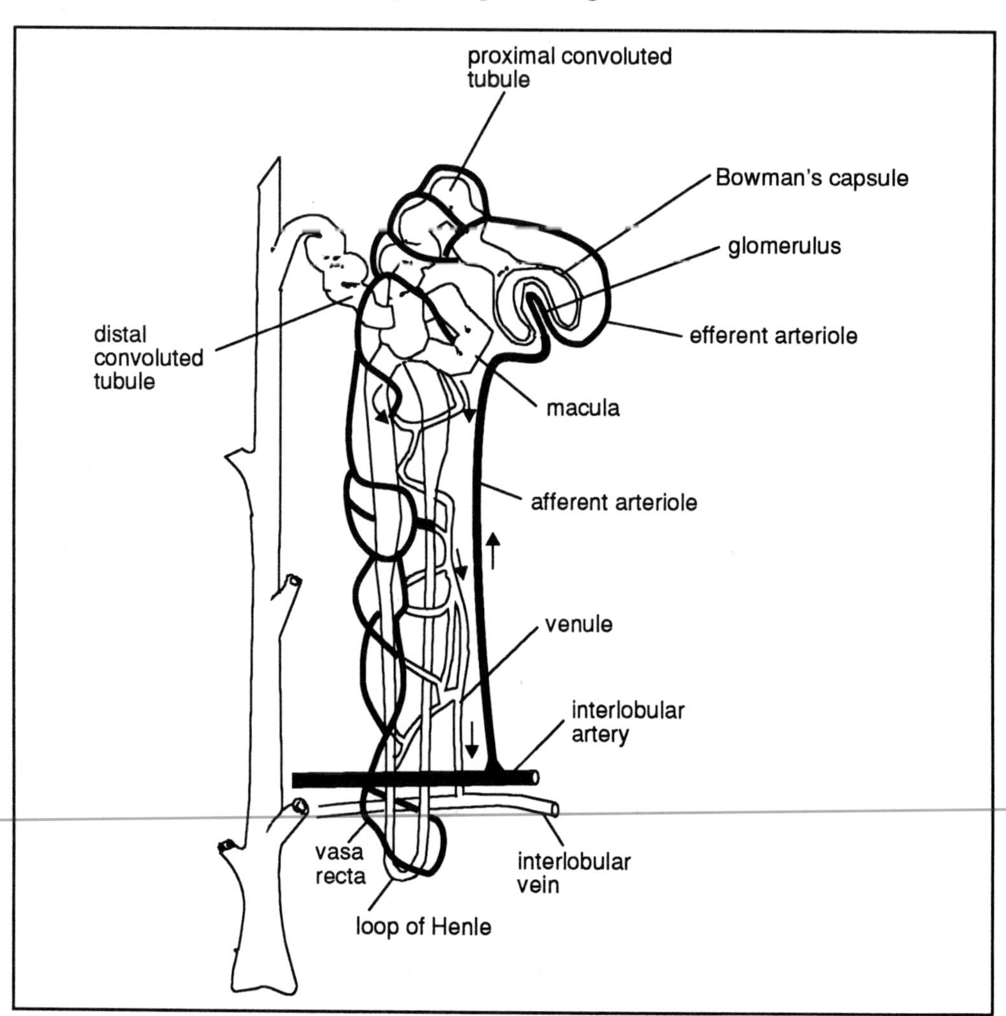

Figure 7.4 The blood supply to a nephron (see text for details).

There are, therefore, two major capillary networks supplying each nephron.

- the glomerulus;
- the peritubular network made up of: a) the peritubular capillaries which surround the proximal and distal convoluted tubules. b) the vasa recta which follows the loop of Henle as it passes through the medullary part of the kidney.

The renal tubule emerges from the Bowman's capsule on the side opposite to the glomerulus to form the first part of the tubule, the proximal convoluted (coiled) tubule. This region of the nephron is lined with cuboid epithelium which has a brush border of microvilli indicating that the epithelium is adapted for absorption.

loop of Henle

The next part of the tubule forms the loop of Henle which runs for varying distances into the medulla before returning to the cortex near to the renal corpuscle from which it arose.

The loop of Henle plays an important role in the formation of a concentrated (hypertonic) urine.

The final portion of the tubule is the distal convoluted tubule which is shorter than the proximal convoluted tubule. The distal convoluted tubules empty into collecting tubules which run through the medulla, uniting with tubules from other nephrons to form collecting ducts. Each collecting duct collects urine from several kidney tubules and transports it through the medulla of the kidney to the top of the pyramids to open into the pelvis.

Look at Figure 7.5 in which we have removed the blood vessels from the nephron so that you can see its structure more clearly.

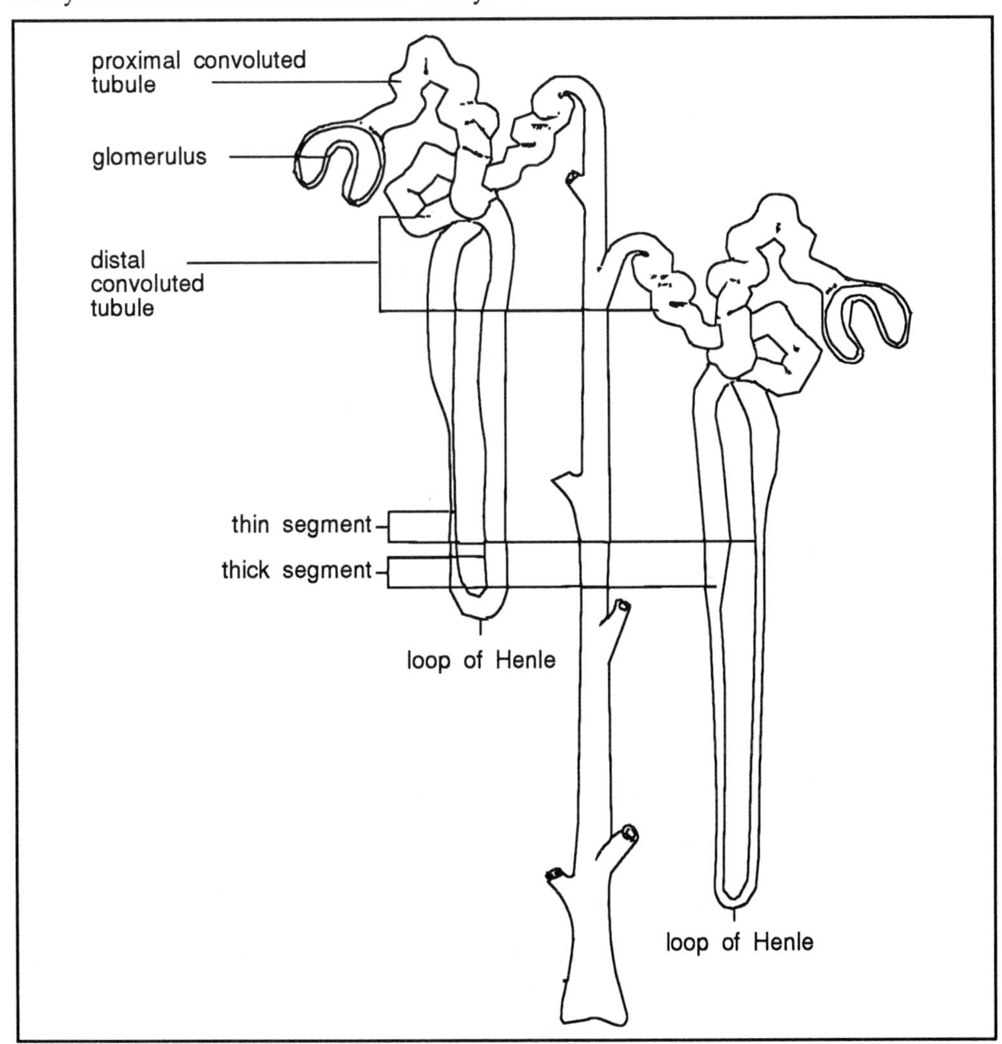

Figure 7.5 Basic structure of a nephron.

juxtaglomerular apparatus

Notice that, where the distal tubule comes up from the loop of Henle, it lies very close to the afferent arteriole supplying the glomerulus. This area is called the juxtaglomerular apparatus and is shown in more detail in Figure 7.6.

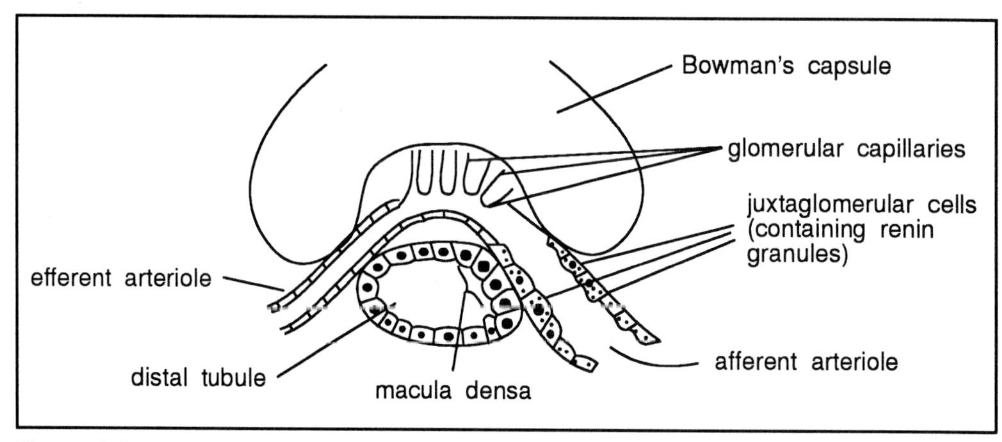

Figure 7.6 Structure of the juxtaglomerular apparatus.

At this point of contact the tubular cells are increased in number to form the macula densa. Also the smooth muscle cells of the afferent arteriole are swollen and filled with granules which are the inactive storage form of the enzyme renin. The juxtaglomerular apparatus is concerned with both the autoregulation of renal blood flow, and in the regulation of sodium reabsorption.

7.7 Functions of the renal tubule

The nephron and its associated capillaries form the functional unit of the kidney. Since the kidney is an organ associated with homeostasis, in order to keep the composition of the body fluids constant, it must:

- remove waste products (eg urea, uric acid and creatinine);
- conserve useful substances (eg glucose and amino acids);
- regulate the balance of water and salts (electrolyte balance);
- regulate acid base balance.

These kidney functions are carried out by the nephrons and we can explain the functioning of the kidney as a whole by explaining how a single nephron functions.

There are three basic components of renal tubular function. These are illustrated in Figure 7.7.

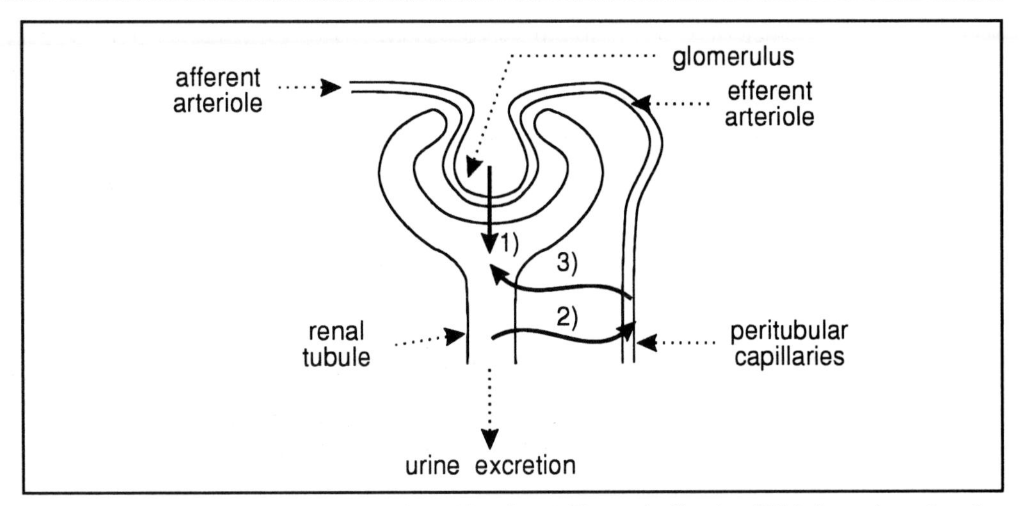

Figure 7.7 The three basic components of renal function. 1 Glomerular filtration. 2 Tubular reabsorption. 3 Tubular secretion.

7.7.1 Filtration

This produces an ultrafiltrate of plasma which is protein free but contains water and dissolved solutes in the same proportions as in plasma.

7.7.2 Reabsorption

Solutes and water are partially or completely reabsorbed from various parts of the renal tubule. Essential materials are reabsorbed quickly into the peritubular circulation before they are lost. Remaining in the tubules are waste products and excess materials which have not been completely reabsorbed into the circulation.

7.7.3 Secretion

Some materials are taken from the circulation and secreted into the tubular fluid to concentrate them in the final urine product.

By means of the above three processes, a filtrate of blood is processed by the renal tubules and converted into urine by the time it reaches the renal pelvis. The changes in the composition of the fluid in the nephrons are summarised in Table 7.2.

	Blood	Filtrate	Urine
protein	80	0	0
amino acids	0.5	0.5	0
glucose	1.0	1.0	0
salts	7.2	7.2	15
urea	0.3	0.3	20

Table 7.2 Concentration of components in g l^{-1}.

Let us now look at each of these processes in turn.

7.8 Glomerular filtration

The blood capillaries are freely permeable to water and solutes of low molecular weight compounds. They are relatively impermeable to large molecules, the most important of which are the plasma proteins. The glomerulus behaves like any other capillary in allowing substances which are small enough to pass from the blood into the Bowman's capsule. Evidence that glomerular filtration occurs and is the first stage in urine formation is shown by three criteria:

- the fluid in the Bowman's capsule is protein free (the presence of proteins in the urine is an indication of renal disease);
- the fluid contains all the low molecular weight compounds in virtually the same concentration as the plasma;
- the glomerular blood pressure is high enough to account for the large volume of filtrate that is normally produced.

Π Bearing in mind that the membranes of the glomerular capillaries and the Bowman's capsule are selectively permeable, which of the following blood constituents would you expect to be filtered and appear in the glomerular filtrate: plasma proteins, red blood cells, water, urea, creatinine, uric acid, glucose, amino acids, inorganic salts?

You will find all these substances in the urine except plasma proteins and red blood cells, which are too large to pass the membranes of the glomerular capillaries and Bowman's capsule.

7.8.1 Glomerular filtration rate (GFR)

GFR

The total amount of fluid formed by glomerular filtration in the nephrons of both kidneys each minute is known as the glomerular filtration rate (GFR). In an average sized man the GFR is 125 ml min^{-1}. This value may vary slightly, for example, GFR is lower at night, and is about 10 per cent less in females than in males.

Π Assuming the GFR to be 125 ml min^{-1} how much fluid is filtered by the kidneys in 24 hours?

180 l day^{-1} is filtered which is four times the volume of our total body water (45 l). Also as we only produce about 1-2 l urine per day, it is obvious that most (99%) of the filtered fluid is reabsorbed from the renal tubules.

Glomerular filtration results from pressure differences between the capillary blood and the capsule cavity. The blood pressure in the glomerular capillaries is much higher (about 70 mm Hg) than in other capillaries of the body (about 20 mm Hg). (Note that we have used the same convention to specify pressure as we used when we were discussing respiration. Pressure differences are expressed relative to atmospheric pressure. 760 mm Hg = 1 atmosphere = 101 325 N s m^{-2}).

Π Look again at Figure 7.3. Can you explain why a rise in blood pressure occurs in the glomerulus?

It may not have been too obvious to you but the diameter of the efferent arteriole, draining blood from the glomerulus is narrower than that of the afferent arteriole, entering the capillaries. As blood leaves the glomerulus a bottle neck is produced by the narrower tube and this causes an increase in pressure. This resulting hydrostatic pressure forces fluid through the walls of the capillaries and the Bowman's capsule and into the space of the capsule. This filtration under pressure is called ultrafiltration.

ultrafiltration

The osmotic pressure due to the plasma proteins in the glomerular capillaries is normally 30 mm Hg. This pressure will oppose the blood pressure and tends to prevent movement of smaller molecules through the membrane. The pressure of fluid within the Bowman's capsule (about 15 mm Hg) will also oppose filtration.

Therefore, the pressure tending to force fluid out of the glomerules is the hydrostatic pressure (70 mm Hg) while the opposing pressure (45 mm Hg) is tending to move fluid in the reverse direction. This opposing pressure is due to the osmotic pressure resulting from dissolved molecules too large to pass the capillary wall (30 mm Hg) plus the hydrostatic pressure exerted by the fluid in the Bowman's capsule (15 mm Hg). The difference (25 mm Hg) between the outward pressure of 70 mm Hg and the inward pressure of 45 mm Hg is the net pressure pushing fluid into the Bowman's capsule. This is called the net filtration pressure (Figure 7.8).

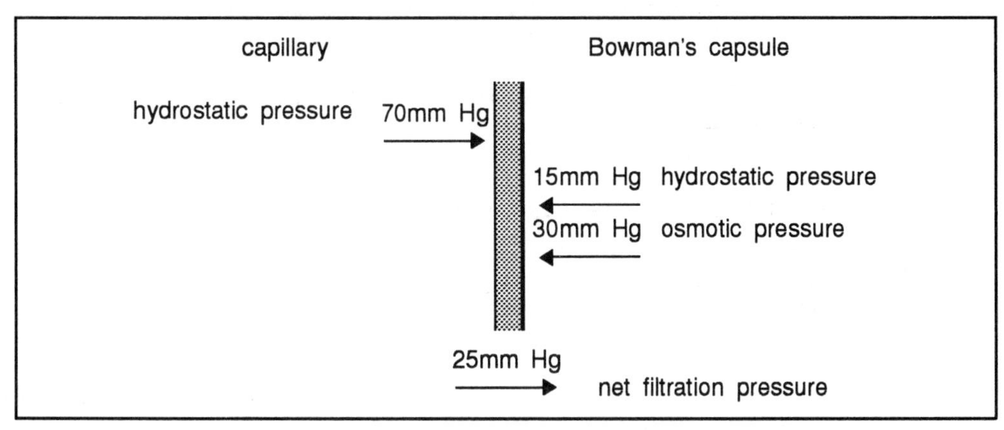

Figure 7.8 The pressures involved in glomerular filtration.

Π Study the diagram carefully. Can you suggest any physiological changes, disease processes or any other factors that can affect glomerular filtration rate (GFR)?

Changes in any of the pressures involved in filtration can alter GFR. You might think of:

- changes in renal blood flow.

  Renal blood flow is reduced during exercise and hypoxia (oxygen lack) and may increase during certain bacterial infections. An increased renal blood flow will lead to an increased GFR.

- changes in glomerular capillary blood pressure.

  This may be due to changes in systemic arterial blood pressure or constriction of the afferent or efferent glomerular arterioles. If the efferent vessels constrict, then glomerular pressure will rise and GFR will increase. If the afferent vessels constrict then glomerular pressure will fall and GFR will be reduced.

- changes in the pressure within the Bowman's capsule.

  This may be due to obstruction of the ureters or oedema (water retention) in the kidney. The kidney is located inside a tight renal capsule and fluid accumulation will increase pressure on the nephrons and therefore the fluid within them and reduce GFR.

- changes in plasma protein concentration.

  If a person is dehydrated the osmotic pressure of the blood increases and therefore filtration pressure is reduced. The opposite is the case if plasma protein concentration falls (hypoproteinaemia).

- renal diseases.

  There is increased permeability of the glomerular filter as a consequence of which protein is filtered and lost in the urine. This reduces the osmotic pressure difference between blood and filtrate. As a result the net filtration pressure and the GFR increase. In some diseases the glomeruli may be destroyed with or without destruction of the tubules. In such cases there is a reduction in the total area of the glomerular capillary bed so GFR will be reduced.

The examples given illustrate that any change in the three pressures contributing to the final filtration pressure can cause a change in GFR. Because the GFR is subject to changes in systemic blood pressure it is vital that renal blood flow is kept constant to ensure proper kidney function and in fact the kidney is able to autoregulate its own blood supply.

7.8.2 Autoregulation of renal blood flow

If the systemic blood pressure were to rise then glomerular filtration would increase. Conversely a fall in blood pressure would reduce glomerular filtration. Short term increases in blood pressure have less serious effects on kidney function since they cause increased glomerular filtration and renal excretion of water, which helps to reduce the body fluid volume and correct the raised pressure. However, a fall in blood pressure can have more serious effects. The body must continually eliminate nitrogenous wastes and excess electrolytes and if the blood pressure falls, glomerular filtration is decreased, fewer waste products are filtered and therefore accumulate in the body. However, mean arterial pressure can drop as low as 70 mm Hg before urea retention becomes a problem. The kidneys function normally over a wide range of blood pressures and must therefore be able to control their glomerular filtration rates. This process is called autoregulation.

autoregulation

For example, if blood pressure rises, the afferent glomerular arterioles are stretched to a point where they reflexly constrict to increase the resistance to blood flow into the glomerular capillaries and therefore prevent filtration pressure rising. By this mechanism the GFR can be maintained.

If the blood pressure falls, the filtration pressure is reduced and fluid flow rate slows down. The juxtaglomerular apparatus detects any changes in the flow rate within the distal tubules and if this becomes too low the renin-angiotensin system is activated (see 7.13.4). This leads to constriction of the efferent arteriole, increasing the resistance to the outflow of blood from the glomerulus and therefore maintaining filtration pressure and glomerular filtration rate.

Autoregulation of renal blood flow by the kidney ensures that glomerular filtration pressure and GFR is maintained at a similar value even when there are quite large changes in systemic blood pressure.

7.8.3 Measurement of glomerular filtration rate

The GFR of a patient may be measured by using the theoretical concept of clearance.

clearance

We need to define the term clearance. Let us say that x amount of a substance is passed from blood into urine every minute. Let us also assume that x amount of substance is found in volume y of the blood. Then the clearance value for the substance is y min^{-1}. It has the units of volume per unit time. In other words, clearance is defined as the blood volume that contains the quantity of a substance which is passed into the urine per minute.

inulin

Measurement of clearance value is a useful indicator of kidney function. To determine the GFR a substance called inulin is used. Inulin is filtered at the glomerulus, but is neither reabsorbed nor secreted in the rest of the renal tubule. The inulin in the urine has therefore only been filtered at the glomerulus and its concentration will be a measure of the GFR.

Inulin is infused (injected into a vein) into the patient and samples are collected to obtain values for inulin concentration in both plasma and urine.

By using the following formula the inulin clearance can be calculated:

$$Cin = \frac{Uin \times V}{Pin}$$

Cin = inulin clearance (ml min^{-1}); Uin = concentration of inulin in the urine (mg ml^{-1}); V = rate of urine production (ml min^{-1}); Pin = concentration of inulin in the plasma (mg ml).

Π Calculate the clearance of inulin from the following figures:

rate of urine production = 1.1 ml min^{-1}; concentration of inulin in the urine = 29 mg ml^{-1}; concentration of inulin in the plasma = 0.25 mg ml^{-1}.

If you have calculated the inulin clearance (and thus GFR) correctly your answer will be 128 ml min^{-1}.

The clearance of creatinine is routinely used clinically to give an approximate indication of the GFR. Creatinine is a normal constituent in the blood and therefore, unlike inulin does not have to be administered.

However, some tubular secretion of creatinine takes place and often the result obtained is slightly high.

Π Explain the significance of: 1) a substance having a clearance value higher than that of inulin (eg penicillin) and 2) one with a clearance value lower than inulin (eg urea)?

The sorts of explanation you should have given are:

1) any substance with a clearance value higher than inulin is filtered by the glomeruli and is also secreted by the kidney tubules into the urine.

2) one with a lower clearance value must be reabsorbed to some extent by the kidney tubules. Urea is reabsorbed to a small extent even though it is a waste product.

7.8.4 Estimation of renal blood flow

The clearance concept can also be used to estimate renal blood flow, but for this the substance used must be completely extracted from the blood as it passes through the kidney. It must be filtered at the glomeruli and secreted into the urine by the tubules with no reabsorption. Such a substance is para-aminohippuric acid (PAH). The clearance of PAH can be calculated, as above, from the blood concentration and the amount excreted in the urine per minute. This clearance value will equal the blood flow through the kidneys, since this volume of blood has actually passed through the kidneys and has been completely cleared of the substance. The PAH clearance (renal blood flow) is about 1200 ml min^{-1}.

SAQ 7.2

The rate of blood flow to both human kidneys is about 1.2 l min^{-1}). Approximately one-fifth of the plasma is filtered off in the glomeruli in a normal adult. Assuming that 50 per cent of the blood volume is plasma, which of the following represents the volume of plasma filtered during a period of one hour?

1) 3300 ml

2) 3750 ml

3) 7200 ml

4) 750 ml

7.9 Tubular reabsorption

Apart from the blood cells and substances of high molecular weight such as plasma protein, the chemical composition of the glomerular filtrate is almost the same as the plasma. There is almost complete transfer of materials from the blood in the glomeruli to the Bowman's capsule, but the actual composition of the urine is very different. This indicates that the filtrate must be modified considerably while passing along the nephrons. The volume of urine excreted is also much less than the volume of filtrate produced in a given time. The first modification of the tubular fluid is reabsorption.

Reabsorption of individual substances is at a rate just sufficient to maintain normal concentrations in the blood. Any excesses stay in the nephron and are later excreted.

7.10 Tubular secretion

Some compounds are secreted from the blood into the tubular fluid by a mechanism called tubular secretion. Secretory transport may be either active or passive. The most important secretory processes are those for hydrogen and potassium ions. The majority of substances secreted by the tubules are foreign substances taken for medicinal purposes.

Let us now examine in more detail reabsorption and secretion in the various parts of the nephron.

7.11 Proximal convoluted tubular function

The fluid that is filtered at the glomerulus contains the water and solutes that have passed through the membrane in the Mapighian (renal) corpuscle. As this fluid passes down the proximal convoluted tubules, water and selected solutes are readsorbed. After passing through the proximal convoluted tubule approximately 75 per cent of the filtered solute and 75 per cent of the filtered water have been removed.

The substances that are gained by or lost from the tubular fluid are shown in Figure 7.9.

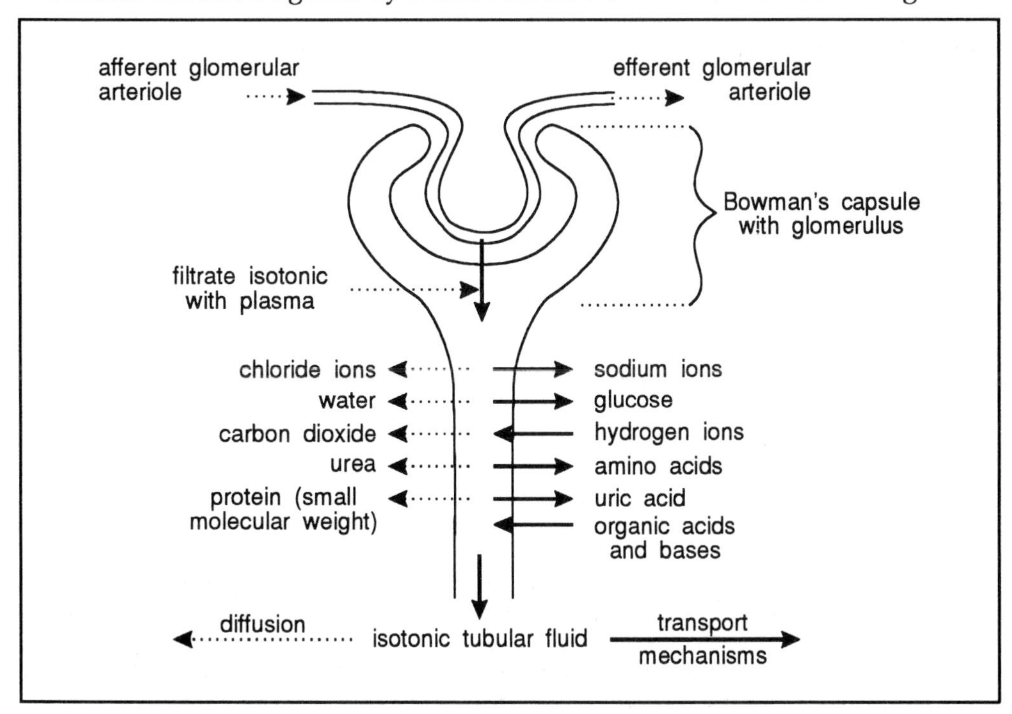

Figure 7.9 Movement of substances into and out of the proximal convoluted tubule.

7.11.1 Glucose reabsorption

glycosuria

The plasma concentration of glucose is 70-100mg/100ml, yet normally no glucose appears in the urine. The filtered glucose is actively reabsorbed from the proximal convoluted tubule. If the blood glucose levels become excessively high (eg 180-200mg/100ml) the reabsorption mechanisms become overloaded and glucose appears in the urine (glycosuria). Glucose has a transport maximum. Above this concentration (180mg/100ml) (renal threshold) the excess glucose that is filtered cannot be transported back into the plasma and thus appears in the urine. This may happen in patients suffering from diabetes mellitus. These patients often have abnormally high blood glucose levels and glucose may be found in the urine.

7.11.2 Amino acids

Amino acids are also actively absorbed from the proximal tubular fluid and this process also has a threshold above which they are not completely reabsorbed and appear in the urine.

7.11.3 Proteins

Proteins are normally too large to be readily filtered but those that do enter the tubular fluid are reabsorbed. Proteins are taken into the proximal tubule cell by pinocytosis. They are thought to be broken down within the tubule cells and passed into the peritubular capillaries as amino acids.

7.11.4 Sodium ions

Sodium ions are actively transported out of the tubular fluid and chloride ions follow, mainly by passive diffusion. Sodium ions have a positive charge and so negatively charged chloride ions follow along the electrical gradient generated by sodium ion transport.

7.11.5 Water

The proximal convoluted tubule is freely permeable to water and this is passively reabsorbed to restore the osmotic balance caused by the active reabsorption of solutes. As stated before, 75 to 80 per cent of the solutes and water are reabsorbed from the proximal tubule.

7.11.6 Urea

Normally 40-50 per cent of the filtered urea and 90 per cent of the filtered uric acid is reabsorbed.

7.11.7 Bicarbonate ions

90 per cent of the bicarbonate ions are reabsorbed from the proximal tubule and hydrogen ions are secreted, the movement of these ions will be discussed later when the role of the kidney in the regulation of blood and body fluid pH is considered.

As a result of the reabsorption of most of the important solutes and water, the volume of fluid leaving the proximal tubule is 5 ml min^{-1} (compared with 125ml min^{-1} filtered). This isotonic fluid then enters the loop of Henle which is the part of the kidney tubule specialised for the concentration of urine.

7.12 Mechanism of urine concentration

anti-diuretic hormone (ADH)

During times of low water intake or dehydration, the kidneys must continue to eliminate wastes and excess electrolytes although at the same time they must conserve water. This is achieved by increasing the amount of water that is reabsorbed into the circulation to cause the excretion of a more concentrated urine. Anti-diuretic hormone (ADH) from the posterior pituitary gland is important in regulating the process of urine concentration. Water leaves the distal tubules and collecting ducts by osmosis along an osmotic gradient. However, the walls of the renal tubules must be permeable to water or it cannot move outwards. When ADH is present in the circulating blood it increases the permeability of the walls of the distal tubule and collecting duct to water.

In order for water to move through the tubular wall an osmotic gradient must be established between the interstitial fluid surrounding the tubules and the fluid inside the tubules.

The close proximity of the loop of Henle and the vasa recta blood vessels create and maintain such an osmotic gradient around the renal tubules. The process which establishes the osmotic gradient are called countercurrent mechanisms and are shown in Figure 7.10.

7.12.1 The role of the counter current mechanisms in the concentration of urine

The loop of Henle plays a crucial role in concentrating the urine. The urine that leaves the proximal convoluted tubule has a concentration of 300mOsmol l^{-1} and it passes from an iso-osmotic region to one that becomes hyperosmotic to the body fluids. The permeability of the walls of the descending limb of the loop of Henle to ions and urea is very low, but is very high for water. Water passes out of the tubules into the more concentrated surrounding tissue, thereby concentrating the fluid in the tubules. As the fluid flows down the descending limb towards the apex of the loop it becomes more and more concentrated and by the time it reaches the hairpin bend of the loop, its concentration can have reached 1200mOsmol l^{-1}, similar to that of the surrounding tissue.

As fluid passes up the thin ascending limb Na^+ and Cl^- move out of the tubules by passive diffusion. The thin ascending tubule is impermeable to water so the urine becomes less concentrated by the loss of NaCl. The thick ascending limb is also permeable to water, but active chloride transport occurs here. Transport of Cl^- is followed passively by an equal quantity of Na^+.

The urine entering the distal convoluted tubule (DCT) is slightly hypo-osmotic to the body fluids, but it is reduced in volume owing to the movements of water and ions that have occurred in the loop of Henle. In the DCT active transport of Na^+ occurs, followed passively by Cl^-. Water follows by osmosis. Also bicarbonate ions are reabsorbed, while K^+, H^+ and NH_4+ are transported into the tubule lumen.

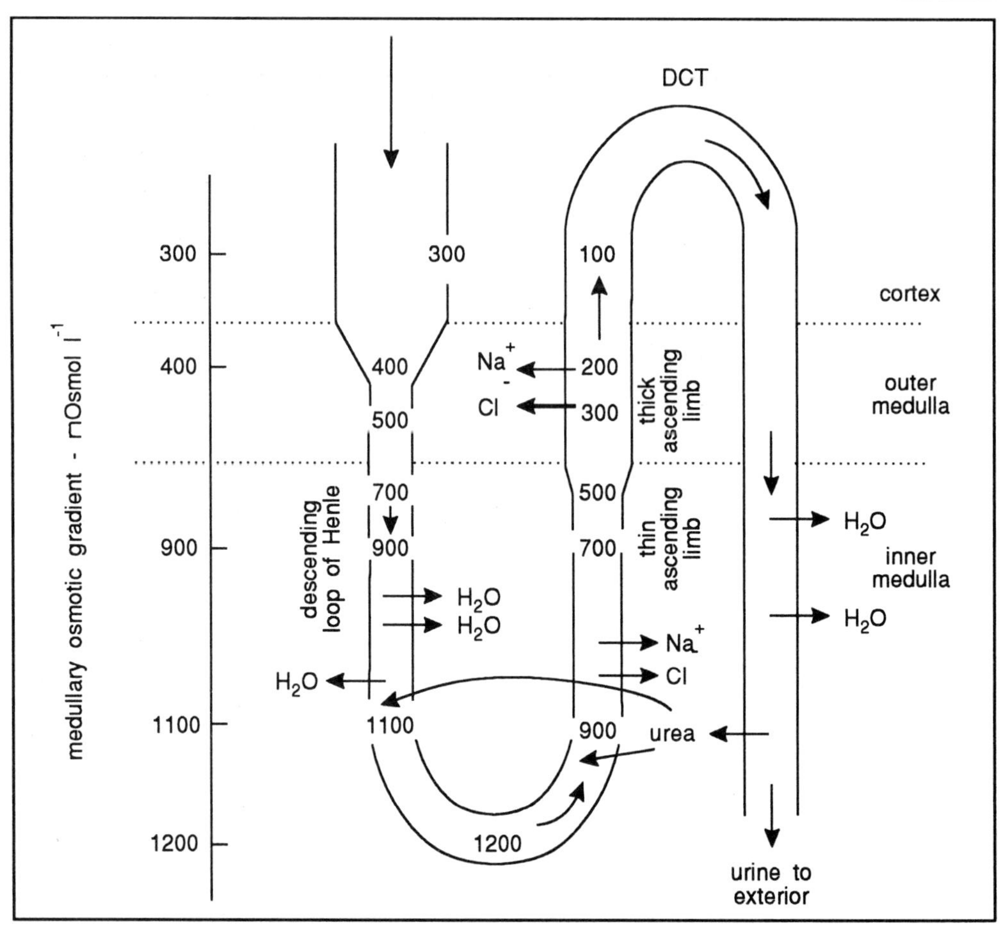

Figure 7.10 Schematic view of a nephron showing the loop of Henle, distal convoluted tubule (DCT) and collecting duct. The values on the left are the osmolarity of the tissue; there is a cortico - medullary gradient, 300 mOsmol l^{-1} in the cortex increasing to 1200 mOsmol l^{-1} in the inner medulla. The fluid within the loop of Henle has about the same osmolarity as the surrounding tissue. Movements of Na^+, Cl^-, urea and water occur as indicated.

The fluid leaving the DCT passes along the collecting tubules and ducts. These are permeable to water, a permeability that can be modified by the hormone ADH (anti-diuretic hormone) depending on the degree of hydration of the body fluids. Because the collecting ducts pass through the hyperosmotic inner medulla more water can be removed to produce a markedly hyperosmotic urine. Actually the loop of Henle is really increasing the solute concentration of the kidney tissue creating a gradient of osmolarity from the cortex to the medulla. Its characteristic U shape provides a countercurrent system which establishes and maintains this high concentration. Without this, water could not be removed from the tubular urine in the loop of Henle or in the collecting ducts, nor would ion diffusion occur from the thin loop of Henle.

Countercurrent multiplication in the loop does not itself concentrate the urine but rather creates conditions in the medulla under which a concentrated urine can be formed.

The effectiveness of the kidney in retaining water can be realised from the fact that of the 120 ml or so of fluid filtered per minute only about 1 ml reaches the ureter. Over 99 per cent of the fluid is reabsorbed in the kidney tubule.

Of crucial importance in the formation of a concentrated urine is the removal of water from the medullary regions. If this did not occur, the medullary osmotic gradients would not exist. Also, NaCl cannot accumulate without limit. Removal of water and excess NaCl is achieved by blood vessels known as the vasa recta, the capillary network associated with the loop of Henle. They operate as countercurrent diffusion exchangers and remove water from the medulla and ensure that the salt concentrations in the medulla are maintained. The vasa recta, therefore, ensure that the medullary osmotic gradient established by the activity of the loop of Henle is maintained. How concentrated the urine finally produced is depends ultimately on the permeability of the collecting ducts, which is controlled by the hormone ADH.

7.12.2 Control of ADH release

role of hypothalamus ADH release

The secretion of anti-diuretic hormone depends on the osmotic pressure of the interstitial fluid that bathes certain cells in the hypothalamus known as osmoreceptors. When the water content of the blood falls, as in dehydration, the interstitial fluid osmotic pressure rises and the receptor cells trigger the release of ADH from the posterior pituitary into the blood stream. The ADH is transported to the kidney tubules where it increases the permeability of the collecting ducts so that water reabsorption is increased. As the reabsorbed water returns to the circulation via the vasa recta, the osmotic pressure of the interstitial fluid falls, removing the stimulation of the osmoreceptors and so ADH secretion decreases.

The feedback mechanism for ADH regulation is shown in Figure 7.11.

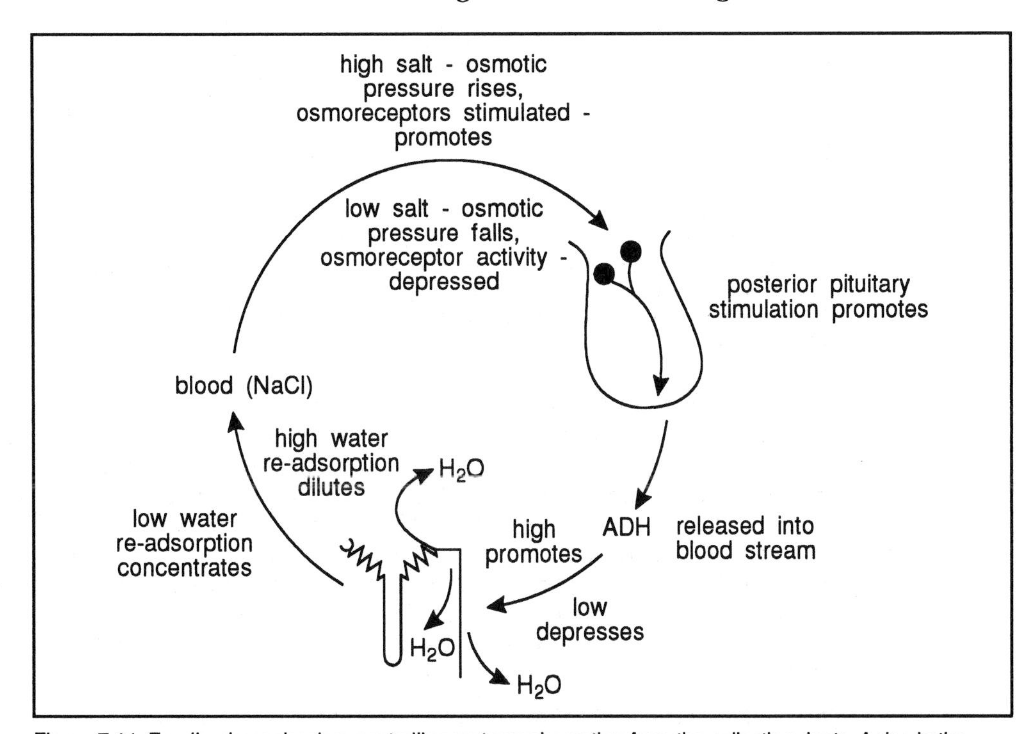

Figure 7.11 Feedback mechanism controlling water reabsorption from the collecting ducts. A rise in the osmotic pressure of the blood and tissue fluids may be brought about by ingestion of an excessive quantity of salt but is might equally result from dehydration of the tissue caused by excessive sweating or failure to drink water.

Π What would be the effect on the volume and composition of urine of drinking one litre of water? Can you suggest why?

diuresis

The water would reduce the osmotic pressure of the interstitial fluid (ie it becomes more dilute) and this would lead to an inhibition of ADH secretion from the posterior pituitary. The collecting ducts become less permeable to water, so less would be reabsorbed. The result would be diuresis which is the production of large quantities of dilute urine. The excess water is, therefore, eliminated. A number of factors which influence ADH secretion are shown in Table 7.3.

Factors stimulating ADH secretion	Factors inhibiting ADH secretion
increased osmotic pressure of interstitial fluid	reduced osmotic pressure of interstitial fluid
reduced extracellular fluid volume	increased extracellular fluid volume
morphine, nicotine	alcohol
barbiturates	
pain, stress	
exercise	

Table 7.3 Factors affecting ADH secretion.

You will see that a reduction in extracellular fluid volume, for example, by bleeding stimulates ADH secretion. This is an attempt to preserve blood volume and as a consequence maintain blood pressure. This mechanism does not play much part in day-to-day life, but is used in emergencies. Alcohol produces its characteristics diuresis by inhibiting the release of ADH.

ADH = vasopressin

ADH also has the effect of constricting the arterioles and can therefore influence blood pressure. This vasoconstrictor action gives ADH its alternative name vasopressin.

You will recall that of a glomerular filtration rate of about 120 ml min^{-1}, 100 ml min^{-1} is reabsorbed from the proximal tubule. The remaining 20 ml min^{-1} of the fluid enters the loop of Henle and is concentrated by the mechanisms described previously. If the body is in a state of water balance the rate of urine production is 1 ml min^{-1} and so 19 ml min^{-1} of the fluid is reabsorbed from the distal tubules and collecting ducts. It is this process which is under the influence of ADH.

The effect of ADH on urine production is illustrated in Table 7.4.

	GFR (ml min^{-1})	Tubular (ml min^{-1}) reabsorption	Urine (ml min^{-1})	Urine per day (ml)
Hot day, sweating high level ADH	120	119.75	0.25	375
Normal hydration	120	119.00	1.00	1500
No ADH [1]	120	105.00	15.00	22500

Table 7.4 Effects of ADH on urine production (1) The failure to produce ADH is symptomatic of some disease conditions such as diabetes insipidus.

7.13 Extracellular fluid volume

As well as producing a concentrated urine to eliminate waste products from the body and at the same time conserve water, the kidneys play a significant role in the regulation of extracellular fluid volume.

An important point to remember is that the excretion of large quantities of sodium ions in the urine always results in the excretion of large quantities of water. A crucial factor in the regulation of extracellular fluid volume is that water is only reabsorbed if sodium ions are reabsorbed first. In contrast, large quantities of water can be excreted even though the urine is virtually free of sodium ions, provided ADH secretion is reduced or inhibited completely.

7.13.1 Regulation of extracellular fluid volume

An increase in extracellular fluid volume leads to the excretion of excess sodium ions and water.

A decrease in extracellular fluid volume is prevented by a reduced urinary excretion of sodium ions and a minimum of water excretion. Thus the most important factor in the regulation of extracellular fluid volume is the control of renal sodium excretion.

Sodium ions are filtered at the glomerulus and are actively reabsorbed from the tubules but they are not secreted. Therefore renal regulation of sodium depends on the control of:

- glomerular filtration rate;
- tubular sodium reabsorption.

For example, sodium excretion is increased by increasing the glomerular filtration rate or by reducing sodium reabsorption or by a combination of both. Let us now consider these possibilities.

7.13.2 Control of glomerular filtration rate

Glomerular filtration rate is controlled primarily by changes in glomerular capillary pressure.

- A reduced arterial pressure reduces glomerular filtration rate by reducing glomerular capillary pressure, although autoregulation (Section 7.8.2) and the renin-angiotensin system (Section 7.13.4) will tend to minimise this.
- Excessive water and sodium loss as in diarrhoea or haemorrhage results in a reduction in arterial pressure and glomerular filtration rate. Therefore, the amount of sodium filtered and excreted is reduced and further loss from the body is prevented.

7.13.3 The control of tubular sodium reabsorption

1500 g sodium chloride filtered daily

Sodium chloride is taken up into the body in the diet and is lost from the body in the urine and in sweat. Under normal conditions the sodium intake exceeds the loss by sweating and the surplus is excreted in the urine. There is an enormous turnover of sodium chloride in the kidney. A daily glomerular filtration of 180 l of fluid (which contains 0.9 per cent sodium chloride) means that about 1500 grams of sodium chloride are filtered daily, most of which is reabsorbed. About 90 per cent of the cations in the extracellular fluid are sodium. Even moderate changes in sodium ion concentration affect such processes as transmission of nerve impulses, function of the brain, strength of cardiac pumping and glandular secretions. Obviously it is extremely important that sodium ion concentration is regulated very precisely.

7.13.4 Hormonal control of sodium reabsorption

role of aldosterone

ADH is not the only hormone that influences kidney function. The regulation of sodium ion concentration in the extracellular fluid is achieved mainly by the hormone aldosterone produced by the adrenal cortex. Aldosterone stimulates distal tubular reabsorption of sodium and in its absence as much as 25g of salt per day may be excreted.

Aldosterone secretion is controlled by reflexes initiated by the juxtaglomerular apparatus of the kidneys. The feedback mechanism regulating aldosterone secretion is shown in Figure 7.12.

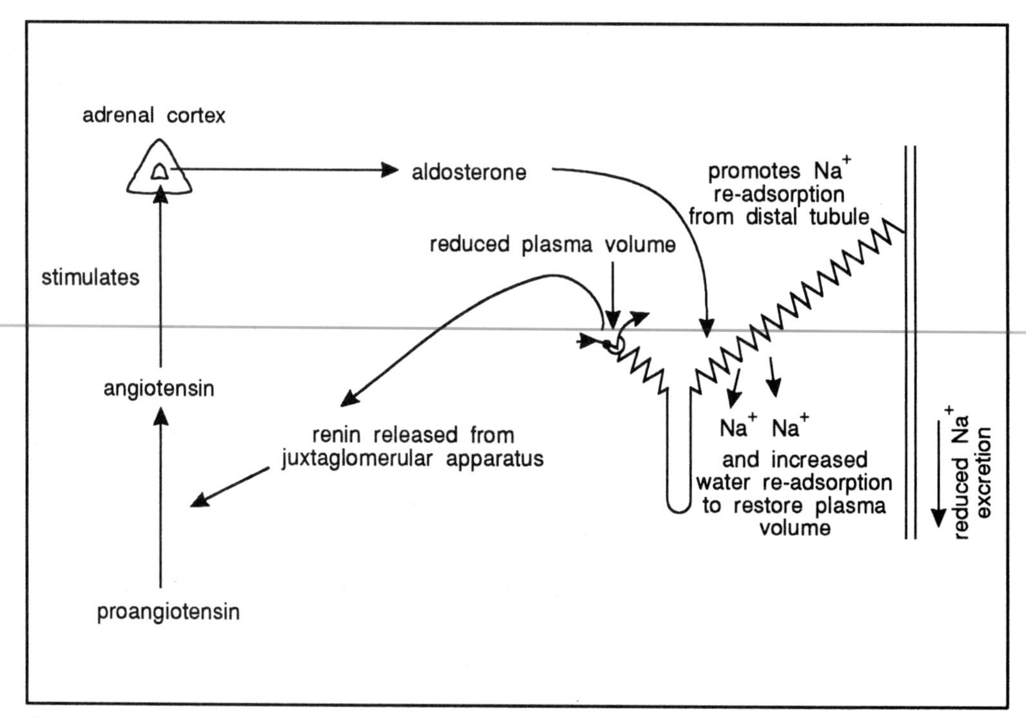

Figure 7.12 Feedback mechanism regulating aldosterone secretion.

A reduction in the sodium concentration within the distal tubule will stimulate the secretion of renin by the juxtaglomerular cells (see Figure 7.6), which are able to detect changes in sodium ion concentration within the distal tubule. This increased renin secretion leads to an increase in aldosterone secretion from the adrenal cortex. The aldosterone acts on the distal tubular cells and promotes the transport of sodium from the tubular fluid back into the circulation. As a consequence of the sodium retention, water is absorbed thereby increasing the blood and fluid volume. Individuals with abnormally high levels of aldosterone secretion show puffy swollen features from the increased fluid volume. Some of the more recently developed drugs used to control high blood pressure aim to reduce renin production. This reduces aldosterone production which reduces water retention and therefore decreases circulatory volume.

The renal tubules are able to regulate the concentration of other ions in the extracellular fluid as well as sodium eg potassium, calcium, magnesium, phosphate, chloride and bicarbonate.

SAQ 7.3

Test your understanding of what we have discussed so far by answering the following questions.

1) a) What is the effect on ADH secretion of an injection of hypertonic saline into the carotid artery?

   b) What is the main pressure responsible for water reabsorption from the collecting tubule?

2) Indicate which of the following are true and which are false. Give a reason for your choice.

   a) Blood flow in all organs varies directly with arterial pressure.

   b) Glomerular filtration rate is directly proportional to systemic arterial pressure.

   c) Glomerular filtration rate is the main factor determining the rate of urine production.

   d) Blood in the efferent arteriole is more viscous than blood in the afferent.

   e) At the tip of the loop of Henle in the medulla, the osmolarity of the tubular contents is several times that of the glomerular filtrate.

   f) The fluid entering the distal convoluted tubule is hypotonic with respect to plasma.

   g) The permeability of the collecting tubule to water is controlled by aldosterone.

7.14 Disorders of fluid and electrolyte balance

7.14.1 Fluid deficit

effects on haematocrit

An increase in the osmotic pressure of the extracellular fluid compartment causes water to move out of the cells disrupting normal cellular activity. Prolonged fluid deficit results in a fall in blood pressure and a rise in the haematocrit (= volume of red blood cells/volume of blood). Blood flow through the kidneys is reduced and urinary output falls below an obligatory amount. This leads to the accumulation of metabolic waste products. To offset the fluid deficit, there is an increase in secretion of ADH and an increased absorption of water by the renal tubules, so minimising the loss of body water.

Π From your own experiences and from the material considered so far in this text, make a list of any factors that could cause fluid deficit.

Fluid deficit may be caused by:

excessive loss of fluid

- from the gastro-intestinal tract through vomiting or diarrhoea;

perspiration

- by excessive perspiration due, for example, to exposure to high environmental temperatures or a rapid breathing rate;
- as a result of haemorrhage or loss of plasma in severe burns;

diabetes insipidus

- a deficiency of ADH secretion as in diabetes insipidus. As a result of this little or no water is reabsorbed from the distal tubules and a large volume (as high as 25 l) of urine is produced each day;

diabetes mellitus polyuria

- an excess of solute demands more water for its elimination from the kidney. For example the accumulation of glucose in the blood in diabetes mellitus results in polyuria (excessive output of urine) causing too great a water loss;
- chronic renal disease may impair the renal ability to concentrate wastes.

insufficient fluid intake

deficiency in electrolytes

deficiency in aldosterone secretion

Addison's disease

A deficiency in aldosterone secretion (Addison's disease) leads to an inadequate reabsorption of sodium by the renal tubules. The osmotic pressure of the extracellular fluid is reduced, promoting a reduction in ADH secretion which leads to an increase in the amount of water lost from the renal tubules.

The necessity to restore normal hydration and electrolyte concentration may be urgent in order to maintain metabolism, circulation and renal function. Intravenous infusion, using a solution based on a patient's need, may be used to re-establish a satisfactory balance quickly.

7.14.2 Fluid excess

oedema

Oedema is the accumulation of an excessive amount of fluid in the interstitial spaces. It implies a positive fluid balance in which the fluid intake has exceeded the output and there is an increase in total body water and sodium.

Π A number of physiological factors can cause oedema. Try to list some.

You may have thought of the following:

- an increase in venous pressure such as may be found in heart failure;
- obstruction of lymphatic drainage;
- a deficiency in blood proteins (hypoproteinaemia) which leads to an increased formation of tissue fluid;
- an increased capillary permeability (as in inflammation, burns and hypersensitivity reactions);
- renal insufficiency due to renal disease or reduced blood flow through the kidneys (eg heart failure or shock);
- excessive aldosterone secretion which leads to increased absorption of sodium by the renal tubules. The osmotic pressure of the extracellular fluid rises, ADH release is increased and more water is reabsorbed by the kidneys;
- immobilisation may lead to generalised oedema, circulatory problems such as hypotension and thrombus formation.

diuretic

Oedema may be controlled by either one or a combination of several treatments. These include a diet in which the salt intake is restricted and the use of specific diuretic drugs.

7.15 Diuretics

Drugs which increase urine flow are called diuretics. Diuretics may be administered to patients suffering from oedema in an attempt to remove the excess fluid. Some diuretics have additional uses, for example, they lower blood pressure in hypertension.

Diuretics which may be used include:

- non-reabsorbable sugars, such as mannitol. These substances are filtered at the glomerulus but are not reabsorbed by the tubules. Their presence in the tubules increases the osmotic pressure of the intratubular fluid and therefore opposes water reabsorption. This results in an increase in urine flow: osmotic diuresis;
- Xanthines, for example caffeine in coffee and theophylline in tea causes dilation of the afferent glomerular arteriole, thereby increasing glomerular filtration pressure and glomerular filtration rate;

- Spironolactone acts by antagonising the actions of aldosterone. It will therefore decrease sodium reabsorption in the distal tubule and increase sodium and water loss;

- Organic mercury salts (Mercuhydrin), and chlorothiazide (Diuril) act in the tubular cells by preventing the action of carriers and enzymes involved in the active reabsorption of sodium and chloride. As a result, larger quantities than usual of these substances remain in the tubules and their increased osmotic pressure opposes the reabsorption of water, increasing urine flow;

- Furosemide (Lasix) is a powerful, rapidly acting diuretic which may lead to severe dehydration unless its adminstration is carefully controlled;

- Acetazolamide (Diamox) also increases sodium excretion and therefore water excretion.

7.16 Endocrine functions of the kidney

erythropoietin

renal erythropoetic factor

As well as in renin production (see Section 7.13.4) the kidney is involved in the production of the hormone erythropoietin. This hormone is formed by the interaction of a substance secreted by the kidneys and a specific plasma protein (a globulin). This renal factor is called renal erythropoietic factor (REF) or erythrogenin. Loss of blood or hypoxia stimulates the formation of erythropoietin which acts on the bone marrow to promote the formation of red blood cells (erythropoesis). The biotechnological production of erythropoietin has enabled treatment of some of the symptoms of kidney failure. (A case study discussing the production of erythropoietin is included in the BIOTOL text 'Biotechnology Innovations in Health Care').

7.17 Abnormal kidney function

Almost any type of kidney damage decreases the ability of the kidney to clear the blood of waste products. Therefore kidney abnormalities usually cause an accumulation of waste metabolic products in the body fluids as well as poor regulation of the electrolyte and water composition of the fluids.

glomerulo-nephritis

Several types of kidney damage can cause the kidneys to stop working suddenly and completely. This is called kidney shutdown and may be caused by heavy metal poisoning, blocking of the kidney tubules with haemoglobin following a transfusion reaction, or by certain antibiotics. Diseases may also cause kidney shutdown. A very common kidney disease is glomerulonephritis which is caused by toxins produced by certain bacteria. The glomeruli become inflamed, swollen and enlarged with blood. Blood flow through the glomeruli almost ceases and also the glomerular membranes become extremely permeable allowing even large molecules such as proteins and red blood cells into the renal tubules. The disease permanently destroys some of nephrons and repeated bouts of glomerular nephritis may result in the destruction of a large number of nephrons leading eventually to oedema, coma and death.

There are usually few symptoms of kidney failure until the glomerular filtration rate has fallen from 120 ml min^{-1} to below 30 ml min^{-1}. Below this dietary restrictions are needed. A diet consisting of carbohydrate and fat is given with the protein content being reduced to the barest minimum. This is to prevent the accumulation of acid and urea from the breakdown of surplus amino acids.

When the glomerular filtration rate falls to below 5 ml min^{-1} control by diet alone is not effective and an artificial kidney (dialysis treatment) or a kidney transplant is needed to maintain life.

7.18 The control of acid base balance

The body continuously produces hydrogen ions which if allowed to accumulate would change the pH of the body fluids and make them more acid. Despite this the pH of the body fluids remains remarkably stable. In this section we will consider the importance and necessity of the homeostatic mechanisms that operate to maintain a normal body pH. The process of acid base regulation is concerned with the regulation of hydrogen ion concentration in the body fluids: any excess must be removed from the body. We will look at how the hydrogen ions are produced and the means by which they are eliminated by both the lungs and the kidneys. The mechanisms of action of the body buffer systems, which will oppose any changes in hydrogen ion concentration will also be discussed. Finally the causes and consequences of acid base imbalance will be examined.

7.18.1 Introduction to the regulation of acid base balance

The term regulation of acid base balance concerns the regulation of hydrogen ion concentration in the body fluids. Hydrogen ion concentration must be regulated precisely for several reasons:

- it influences the activity and structure of protein molecules many of which are enzymes. Each enzyme operates optimally at a particular pH and changes in hydrogen ion concentration will result in changes in enzyme activity;
- it influences the distribution of other ions between the extracellular fluid and the intracellular fluid;
- it influences the activity of drugs and hormones, in many cases affecting the binding of these substances to plasma proteins.

From the examples given you can see that any deviation of the hydrogen ion concentration from normal will produce disturbances in cellular activity.

Normally body fluids are slightly alkaline having a pH of 7.35 to 7.45. The normal pH is kept within this narrow range because small variations in either direction are incompatible with normal cellular activity and may be life threatening.

7.18.2 Acids, bases and buffers

An acid is a substance which in solution will dissociate to release hydrogen ions (H^+) in a process called dissociation. For example, when hydrochloric acid (HCl) is added to water the following dissociation reaction occurs:

$$HCl \rightarrow H^+ + Cl^-$$

dissociation

Because hydrochloric acid dissociates completely it is called a strong acid.

An acid that only partially dissociates is called a weak acid, for example, a molecule of carbonic acid (H_2CO_3) dissociates into a hydrogen ion and a bicarbonate ion (HCO_3^-):

$$H_2CO_3 \rightarrow H^+ + HCO_3^-$$

A strong acid will have a lower pH than a weak acid because it is more highly dissociated.

Substances which accept or take up hydrogen ions when in solution are called bases.

Π Which of the following represents a substance acting as an acid and which a substance acting as a base.

1) $H_2CO_3 \rightarrow H^+ + HCO_3^-$

2) $HCO_3^- + H^+ \rightarrow H_2CO_3$

3) $NH_3 + H^+ \rightarrow NH_4^+$

4) $H^+ + OH^- \rightarrow H_2O$

Your answers should read:

1) H_2CO_3 (carbonic acid) acts as an acid.

2) HCO_3^- (bicarbonate) acts as a base.

3) NH_3 (ammonia) acts as a base.

4) OH^- (hydroxyl) acts as a base.

carbonic acid

carbonic anhydrase

Chemical processes (ie metabolism) in cells produce relatively large amounts of acids, but the body possesses control mechanisms that maintain the required normal pH of about 7.4. The chief acid resulting from metabolism is carbonic acid which is formed when carbon dioxide (CO_2) combines with water. This combination is accelerated by the enzyme carbonic anhydrase within the cells:

$$CO_2 + H_2O \xrightarrow[\text{anhydrase}]{\text{carbonic}} \underset{\text{carbonic acid}}{H_2CO_3}$$

As well as carbonic acid, cellular metabolism produces stronger acids such as sulphuric, phosphoric, hydrochloric, lactic and other organic acids.

You are probably familiar with cramp which develops when you have overworked muscles. This is because lactic acid is produced as an end product of the anaerobic metabolism (glycolysis) taking place in the muscles under these conditions.

The body has several mechanisms for removing acids. In the case of the lactic acid build up described above, two main mechanisms come into play. Some of the lactic acid may be transported away in blood while the remainder may be oxidised (via pyruvate and the tricarboxylic acid cycle) to form carbon dioxide and water. Remember that carbon dioxide can readily form carbonic acid. Carbonic acid produced by cells from the oxidation of organic substrates is transported via the blood stream. It is removed by the lungs in the form of carbon dioxide, other acids are excreted by the kidneys.

The normal pH of the body fluids is maintained by acid base buffer systems.

7.18.3 Buffer systems

A buffer is a chemical system that prevents a rapid or excessive change in pH when an acid or a base is added to a solution containing the buffer.

We will not explain the principles of buffering in detail here, since it is assumed that the reader has encountered the concept of buffers previously. (A description of this is given in the BIOTOL text 'Molecular Fabric of Cells'). Here we will mainly explain the role of the kidneys in controlling the buffering capacity of the body.

7.18.4 Bicarbonate buffer system

This is the main buffer system of the plasma. The main components of this system are carbonic acid and biocarbonate.

We can represent this system as:

$$H_2CO_3 \rightleftarrows H^+ + HCO_3^-$$

Addition of acid (ie H^+) to this system pushes the reaction towards the left, thus leading to the removal of H^+ ions and the formation of undissiociated H_2CO_3.

Π What would be the consequence of adding a strong base (for example NaOH) to the carbonic acid: biocarbonate system?

You should have concluded that the addition of a strong base effectively adds OH^- ions. These will combine with the H^+ ions to form water. This in turn will cause the carbonic acid to dissociate to re-establish the H^+ ion concentration. Thus:

$$H_2CO_3 \rightleftharpoons [H^+] + HCO_3^-$$

$$[H^+] + OH^- \rightarrow H_2O$$

Thus the carbonic acid: biocarbonate system resists changes in pH when either acid or base is added (that is it acts as buffer).

Henderson-Hasselbalch equation

The greatest buffering capacity of this system is established when the concentrations of carbonic acid and biocarbonate are equal. We will not go into details here but this is predicted by the Henderson-Hasselbalch equations which relates pH to the dissociation constant of the acid (Ka) and the relative concentrations of the acid and its conjugate base. Thus:

$pH = pKa + \log \frac{[A^-]}{[HA]}$ where [HA] = concentration of acid; $[A^-]$ = concentration of the conjugate base; pKa = -log (dissociation constant).

If you substitute a series of values for the ratio of $\frac{[A^-]}{[HA]}$ into this equation, you will find that pH is only changed slightly when the ratio of $[A^-]/[HA]$ is close to 1. When the ratio of $[A^-]/[HA]$ is large (eg 10-100) or small (eg 0.001-0.01) then a small change in either $[A^-]$ or [HA] leads to a large change in pH. Thus such systems display greatest buffering capacity close to their pKa values.

SAQ 7.4

Calculate the ratio of $[HCO_3^-]/[H_2CO_3]$ if the pKa of carbonic acid 6.4 and the pH is a) 5.4, b) 8.4.

7.18.5 Phosphate buffer system

Phosphate are also present in biological systems.

Phosphate can dissociate in the following ways.

$$H_3PO_4 \rightleftharpoons H^+ + H_2PO_4^- \rightleftharpoons H^+ + HPO_4^{2-} \rightleftharpoons H^+ + PO_4^{3-}$$

(phosphoric acid) (dihydrogen phosphate) (hydrogen phosphate) (phosphate)

Thus if we begin at a very low pH, the phosphate is mainly in the form H_3PO_4 if we add alkali (that is remove H^+ ions by the reaction $H^+ + OH^- \rightleftharpoons H_2O$), then H_3PO_4 will dissociate to form $H^+ + H_2PO_4^-$, thus resisting the change in pH. If we continue to add OH^- ions, H_3PO_4 will continue to dissociate until it is all in the form $H_2PO_4^-$. If we continue to add OH^- ions, then $H_2PO_4^-$ will dissociate to form HPO_4^{2-} and so on until all of the $H_2PO_4^-$ is converted to HPO_4^{2-}. If we continue to add OH^- ions then the HPO_4^{2-} dissociate to form PO_4^{3-}.

The mid points of these three dissociations have pKa values of 2.1, 7.2, 12.3. Thus the phosphate buffer system has strong buffering capacity around pH2.1 because of the $H_3PO_4 \rightleftharpoons H^+ + H_2PO_4^-$ dissociation.

Π Which phosphate dissociation strongly resist pH changes around pH12.3?

You should have concluded that the dissociation relevant at pH12.3 is:

$$HPO_4^{2-} \rightleftharpoons H^+ + PO_4^{3-}$$

If acid (H^+) is added to this system, the equilibrium move to the left thus reducing the H^+ ion concentration. On the other hand addition of a base (OH^-) results in H^+ being removed ($H^+ + OH^- \rightleftharpoons H_2O$). This leads to further dissociation of HPO_4^{2-} thereby restoring the H^+ ion concentration.

Π Which dissociation of phosphate is mainly involved in buffering in living systems?

Typically the pH of animal tissues is around pH7.4. Thus the phosphate dissociation at this pH is:

$$H_2PO_4^- \rightleftarrows H^+ + HPO_4^{2-}$$

Phosphate buffering is particularly important in the kidneys. The production of acids would lower the pH. However, the H^+ ions combine with HPO_4^{2-} ions to form $H_2PO_4^-$. In this way the pH is maintained at an appropriate level. The $H_2PO_4^-$ ions produced are excreted in the urine.

Π In SAQ 7.4 we quoted the pKa value of carbonic acid as 6.4. We should have quoted two values. Why?

The answer is carbonic acid can undergo two dissociations. Thus:

$$H_2CO_3 \rightleftarrows H^+ + HCO_3^- \rightleftarrows H^+ + CO_3^{2-}$$

The pKa of the first dissociation is 6.4, whilst that of the second is 10.3. At physiological pHs (pH7.4) we are mainly dealing with the first dissociation in this sequence.

7.18.6 Proteins as buffers

Plasma and tissue proteins also act as buffers. They are buffers because they contain chemical groups (NH_2, NH_3^+, COO^-, COOH) which will accept or donate hydrogen ions. The protein haemoglobin within the red blood cells also acts as a buffer and this will be discussed later.

In summary, the buffer systems of the body can hold the pH of the body fluids within very narrow limits because they take up or release hydrogen ions, becoming bound to or being dissociated from them thereby preventing them from changing the free hydrogen ion concentration, which would change the pH. The hydrogen ion concentration of the body can not only be regulated by the buffering systems but the physiological buffer systems can actually remove H^+ ions from the body. This takes place in the kidneys.

7.18.7 Production of hydrogen ions in the body

There are three major sources of hydrogen ions in the body:

1) Carbon dioxide (CO_2) produced by the tissues of the body during respiration. The CO_2 combines with water in the red blood cells to form carbonic acid (H_2CO_3). This dissociates immediately to release hydrogen ions as shown in the following reaction:

$$CO_2 + H_2O \xrightarrow[\text{carbonic anhydrase}]{} H_2CO_3 \longrightarrow H^+ + HCO_3^-$$

The first reaction is speeded up by the enzyme carbonic anhydrase found in the red blood cells. Most of the hydrogen ions are buffered by combination with haemoglobin. Carbon dioxide is the major generator of hydrogen ions.

2) The phosphorus and sulphur groups present in many biological molecules. On metabolism, phosphoric and sulphuric acids are produced which are released into the extracellular fluid where they dissociate to release hydrogen ions.

3) Organic acids such as fatty acids and lactic acid. These are end products of metabolism and also release hydrogen ions by dissociation.

When food is oxidised in the body to produce energy, its final products after metabolism will be acidic, basic or neutral. Acid forming foods include meat, eggs, fish and seafood whereas most fruits and vegetables are basic substances. Butter and milk are nearly neutral.

In normal metabolism, a general mixed diet results in the production of a large excess of acid, primarily in the form of carbon dioxide which is the major generator of hydrogen ions in the body.

alkali reserve

Let us use a simple example to illustrate the problem. The 3000 kilojoule diet of an average adult will produce about 480 l of carbon dioxide each day. This is equivalent to about 2 l of concentrated hydrochloric acid as far as H^+ production is concerned. Therefore, the body must excrete large amounts of acid and at the same time conserve bases to maintain a proper balance of pH. Alkali reserve is the name given to the base (usually sodium bicarbonate) which removes the excess hydrogen ions. In actual fact it is the base available in excess of normal body needs to handle hydrogen ions. But does this mean that the level of bicarbonate in the body will drop? The answer is no, and we will see later in the text how the body maintains this level.

Most of the acid produced by metabolism is excreted as carbon dioxide by the lungs and as inorganic and organic acids by the kidneys. It may seem a little odd to describe acid produced by metabolism is excreted by carbon dioxide since carbon dioxide alone is not an acid. You will however learn in the next section exactly what we mean by this statement. Therefore, respiratory regulation and kidney regulation are very important control mechanisms in the maintenance of acid base balance.

Before we examine how the respiratory system regulates acid base balance let us look briefly at how the blood is involved in maintaining hydrogen ion concentration.

7.18.7 The role of the red blood cell in pH regulation

Carbon dioxide, produced as a waste product by the cells, diffuses into the red blood cells and combines with water to form carbonic acid. This dissociates and the resulting hydrogen ions are buffered by the:

haemoglobin (Hb)/potassium haemoglobin (KHb) buffer system to form acid haemoglobin (HHb) and potassium bicarbonate ($KHCO_3$). The reactions are as follows:

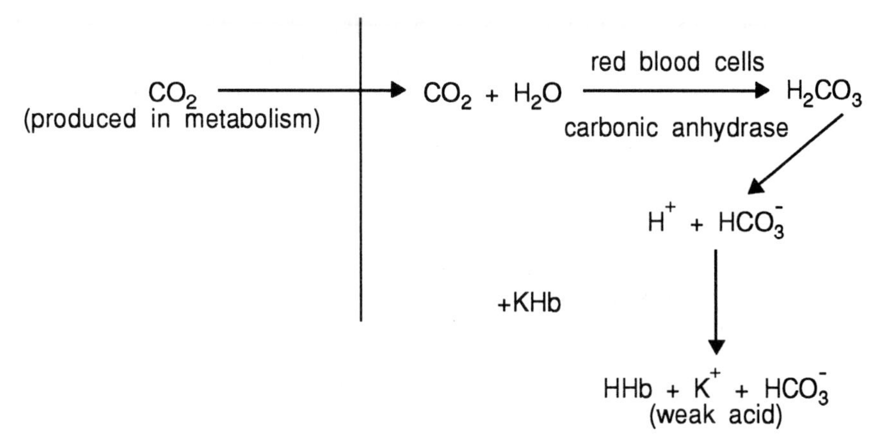

When the acid haemoglobin (HHb) combines with oxygen in the lungs it becomes a stronger acid ($HHbO_2$) and dissociates. The hydrogen ions are taken up by bicarbonate. The result of this is the formation of oxyhaemoglobin and carbonic acid:

$$HHb + O_2 \rightarrow \underset{\text{(stronger acid)}}{HHbO_2} \longrightarrow \underset{\text{(oxyhaemoglobin)}}{HbO_2} + H^+ + HCO_3^- \rightarrow H_2CO_3$$

The carbonic acid breaks down into carbon dioxide and water:

$$\underset{\text{carbonic acid}}{H_2CO_3} \longrightarrow CO_2 + H_2O$$

and the carbon dioxide diffuses out of the blood and into the lungs and is breathed out. It is as bicarbonate ions that most of the carbon dioxide is carried to the lungs. The loss of carbon dioxide through the lungs removes potential hydrogen ions, a build up of which would reduce blood pH.

7.18.8 Respiratory regulation of pH

influence of CO_2 and H^+ on breathing

The nerve cells of the respiratory centre in the brain stem are very sensitive to the concentration of carbon dioxide and hydrogen ion in the body fluids. An increase in either stimulates the centre to produce an increase in the rate and depth of breathing so that more carbon dioxide is lost from the body. Conversely a decrease in the normal concentration of carbon dioxide leads to slower, shallower breathing so that carbon dioxide is retained to form carbonic acid which will lead to an increase in hydrogen ion concentration and a reduction in pH.

Π What would be the effect on body fluid pH of any condition that reduces the ability of the lungs to eliminate carbon dioxide?

The pH would fall because carbon dioxide would accumulate in the body where it will combine with water to form carbonic acid which ionises to release hydrogen ions. Because there is so much carbon dioxide, the buffer systems would become saturated and the pH of the body fluids would fall. If it falls below normal levels the resulting condition is called acidosis. On the other hand, increased ventilation may cause excessive loss of carbon dioxide, a reduction in carbonic acid and therefore a decrease in hydrogen ion concentration leading to an increase in the pH of the body fluids. This condition is called alkalosis.

acidosis

alkalosis

The lungs, however, cannot restore any loss of bicarbonate reserves (alkali reserve) because each time a hydrogen ion is removed via carbon dioxide removal, a bicarbonate ion is lost as well. The kidney, however, has the ability to eliminate hydrogen ions and at the same time retain bicarbonate.

7.18.9 The regulation of pH by the kidney

The kidneys play an important role in maintaining acid base balance by excreting excess hydrogen ions and forming bicarbonate which is returned to the circulation to form a reserve of alkaline ions (ie the alkali reserve).

The cells of the distal renal tubules are sensitive to changes in pH of the tubular fluid. When there is a fall in the normal pH of the body fluids the kidneys excrete hydrogen ions and bicarbonate is formed and retained. Conversely when the pH of the body is raised above normal (ie becomes more alkaline) hydrogen ions are retained and bases are excreted. The kidneys therefore excrete a variable amount of hydrogen ion and produce more or less bicarbonate depending on the adjustments needed to maintain pH.

The elimination of hydrogen ions and the formation of bicarbonate involve a number of processes within the kidneys.

1) Within the tubular cells carbon dioxide and water form carbonic acid which ionises to release hydrogen ions and bicarbonate ions. The hydrogen ion moves out into the renal tubule and is exchanged for a sodium ion. The reactions are shown in Table 7.5.

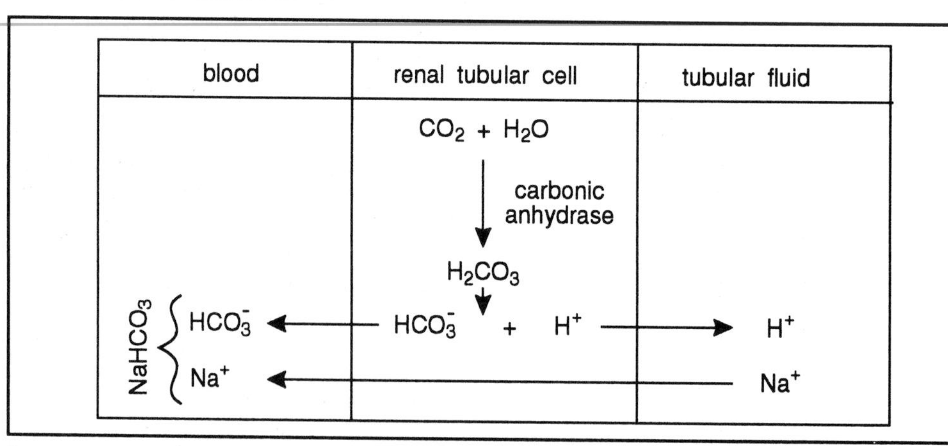

Table 7.5 Renal elimination of hydrogen ions.

Notice that the excretion of hydrogen is accompanied by the reabsorption of a bicarbonate ion in the blood. Sodium bicarbonate forms the alkali reserve and is part of an important buffer system.

2) The hydrogen ions in the tubular fluid must now be buffered. They combine in the tubular fluid with disodium hydrogen phosphate (Na_2HPO_4) which has been filtered from the blood at the glomerulus. This is converted to dihydrogen sodium phosphate (NaH_2PO_4) by accepting the hydrogen ion and releasing a sodium ion. The sodium dihydrogen phosphate is excreted in the urine. The reactions involved are shown in Table 7.6.

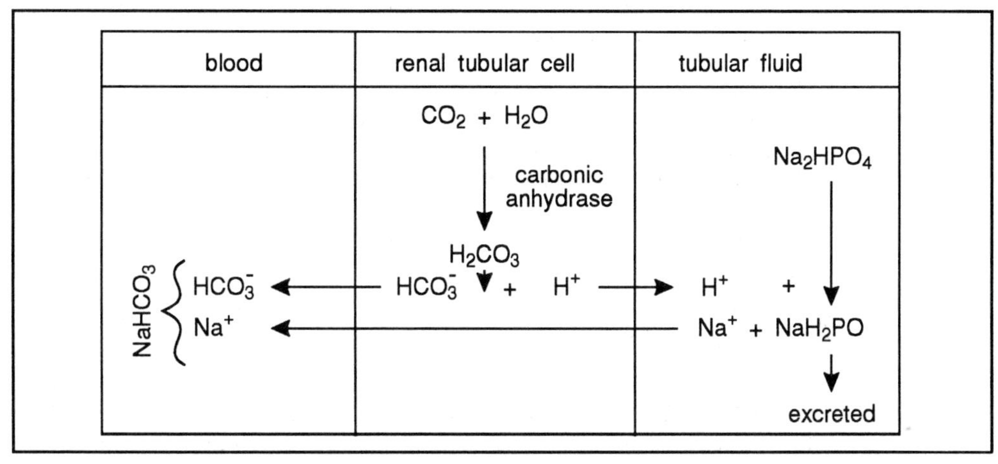

Table 7.6 Renal elimination of acid phosphate.

You will notice that again sodium bicarbonate is conserved.

3) If the acid content of the blood is high as occurs in acidosis, the acidity of the tubular fluid stimulates the production of ammonia (NH_3) in the tubular cells. Ammonia diffuses into the tubular fluid where it combines with secreted hydrogen ions to form ammonium ions (NH_4^+).

The ammonium ion combines with a chloride ion (Cl^-) to form ammonium chloride which is excreted in the urine. When there is a large load of hydrogen ions to be excreted, most of it is excreted as ammonium salts. Under lower H^+ loadings, the carbonate and phosphate systems predominate. Notice that with the ammonium system, sodium ions are again reabsorbed and form sodium bicarbonate in the blood. The reactions involved are shown in Table 7.7.

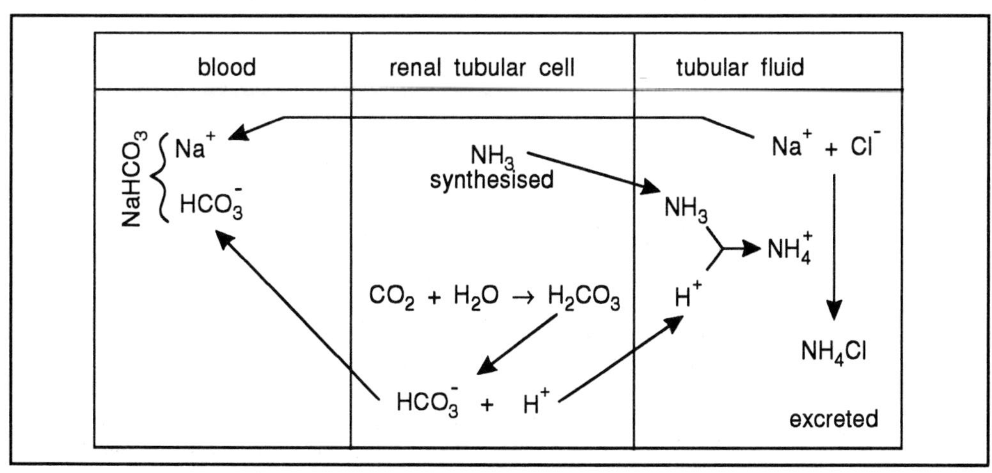

Table 7.7 Renal elimination of ammonia.

limiting pH

The amount of hydrogen ions that can be secreted into the tubular fluid depends on the pH of the urine within the tubule. If the urine pH falls to 4.5 or less the hydrogen transport mechanism cannot function. This is called the limiting pH. Normally the pH of urine is about 6.0. However, when excess acids are excreted, for example in acidosis, the pH can fall as low as 4.5. On the other hand urine pH may go as high as 8.0 when excess bases are excreted.

In summary, to preserve the normal pH, the kidneys secrete hydrogen ions into the urine in exchange for sodium ions. They acidify alkaline phosphate, form and retain sodium bicarbonate and also produce an ammonium salt. By losing hydrogen ions and adding bicarbonate, acidosis may be quickly corrected. In this way the kidneys are much more effective in controlling blood pH than the respiratory system. Renal adjustments of the acid base balance, however, while more effective are much slower than the respiratory system and may take several hours or even days.

So far we have examined the processes by which the pH of the body fluids is maintained. In the next section we will consider the factors which may produce disturbances of this pH.

SAQ 7.5

Now that you have completed this section, answer the following true/false questions.

1) When there is a large load of hydrogen ions to be excreted, most of it appears in the urine in the form of ammonium salts.

2) Potassium is normally reabsorbed from tubular fluid in exchange for hydrogen ions.

3) Hydrogen ion are secreted into the urine by cells lining the tubules.

4) The hydrogen ions reacts with NaH_2PO_4 in the tubular fluid to give Na_2HPO_4.

5) Carbonic acid is present in the urine in very high concentrations compared with plasma.

7.18.10 Alterations of acid base balance

acidosis and alkalosis

If the pH of body fluids moves beyond normal limits (pH 7.35 to 7.45) the resulting conditions will be either acidosis (pH below 7.35) or alkalosis (pH above 7.45). Either condition is life threatening unless remedied without delay. Acidosis and alkalosis may be classified as either respiratory or metabolic according to the causal source.

7.18.11 Acidosis

When the hydrogen ion concentration is increased in the body fluids, three control mechanisms try to restore the normal pH.

Π Can you name these three control mechanisms?

ms;

vity.

If the carbonic acid/bicarbonate ratio can be kept normal by increasing respiratory elimination of carbon dioxide, by increasing elimination of hydrogen ions and by the formation of bicarbonate by the kidney, then the pH can be kept within the normal range. If this mechanism is saturated by excess acid and cannot compensate, an increase in the carbonic acid/bicarbonate ratio develops, pH falls and acidosis develops.

Respiratory acidosis

hypoventilation

The most common cause of respiratory acidosis is the build up of carbon dioxide in the body caused by either inadequate or inefficient elimination by the lungs, for example in hypoventilation where there is low or restricted ventilation.

This hypoventilation may be caused by:

- acute or chronic respiratory disease (pneumonia or emphysema);
- circulatory impairment or failure of the lungs;
- depression of the respiratory centre by drugs or cerebral disease or weakness of the respiratory muscles.

If the carbon dioxide is not removed efficiently by the lungs, the excess hydrogen ions will saturate the buffer and will accumulate in the blood and body fluids causing acidosis.

Π How would the kidney respond to respiratory acidosis?

The kidneys respond to the increased level of carbon dioxide by excreting more hydrogen ions from the tubules in exchange for sodium ions. These combine with bicarbonate, left behind by the hydrogen ions, to form sodium bicarbonate. This is reabsorbed into the blood stream to add to the buffering capacity.

The kidney also increase the formation and excretion of ammonia, which combine with hydrogen ions in the tubular fluid to form ammonium ions.

7.18.12 Metabolic acidosis

Metabolic acidosis occurs as the result of:

- an excessive production or intake of acid;
- the retention of acid;
- a depletion of bicarbonate base (the body's normal alkali reserve).

This condition may be caused by:

- diabetes mellitus, where abnormal amounts of keto acids (organic acids) are produced by fat metabolism in the absence of carbohydrates. These deplete the bicarbonate buffer and the pH of the body fluids falls;
- an overdose or prolonged administration of certain compounds eg aspirin or acetazolamide (a diuretic);
- a reduced urinary output, whether due to renal disease or severe dehydration. The result is retention of acids;
- an abnormal loss of alkaline secretions that are normally reabsorbed eg the loss of intestinal juices in diarrhoea;
- a reduced blood supply to certain tissues results in the accumulation of acids and impairment of normal cell function, eg shock.

respiratory compensation

The body tries to correct metabolic acidosis by increasing the respiration rate to increase carbon dioxide elimination and reduce carbonic acid levels. This is called respiratory compensation. The kidneys also increase hydrogen ion excretion for elimination in the urine.

7.18.13 Alkalosis

Alkalosis describes the condition in which there is an increase in the normal pH to a level, greater than 7.45. Alkalosis may be the result of a lack of carbonic acid or excess bicarbonate but whatever the cause the normal equilibrium of carbonic acid bicarbonate is upset.

7.18.14 Respiratory alkalosis

hyperventilation

Respiratory alkalosis can be caused by hyperventilation since this process leads to carbon dioxide excretion by the lungs in excess of its production by the tissues. This depletes carbonic acid from the buffer system.

Π Make a list of as many circumstances which might produce hyperventilation as you can?

Rapid deep breathing may be caused by anaemia, anxiety, hysteria, or central nervous system disease producing overstimulation of the respiratory centre in the brain. A high fever or oxygen lack (hypoxia) can also cause hyperventilation. Treatment such as oxygen inhalation may be given if the respiratory alkalosis is caused by hypoxia. If the cause is anaemia then a blood transfusion may be administered.

7.18.15 Metabolic alkalosis

In this condition, the increase in pH may develop as the result of:

- an abnormal loss of hydrochloric acid from the stomach in vomiting;
- the excess ingestion of alkaline substances (eg sodium bicarbonate);
- a potassium deficit.

In these conditions plasma concentration of bicarbonate is raised and produces a corresponding increase in pH. Respiration becomes slow and shallow in an effort to retain carbon dioxide and therefore, increase the carbonic acid content of the blood. The kidney tries to compensate by conserving hydrogen ion and increasing secretion of bicarbonate. If the alkalosis is caused by vomiting then the patient is also likely to be dehydrated. This state would lead to a reduced urinary output and reduced renal compensation.

If you review the examples given, you will see that a number of disorders of either the respiratory system or the kidneys as well as metabolism can cause serious disturbances in acid base balance.

Acidosis generally depresses mental activity resulting in coma and death if left untreated, whereas alkalosis causes overexcitation of the nervous system, leading to tetany or convulsions which may also be fatal. A very important point to remember is that the lungs and kidneys work together to achieve compensation of acid base disorders.

7.18.16 Summary diagram

Diagrammatic representation of the mechanisms that operate to maintain normal acid base balance and the factors causing acidosis or alkalosis are shown in Figure 7.13.

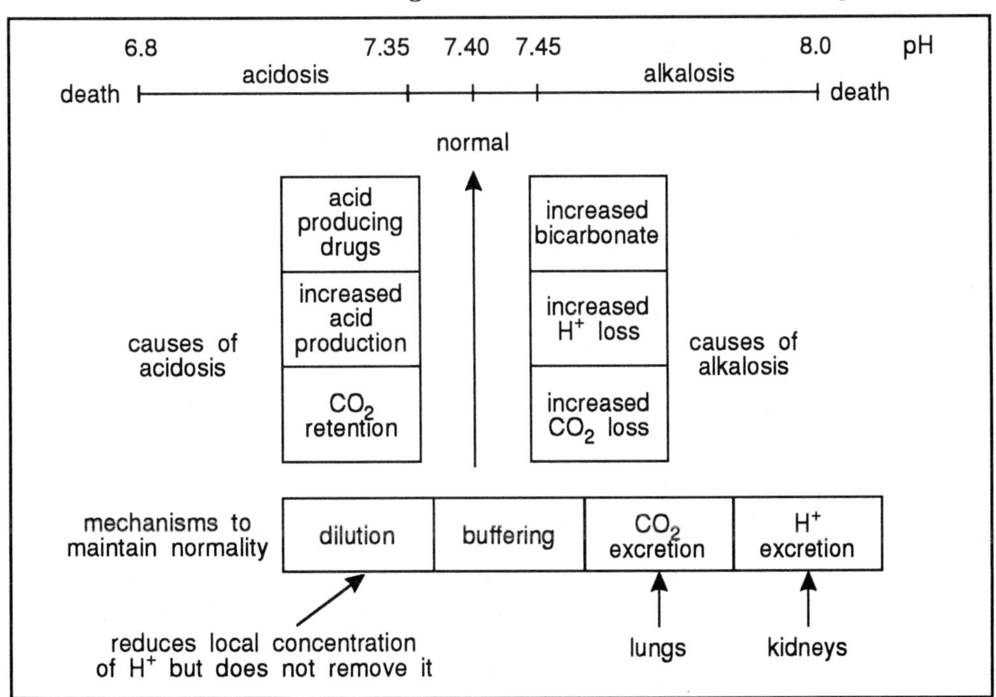

Figure 7.13 The maintenance of acid/base balance.

Summary and objectives

In this chapter we have examined the structure and function of the kidney with particular emphasis on its role in water and electrolyte balances. We began the chapter by considering the nature of excretory products and the general layout of the urinary system. We then examined in depth the structure of the kidney and related this structure to its function. We also explored how the activity of the kidney was regulated. We also examined the consequences of kidney malfunction. In the final part of the chapter we examined the role of the kidney in maintaining acid: base balance within the body and learnt that the lungs were also important in maintaining body pH.

Now that you have completed this chapter you should be able to:

- relate the excretion of a wide variety of excretory products to their appropriate excretory organ;
- list a wide variety of excretory products and describe in outline the composition of urine;
- describe the general layout of the kidney and the general form of nephrons;
- describe the blood supply to and from the kidney with particular emphasis on the blood supply in the glomeruli;
- describe the processes of glomerular filtration, tubular secretion and readsorption;
- calculate a variety of parameters such as glomerular filtration rates (GFR) and clearance rates from supplied data;
- describe how the output of the kidneys is regulated through the control of the blood supply, glomerular filtration rates and readsorption;
- list and explain the origins of a variety of disorders arising from incorrect fluid and electrolyte balances;
- explain how bicarbonate, phosphate and ammonium ions can act as buffers and describe the role of these buffer systems in maintaining pH in the body;
- describe the roles of the kidney and lungs in regulating body pH and explain the consequences of incorrect pH regulation.

Nutrient supply

Nutrient supply

8.1 Introduction

In Chapter 1 you were introduced to different types of cells, specialised in a way that enables them to aid the functioning of the particular tissue, organ, and system to which they belong. In Chapter 4 you learned about the digestion and absorption of the various nutrients required by cells. This chapter will review the supply of cellular nutrients and show why different tissues have different requirements. We will examine the ways by which nutrient supply is controlled or altered in response to different physiological circumstances. In addition, some consequences of abnormal nutrient intake will be discussed.

8.2 Functions of the nutrients

main nutrients

Most people, if asked for the names of the different types of nutrients, would probably suggest carbohydrates, fats (lipids), proteins and vitamins. The often forgotten ones are water, minerals and fibre. You might think it strange to include water and fibre in this list, but a nutrient is usually defined as a substance consumed and necessary for the optimum functioning of the body, rather than something which is necessarily absorbed by the gut or metabolised by cells.

Π Can you think of a way of subdividing these seven types of nutrients into two groups with three or four types in each group?

One way would be to divide them into nutrients metabolised by Man and those which are not. The latter group includes fibre, water, mineral and vitamins. However, this division would not be strictly true, since water and some vitamins do undergo metabolic changes.

Carbohydrates, lipids and proteins can be classified as nutrient fuels. For most adult humans, carbohydrates supply about 65% of their total energy, though in Europe and North America the figure is only about 45%. Lipids usually supply between 10 and 45% of the energy, and the remainder (usually around 10-15%) comes from protein.

8.2.1 Carbohydrates

The chief role of carbohydrates in the diet is to provide energy. After digestion and absorption, dietary carbohydrates are present in the form of simple sugars, mainly glucose, but also fructose and lactose, the latter two being converted to glucose and/or metabolised by the liver. Thus glucose is the only significant carbohydrate fuel available to most body tissues. The metabolic fates of glucose are shown in Figure 8.1.

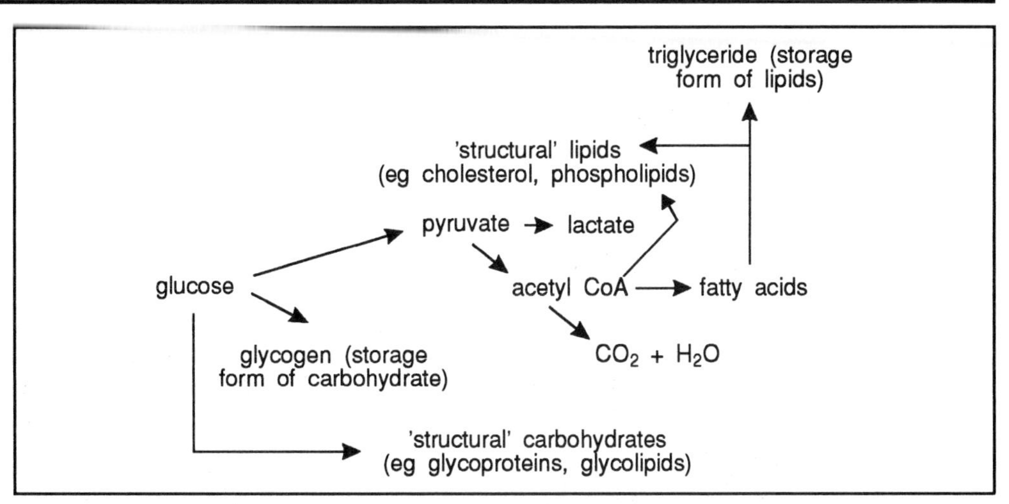

Figure 8.1 Metabolic fate of glucose.

Π Using Figure 8.1, list the storage and structural forms of carbohydrates and lipids.

The potential energy available from carbohydrate-containing foods can be stored as glycogen (a carbohydrate), or as triglyceride (fat), or glucose can be used to make the structural components of cells, such as glycoproteins, glycolipids, cholesterol and phospholipids.

8.2.2 Lipids

fatty acids, ketone bodies, acetoacetate, 2-hydroxy-butyrates

Dietary fat, after conversion to fatty acids, can be catabolised directly to CO_2 and H_2O or, in the liver, to ketone bodies (acetoacetate and 2-hydroxybutyrate) which can be further oxidised to CO_2 and H_2O by other tissues. Alternatively, fatty acids may be re-esterified to triglyceride (for storage), or converted into phospholipids and other structural lipids. Dietary cholesterol can be used in membranes or converted to bile acids (to aid digestion and absorption), to steroid hormones, or to Vitamin D_3. The eicosanoid group of hormone-like substances (prostacyclins, thromboxanes and leukotrienes) are normally synthesised from polyunsaturated fatty acids.

eicosanoids

The pathways of lipid metabolism are described in the BIOTOL text 'Energy Sources for Cells'.

8.2.3 Proteins

metabolic products of amino acid derived nitrogen

Protein is often thought of as a body building food. It is true that the amino acids derived from the digestion of proteins are used for the synthesis of tissue protein, the most obvious being muscle proteins. Amino acids are also used to synthesise the plasma proteins, peptide hormones, digestive enzymes and other secreted proteins. In addition amino acids are converted into non-protein, nitrogen containing molecules such as choline (for phospholipids), purines and pyrimidines (nucleic acid synthesis), thyroid hormones and catecholamines. The carbon skeleton (frame work) of the amino acids can also be used to make fat and some amino acids can be used for glucose production. In most adults, the anabolism of amino acids is exactly balanced by the catabolism of these compounds to products such as urea, uric acid, carbon dioxide and water. Overall, there is net catabolism of the dietary protein and we can, therefore, regard dietary protein as an energy producing nutrient. (Note that the metabolic pathways for amino acid catabolism are described in the BIOTOL text 'Energy Sources for Cells'.

8.2.4 Water

Water has five main functions:

- it is a solvent and suspending medium;
- it participates in metabolic reactions;
- it is a transport medium;
- it lubricates;
- it helps to maintain a constant body temperature due to its high thermal capacity.

8.2.5 Fibre

Fibre is a difficult nutrient to define as it is composed of a number of different substances, including pectin, hemicellulose, cellulose and lignin. Some of the components are fermentable by gut microbes, producing short-chain fatty acids which are absorbed and can be oxidised. However, the major role of fibre is in the regulation of the absorption of glucose, fats and minerals and control of gut motility.

8.2.6 Vitamins

coenzymes

cofactors

Vitamins are essential organic micronutrients. They must be present in the diet, but only in very small amounts. They are involved in a wide range of body functions, including acting as coenzymes or cofactors in the enzymes involved in metabolism. Some are incorporated into molecules that perform hormonal functions. Deficiency in dietary intake of vitamins usually leads to significant physiological disorders often including reduction in growth rate. Since vitamins are required for normal growth, some people would regard vitamins as growth factors. The use of this term should not, however, be confused with the use of the term growth factor which is used to describe molecules produced by one set of cells in the body which regulate the growth of other groups of cells. For example the epithelial cell growth factor is a biologically produced molecule essential for the growth of epithelial cells.

8.2.7 Minerals

The functions of minerals are also many and varied. They help determine the osmotic properties and pH of the body fluids. They are used structurally in bones and teeth or may function as cofactors in metalloprotein or metalloenzymes activity and are also implicitly involved in the control of various cell processes such as nerve conduction, and intracellular signalling.

8.2.8 Other cell nutrients

In addition to the nutrients that have been mentioned above, there are some important compounds that, although not significant components of the diet, are nonetheless important nutrients for some cells. These compounds are produced by cells in one tissue and transported via the blood to another tissue where they are taken up and metabolised. Pyruvate, lactate, acetoacetate and 2-hydroxybutyrate are examples of this group.

This chapter will deal only with the fuel nutrients. Examples of the importance, roles and regulation of the other nutrients occur where applicable in the other chapters.

The following SAQ is intended to test your knowledge of the functions of nutrients.

SAQ 8.1

From the list below, complete the following table, writing in the numbers of the appropriate responses. Note, numbers may be used more than once, and more than one number may be entered for each nutrient.

Nutrient	Functions
Glucose	
Fatty acids	
Cholesterol	
Amino acids	
Fibres	
Vitamins	
Minerals	

1) Formation of triglycerides.
2) Provision of energy.
3) Formation of steroid hormones.
4) Structural component of bone.
5) Formation of proteins.
6) Act as cofactors for enzymes.
7) Act as growth factors.
8) Involved in cell signalling.
9) Formation of glycoproteins.
10) Major constituents of cell membranes.
11) Regulation of gut motility.

8.3 Nutrient requirements of different tissues

mitochondrial content

Different tissues have different nutrient requirements. There is often an obvious link between the nutrient requirements of a tissue and its mitochondrial content. Table 8.1 lists several tissues and their preferred fuels immediately after a meal or at other times and Table 8.2 lists the same tissues, indicating their main functions and their mitochondrial content. Read these tables carefully then attempt the intext activity.

Tissue	Main Fuels Well-fed state	 Other states
Liver	lactate	Fatty acids
Heart muscle	Glucose	Fatty acids, lactate, ketones, pyruvate
Skeletal muscle (white) type IIB	Glucose/fatty acids	Fatty acids, BCAA, ketones (glycogen when contracting)
(red) type I	Glucose/fatty acids	Fatty acids, BCAA ketones (glycogen)
Brain	Glucose	Glucose, ketones
Adipose tissue	Glucose	Fatty acids
Red blood cells	Glucose	Glucose
Eye lens	Glucose	Glucose
Lung	Glucose	Glucose
Skin	Glucose	Glucose
Kidney medulla	Glucose	Glucose
Kidney cortex	Fatty acids/lactate	Fatty acids, lactate

Table 8.1 The preferred fuels of selected tissues and organs. BCAA = branched chain amino acid.

Tissue	Main Functions	Mitochondrial content
Liver	Regulation of supply of nutrients, synthesis of products, detoxification & disposal of waste	Plentiful
Heart muscle	Regular contraction	Numerous
Skeletal muscle (white)	Intermittent contraction high power, short acting	Few
(red)	Contraction, medium power, long lasting	Moderate
Brain	Electrical signalling	Moderate/few
White adipose tissue	Storage/release of fat	Few
Red blood cells	Transport of O_2	None
Eye lens	Focussing light	Few
Lung	Gaseous exchange	Moderate
Skin	Protection, excretion	Few
Kidney medulla	Filtration	Few
Kidney cortex	Reabsorption	Many

Table 8.2 The mitochondrial content and main function of selected tissues and organs.

Π From Tables 8.1 and 8.2 what general rule can you make about the relationship between the fuel used and the mitochondrial content of a tissue? Write down your answer and try the next question.

What do you think is the biochemical reason for this relationship?

relationship between function and mitochondrial content

It is generally true that tissues with few mitochondria requires glucose as their major fuel. Remembering that mitochondria are the site of aerobic catabolism you probably deduced correctly that the tissues which contain only a few mitochondria must obtain ATP largely anaerobically and to do this they need to catabolise glucose to lactate by glycolysis. You probably noticed that it would not be as valid to say that the tissues which use glucose all had few mitochondria because there are several exceptions, for example, brain and lung. Look again at the two tables, particularly at the tissues with few mitochondria. White adipose tissue is primarily concerned with the storage and release of preformed fatty acids, although it does synthesise some itself. It has a relatively low requirement for ATP and thus does not need many mitochondria. Red blood cells carry out relatively little metabolism and can meet their ATP requirements by glycolysis. The eye lens must be transparent to light, and therefore, the cells cannot contain many organelles, hence there are few mitochondria. The lung's function is to exchange O_2 for CO_2. If the lungs consumed much O_2 this would make gas exchange more difficult because there would be a reduction in the concentration gradient across the cells. The skin and kidney medulla have low energy requirements and are poorly supplied with blood (and hence O_2) and so rely on anaerobic ATP generation. The kidney cortex, however, has a high energy demand, is well supplied with blood, and therefore operates aerobically.

white adipose tissue

While obviously the brain does operate aerobically (you lose consciousness quickly in the absence of O_2) it requires glucose because, unlike glucose and ketone bodies, fatty acids cannot readily diffuse from the blood vessels into brain cells.

muscle activity and mitochondria

Heart muscle is also well supplied with O_2, and its work rate is relatively constant, only changing about 4-fold from the rate in a man at rest to one exercising at maximum capacity. Although it utilises glucose after a meal, most of the time it oxidises fatty acids or lactate. In contrast, white skeletal muscle (Type IIb, fast twitch fibres) has a poor blood supply and few mitochondria and so obtains ATP from glucose or from glycogen which is broken down to lactate. These white muscle fibres can only operate at high power output for a short time because the muscle runs out of glycogen or glucose. Red skeletal muscle (Type I and IIa fibres) have a better blood supply and a lower power output, therefore the ATP demand can be met aerobically and activity can be sustained for a long time. We will deal with muscle activity in the next chapter.

regulation of nutrient supply by the liver

A major role of the liver is to regulate the supply of nutrients to other tissues. In particular, it has a central role in controlling the blood glucose concentration. It is important to realise that unlike other organs, most of the liver's blood supply comes via another organ, the gastrointestinal tract. The blood supply of the liver enters the liver mainly via the portal vein, and only a small proportion is arterial blood which it receives via the hepatic artery. The result is that the liver is the first organ exposed to the nutrients which have been absorbed by the gastrointestinal tract. The only exception to this is fat, which enters the circulation from the lymphatic system into the neck veins and, therefore, does not go directly to the liver.

Π Why do you think that it is important that the blood glucose concentration is maintained between 4-10 mmol dm^{-3} in peripheral blood?

If the concentration were to fall too low, the brain would not get glucose quickly enough to supply its needs, and this would lead to unconsciousness. If the concentration were to rise too high, there would be a wasteful loss of glucose into the urine, harmful osmotic effects in the brain, and possibly long term effects due to nonenzymic reactions between glucose and proteins.

regulation of blood glucose by the liver, glycogen breakdown gluconeogenesis

The liver, therefore, regulates the blood glucose concentration by increasing its uptake when the concentration rises, and releasing glucose into the blood when it falls. The balance point is about 8-10 mmol dm^{-3} in the portal vein. Glucose moves freely across the liver cell membrane according to the concentration gradient. Therefore, glucose uptake can only be increased if glucose is metabolised at a faster rate. Most glucose taken up by the liver is either stored as glycogen or converted to fatty acids and then to triglyceride, which is exported from the cell. Glucose released to the blood can either come from the breakdown of glycogen or from gluconeogenesis (production of new glucose) from lactate, alanine or glycerol so at no time does the liver use a significant amount of glucose as fuel itself. When it is necessary to supply glucose to other tissues, the liver uses fatty acids for fuel. When there is plenty of glucose, it uses the lactate which it no longer requires for the production of glucose.

urea synthesis

amino acid deamination

As well as controlling the blood glucose concentration, the liver also regulates the supply of amino acids to other tissues. It is the only organ that has the ability to synthesise urea, which is made from CO_2 and ammonia, the latter being derived by the deamination of amino acids. The liver is able to catabolise amino acids that are not required for anabolism either in itself or in other organs. The carbon skeletons of the amino acids can be used to synthesise glucose or fat. Exceptions to this are the branched chain amino acids ((BCAAs), leucine, isoleucine, valine) which are mainly oxidised by muscle, the amino groups being transferred to pyruvate to form alanine, which is exported from the muscle to the liver.

The liver is also able to synthesise nutrients for other tissues. When the supply of glucose is limited, fatty acids can be converted into ketone bodies (acetoacetate and 2-hydroxybutyrate) which can be used as fuel by several organs (see Table 8.1). When the supply of glucose is plentiful, triglyceride synthesised in the liver can be exported to other tissues. The liver is also a major source of cholesterol.

The transport and uptake of the nutrient is described in the next section, but before reading it, try the following SAQs.

SAQ 8.2

Answer the following questions by circling the true statements.

1) Cells which have very few or no mitochondria are dependent on the aerobic oxidation of glucose.

2) Cells have different activities of particular enzymes according to their function.

3) Muscle cells with a very high power output require a constant supply of fatty acids to provide the energy for this.

4) The brain requires a constant supply of glucose because it is entirely dependent on it for its energy supply.

SAQ 8.3

Complete the sentences below by writing an appropriate word in the blanks. The sentences all relate to the liver's role in regulating the supply of nutrients to other tissues. For questions 2, 4 and 5 answer either high or low.

1) The [ ] is the first organ to receive blood containing nutrients absorbed from the gut.

2) The liver has a [ ] capacity for glucose uptake.

3) The liver has the metabolic pathways of [ ] and [ ] synthesis, which are unique to it.

4) The liver has a [ ] capacity for the oxidation of fatty acids.

5) The liver has a [ ] capacity for the synthesis of glucose.

6) The liver has a high capacity for the storage of glucose as [ ].

8.4 Transport of nutrients

Π Examine the list of blood borne nutrients below. Can you think of what is different about the two groups?

Group A	**Group B**
Glucose	Triglycerides
Lactate	Cholesterol
Pyruvate	
Amino acids	
Glycerol	

One important difference relating to the transport of these nutrients is that the members of Group A are all water soluble, while the members of Group B are not. An important nutrient missing is the fatty acids.

Π Which group should fatty acids belong to?

This is actually a tricky question, since the short chain fatty acids like acetic, propionic and butyric acids are water soluble, whereas the long chain fatty acids such as palmitic, stearic and oleic acid are poorly soluble in water.

Obviously there should be no problem in transporting the water soluble nutrients in the blood, because they can simply circulate in solution. The lipids, however, pose a problem, which is overcome in a similar manner to solving the problem of washing up greasy plates, which is, of course, to use a detergent.

chylomicrons

Globules of a mixture of triglyceride and cholesterol are surrounded by a layer composed of phospholipid and protein, with their hydrophobic (lipid) parts towards the centre, and the hydrophilic parts to the outside, in contact with water, thus solubilising the lipid globule. These structures are called chylomicrons (long chain non esterified fatty acids are more simply solubilised by being bound to serum albumin).

very low, low, intermediate, high density, lipoproteins

These solubilised lipid particles are known as circulating lipoproteins, the associated protein molecules involved are called apoproteins. The composition of the lipoproteins which is dependent on their origin is given in Table 8.3. They are named mainly according to their density, thus DL stands for density lipoproteins and the VL, L, I and H for very low, low, intermediate and high respectively.

	Chylomicrons (CM)	VLDL	IDL	LDL	HDL
origin	gut	liver, gut	blood	blood	liver
lipids % by weight	98	92	85	79	50
cholesterol	9	22	35	47	19
triglyceride	82	52	20	9	3
phospholipids	7	18	20	23	28
major apoproteins	ABCE	BCE	BE	B	A
function	TG transport from gut	TG transport from liver		cholesterol transport	cholesterol transport

Table 8.3 Characteristics of the lipoproteins. TG = triglyceride (see text for further discussion).

Π What are the major differences in lipid composition between LDL and VLDL?

LDL contains much less triglyceride. This is because VLDL loses triglyceride to tissues such as adipose tissue and muscle and is converted into LDL, the IDL being an intermediate form.

The interactions of the lipoproteins with tissues and with each other are very complex. Figures 8.2 and 8.3 outline the major pathways for the supply of triglyceride and cholesterol to different tissues.

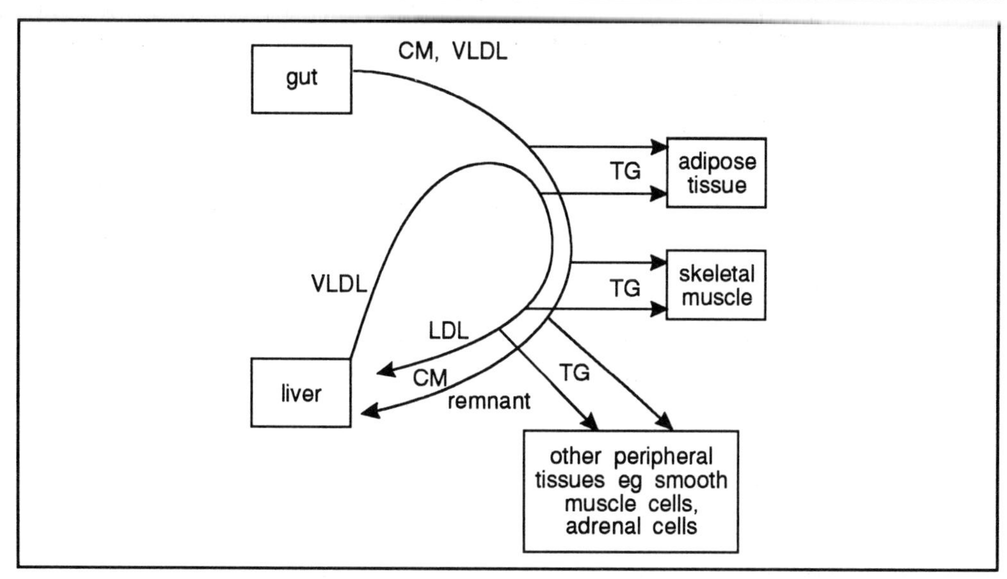

Figure 8.2 Supply of triglyceride to tissues. (Note: TG = triglyceride - see text for other abbreviations).

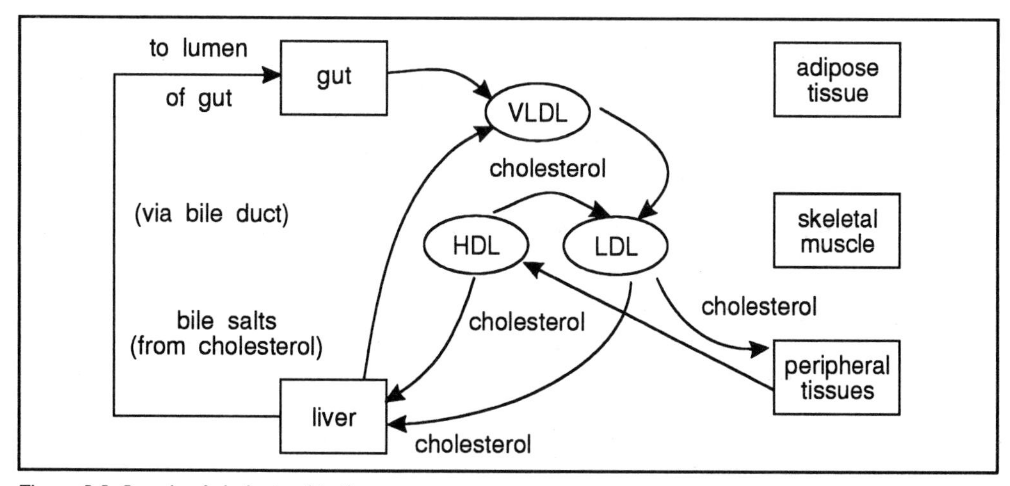

Figure 8.3 Supply of cholesterol to tissues.

Π Since all the tissues are exposed to all of the lipoproteins in the blood, why is it that they vary in their interactions with the lipoproteins? (Examine Table 8.3 again it might give you a clue).

apoprotein receptors

You saw that the lipoproteins varied in their lipid composition, but Table 8.3 also showed that they contained different apoproteins. These are recognised by specific receptors on the cell surface. If the receptors are absent, there is little interaction between the cells and the particular lipoproteins.

Although cells do not have receptors for the uptake of the water soluble nutrients, they often have carrier proteins which facilitate the transport of these substances across the cell membrane. The transport across cell membranes of lactate, pyruvate, and the ketone bodies is by simple facilitated diffusion, and thus the direction depends on the concentration gradient. This is also true for glucose in the liver, brain and red blood cells. In other cells, such as those in adipose tissue and skeletal muscle, although the uptake of glucose is still dependent on facilitated diffusion, the rate of this process depends on the plasma concentration of the hormone insulin, which is stimulatory.

In contrast, the uptake of amino acids by most cells is by active transport, as in the process of absorption in the gut (see Chapter 4).

SAQ 8.4

Indicate which of the following statements are correct.

1) The main role of VLDL is to transport triglyceride from the liver to tissues such as adipose tissue and skeletal muscle.

2) LDL is formed as the result of the loss of triglyceride from VLDL.

3) LDL carries cholesterol to tissues.

4) HDL removes cholesterol from tissues other than the liver.

5) The water soluble nutrients cross cell membranes by active transport.

6) Glucose uptake by cells is dependent on the hormone insulin.

8.5 Changes in nutrient supply

In order to understand the changes that occur in nutrient supply you need to appreciate

- the metabolic interrelationships of carbohydrates, lipids and proteins;
- the irregular, intermittent supply of nutrients to the body;
- the size of the energy stores.

8.5.1 Metabolic interrelationships

If ingested in sufficient amounts, any of the three energy producing nutrients, carbohydrate, fat and protein can provide the body with the necessary energy on a short term basis.

Figure 8.4 shows that much interconversion among these is possible. In particular, carbohydrate is readily converted into fat.

Π Examine Figure 8.4. Why is it not possible to directly convert fatty acids into carbohydrates?

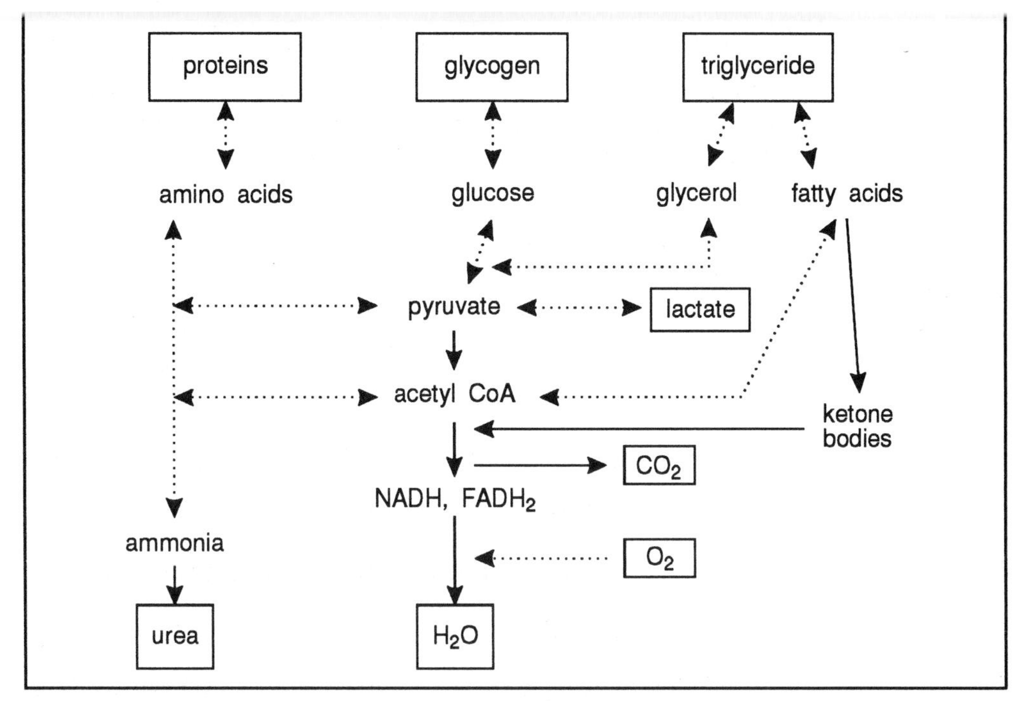

Figure 8.4 Metabolic interrelationships. Note that detailed description of the metabloic pathways involved in these interrelationships are given in the BIOTOL texts 'Principles of Cell Energetics' and 'Energy Sources for Cells'.

Fatty acid oxidation gives rise to acetyl CoA which can enter the TCA cycle. There is no pathway in mammals by which there can be a net synthesis of carbohydrate from acetyl CoA.

glycogenic and ketogenic amino acids

Metabolism of amino acids gives rise to intermediates that can be oxidised or used to make carbohydrate or fat. Those able to be converted to glycogen are termed glycogenic and those producing acetyl CoA are ketogenic. Several amino acids can give rise to both, and only leucine and lysine are purely ketogenic. All of the amino acids giving rise to acetyl CoA directly are essential, that is, their carbon skeletons cannot be synthesised by the body but must be obtained from food. The biochemistry of their interconversion are described in the BIOTOL text 'Energy Sources for Cells'.

8.5.2 Meals and energy stores

Because of the periodic nature of eating, there is no constant supply of nutrients from the gut to other organs. Therefore it is necessary for the body to store nutrients. The stored forms are glycogen, triglyceride and protein.

Table 8.4 shows the typical amounts stored in an average man.

Fuel store	Weight	Energy content	
	(g)	(kJ)	(kJ g^{-1})
Glycogen (in muscle)	350	6000	17
Glycogen (in liver)	90	1500	17
Triglyceride	9000	337 000	37.5
Protein	8800	150 000	17

Table 8.4 Fuel stores in the average man.

Π Try the following calculations:

1) If the energy expenditure is 13000 kJ per day. Calculate how long the different fuel stores would be able to provide this energy.

2) If the requirement for glucose at rest is 300g per day, how long would the supplies of glycogen be able to meet this demand.

importance of gluconeogenesis

You should find that glycogen in liver and muscle would only last for less than 3 and 12 hours respectively, whereas the triglyceride and protein would last for 26 and 12 days. Of course this is rather simplistic because these fuels are not completely available to all tissues, but it does illustrate the greater importance of lipid as a stored fuel, and the limited store of carbohydrate. The second calculation gives a slightly different answer; 28 hours for muscle and 7 hours for liver. The problem here though is the same, the glycogen present in the muscle is not available to other tissues. Only the glycogen in the liver can be released as glucose into the blood. This would suggest that you could only survive 7 hours without eating a meal, because the tissues that require glucose would run out of fuel. Obviously this is nonsense! What we have not taken into account is gluconeogenesis, the synthesis of glucose.

lactate, glycerol, glycogenic amino acids

The liver (and to a much smaller extent the kidney cortex) is able to carry out gluconeogenesis from lactate, glycerol and the glycogenic amino acids. The lactate arises from glycolysis in the tissues which oxidise glucose anaerobically. The glycerol comes from the breakdown of triglyceride in adipose tissue with the release of fatty acids into the blood. The amino acids are produced from the breakdown of protein, mainly in skeletal muscle. Thus the liver recycles the waste product lactate back into glucose, the preferred fuel for several tissues. This is done at the expense of the oxidation of fatty acids. Because the process is not 100% efficient, and because some glucose is oxidised to CO_2 and H_2O, the liver requires other sources of precursors and so it, therefore, uses amino acids. In this way, protein acts as an energy store in the same way as triglyceride.

8.5.3 The fed to fasting change

Every day most people undergo a series of feed-fast cycles. Figure 8.5 shows the disposition of the major fuels in the well fed state. This figure is quite complex. To help you understand it identify the energy store(s) in each organ (eg triglycerides in adipose tissue; triglycerides, glucogen and protein in the liver etc). Then look at the ways these may be metabolised in each of the tissues. We have included many, but all of the interrelationship. The figure will, however, help you to follow the principal ways in which these compounds are cycled between the various organs.

Π This and the following two figures contain a lot of information. One way to remember the information they contain is to examine the figures carefully and read the corresponding text. Then re-draw the figures yourself. We suggest you do this by drawing boxes to represent the organs. Then write into the boxes the appropriate metabolic processes that takes place in each of the organs. Then, finally draw in arrows to represent the transfer of materials between each of the organs.

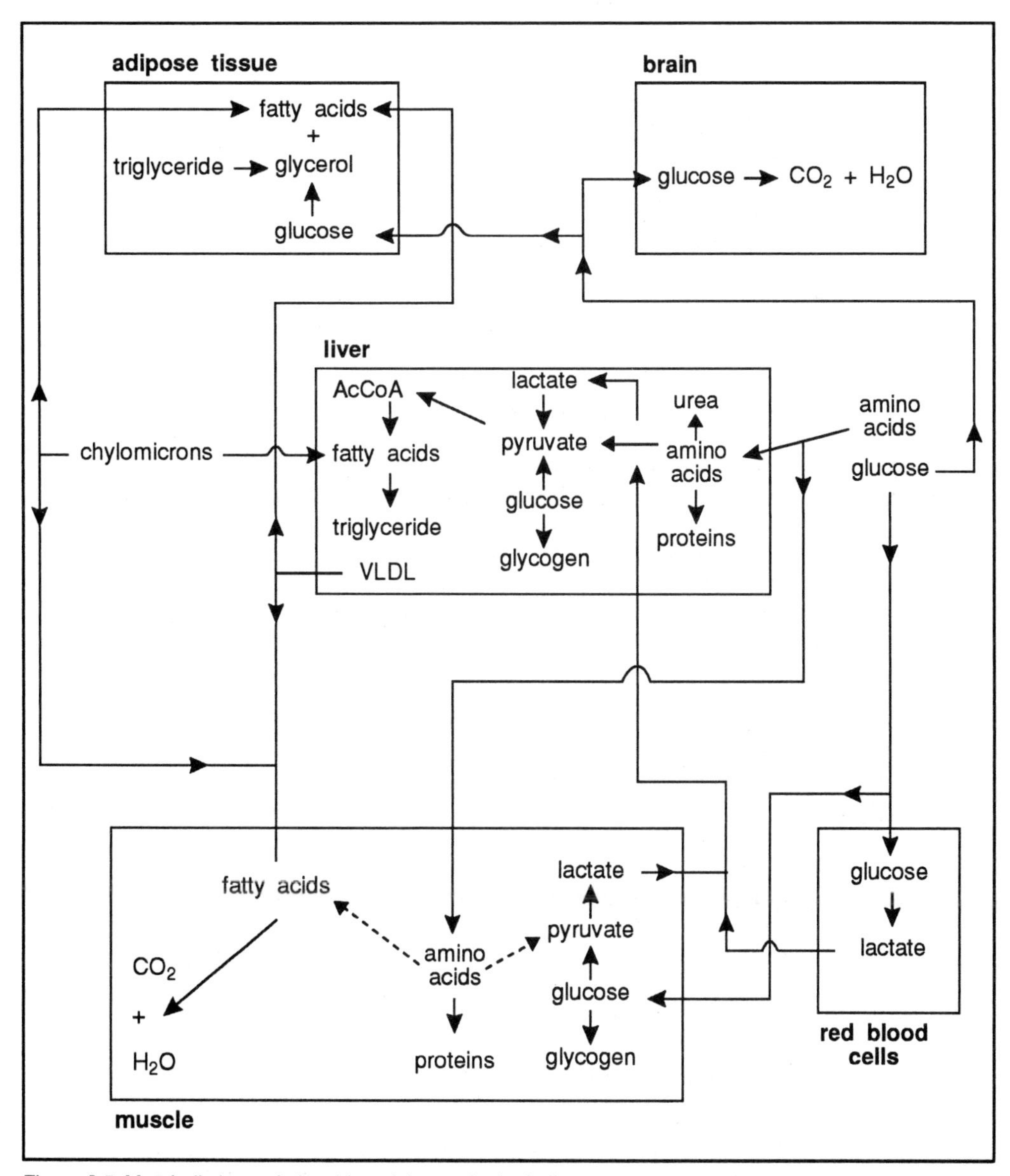

Figure 8.5 Metabolic interrelationships of tissues in the fed state.

Glucose from the gut is converted to glycogen and fat by the liver, or it is completely oxidised by the brain or oxidised to lactate by muscle and red blood cells and other peripheral tissues. Muscle is also able to store glucose as glycogen. Adipose tissue converts glucose to fat, and also stores fat arriving directly from the gastrointestinal (chylomicrons) or from the liver (in VLDL). Muscle can also oxidise fat from these sources.

Amino acids from dietary protein are in part taken up by the liver which uses them for synthesis of either protein or fat. The remainder are available to other tissues, especially for protein synthesis which is increased after a meal. The branched chain amino acids (leucine, isoleucine and valine) are not catabolised by the liver, but are deaminated by skeletal muscle, which may then oxidise them completely, or export them either as the α-ketoacids or as glutamine.

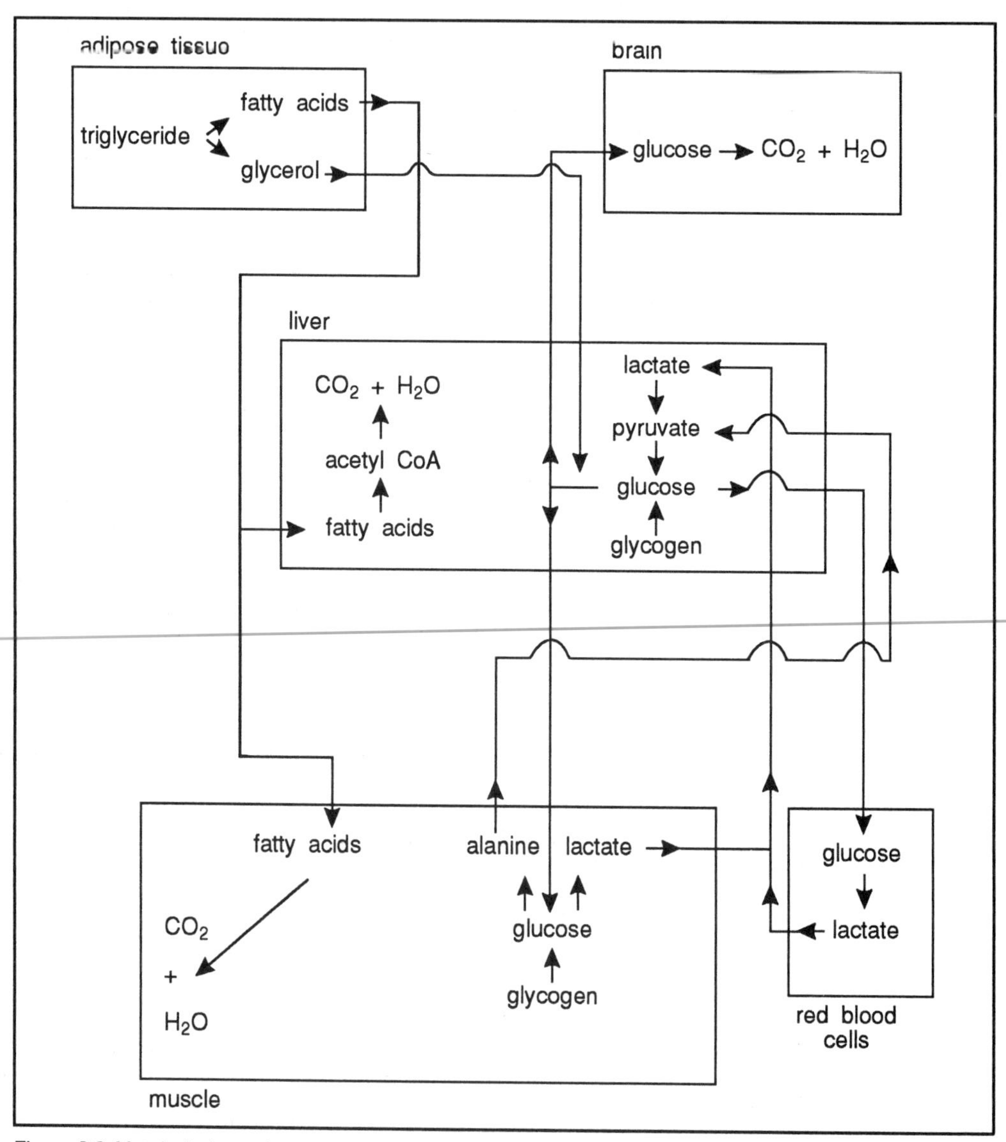

Figure 8.6 Metabolic interrelationships of tissues in the early fasting state.

Figure 8.6 shows what happens during the period of early fasting after fuel stops arriving from the gut. The liver starts to release glucose from glycogen, and the synthesis of glucose via gluconeogenesis starts. Amino acids catabolism is also reduced and some alanine is released by muscle, having been produced there by transamination of pyruvate. Fatty acids and glycerol begins to be released by adipose tissue.

In the fasting state (about 24 hours after the last meal) no fuels arrive from the gut and there is now little glycogen left in the liver (Figure 8.7). Tissues dependent on glucose are supplied because of the increased gluconeogenesis (from lactate, glycerol and alanine) present in the liver. Amino acids from muscle protein become the major source of carbon for net glucose synthesis. Alanine and glutamine are released in large amounts by muscle, the amino acids being largely formed by catabolism of other amino acids. Tissues which are able to use fatty acids instead of glucose (eg kidney cortex, heart and skeletal muscle) do so.

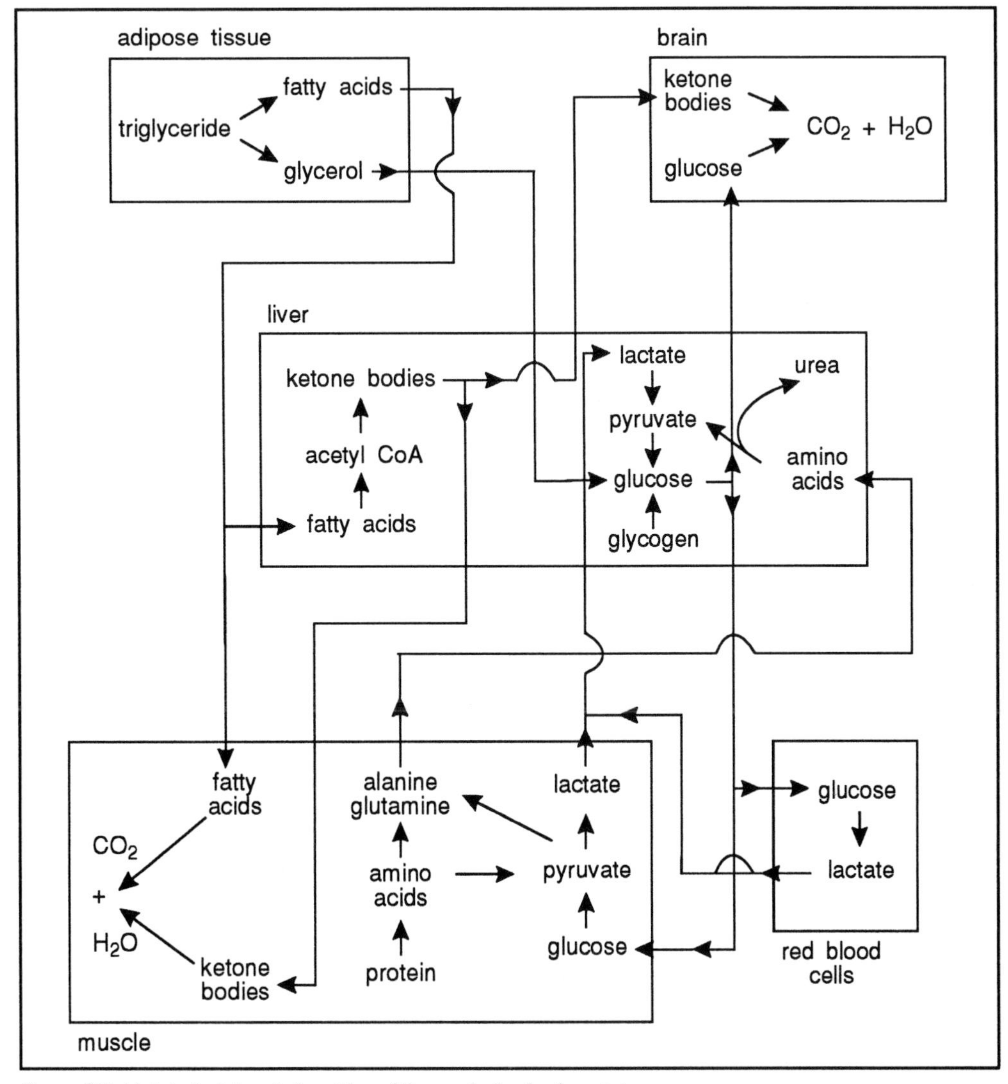

Figure 8.7 Metabolic interrelationships of tissues in the fasting state.

Adipose tissue switches over from storing fat to releasing it as fatty acids and glycerol. The liver takes up the glycerol (for gluconeogenesis) and some of the fatty acids. Oxidation of the fatty acids provides energy, but also ketone bodies are produced which are then released into the blood and used for energy by many tissues. In particular, the brain is able to use ketone bodies to replace glucose, with the result that after a few days they reduce the brains glucose requirement by about 1/3 and after several weeks starvation, ketone bodies may provide about 70% of the brain's total energy requirements. The result of this change is a reduction in the need for gluconeogenesis and hence a reduction in the rate of protein breakdown.

amino acid catabolism

The catabolism of amino acids necessarily results in the problem of disposal of the amino nitrogen. This is achieved by increased urea formation in the liver, and by excretion of ammonia by the kidney. Amino acids, in particular glycine, glutamate, alanine and aspartate can be deaminated by the kidney, the carbon skeletons being used for gluconeogenesis. This pathway becomes significant during starvation.

Π What would you predict would happen soon after refeeding? (ie when fuels start to arrive from the gut).

The obvious answer is a return to the well fed state, but while this is true for fat metabolism, it is not quite correct for glucose. The liver actually keeps carrying out gluconeogenesis for some time in order to replenish the glycogen store, while the glucose from the gut is metabolised by other tissues. There is a gradual change to the well fed state over a few hours.

SAQ 8.5

Enter an arrow to indicate an increase, or a decrease or no change in the processes mentioned, in fasting relative to the well fed state. Use ↑ to signify an increase, ↓ to signify a decrease and → to signify no change.

1) Liver glycolysis.
2) Muscle fatty acid oxidation.
3) Adipose tissue glucose uptake.
4) Brain glucose uptake.
5) Red blood cell glucose uptake.
6) Protein synthesis in muscle.
7) Muscle ketone body oxidation.
8) Muscle glycogen breakdown.
9) Adipose tissue fat synthesis.
10) Liver urea synthesis.

If starvation continues, as long as there is sufficient fat to keep the ketone body concentrations high, protein breakdown will be restricted to the minimum necessary, as little as 3-4 g per day. Eventually, of course, the supply of either fat or protein will have run out and the individual will die.

8.6 Regulatory mechanisms

You are now aware of the great changes which occur in nutrient supply between the well fed and the fasting state. Now we need to consider the mechanisms by which these changes are controlled.

In Chapter 3 you studied an introduction of the function of the endocrine glands which secrete various hormones into the blood. These hormones have diverse effects on their specific target tissues, either stimulating or inhibiting some metabolic process. In this chapter only the major hormones affecting nutrient supply to the tissue will be mentioned.

In Section 8.4 one of the most important hormones, insulin was mentioned. Table 8.5 lists the principal hormones and their modes of action on nutrient supply.

Π After reading Table 8.5, divide the hormones into those you would expect to be increased, remain the same, or be decreased, on moving from the well fed state to the fasting state.

Hormone	Principal Actions
Insulin (secreted by the pancreas)	Increase glucose uptake by cells (not in liver, brain), Stimulates glycogen synthesis from glucose, Stimulates fat synthesis, Inhibits fat breakdown and glycogen breakdown, Stimulates amino acid uptake, Stimulates protein synthesis
Glucagon (secreted by pancreas)	Stimulates glycogen breakdown to glucose (in liver), Stimulates gluconeogenesis (in liver & kidney), Stimulates release of fatty acids from adipose tissue
Adrenalin (secreted by adrenal gland)	Stimulates glycogen breakdown (in muscle, less so in liver), Stimulates gluconeogenesis, Stimulates release of fatty acids from adipose tissue
Growth hormone (somatrophin secreted by pituitary)	Stimulates amino acid uptake, Stimulates protein synthesis, Stimulates gluconeogenesis, Stimulates release of fatty acids from adipose tissue
Cortisol (secreted by adrenal gland)	Stimulates protein breakdown in muscle, Stimulates gluconeogenesis in liver, Decreases peripheral glucose uptake, Stimulates fatty acid release from adipose tissue, Increases the effectiveness of glycogen and adenaline

Table 8.5 Principal hormones regulating nutrient supply.

You are doing well if you have suggested a fall in insulin secretion and a rise in all of the others. You probably had difficulty with growth hormone because some of its effects would fit either the fed or fasted states, and you may be surprised that there are apparently four hormones concerned with the fasting response and only one with the fed state. However, this is not quite true. If growth hormone is present together with insulin, the net effect is to increase protein synthesis but also to reduce fat synthesis.

Table 8.6 lists the same hormones and the factors that stimulate their release.

Hormone	Secretion stimulated by	Notes
Insulin	Increased blood glucose	Occurs on feeding
	Increased growth hormone	
	Increased gut hormone	Occurs on feeding
Glucagon	Decreased blood glucose	Occurs on fasting
	Increased amino acids especially alanine	Alanine released from muscle on fasting
	Exercise	
Adrenaline	Stress	
	Exercise	
Growth hormone	Decreased blood glucose	Occurs on fasting
	Increased amino acids especially arginine	Occurs on feeding
	Stress	
	Exercise	
	Increased glucagon	
	Increased insulin	
	Increased cortisol	
Cortisol	Stress	

Table 8.6 Control of hormone secretion.

Π Does Table 8.6 confirm your ideas about the various roles of hormones in the well fed and fasting states?

The fall in blood glucose concentration, such as occurs when the supply of glucose from the gut is reduced, causes both the stimulation of the pancreas to secrete glucagon and to reduce the secretion of insulin and the pituitary to secrete growth hormone.

Although in the short term there may be stress resulting in the increased secretion of adrenaline and cortisol, in the long term the secretion of these are decreased. If you refer back to Table 8.5 you can see that these events will bring about the necessary changes in nutrient supply.

8.7 Changes in response to either stress or exercise

8.7.1 Stress

The body normally tries to maintain the fuel supplies between certain values, such that if the supply of glucose increases, mechanisms operate to reduce the rise in glucose concentration.

adrenaline effects in glycogen breakdown and gluconeogenesis

However, some stressful situations such as injury or infection can bring about a general adaption syndrome in which the blood glucose, fatty acid levels, and protein turnover are all increased above normal. The rise in adrenaline secretion results in increased glycogen breakdown, increased gluconeogenesis and fatty acid release from adipose tissue. These changes increase the availability of fuels to the tissues. Increased cortisol secretion stimulates several mechanisms such as the protein breakdown in skeletal muscle, fatty acid release from adipose tissue, and gluconeogenesis in the liver. The result is a raised blood glucose level and rapid loss of skeletal muscle protein. There is increased synthesis of some proteins such as fibrinogen, complement and immunoglobulins, which are involved in wound repair and defence against infection. The overall metabolic rate is increased. The body thus prioritises short term survival. If the stress is prolonged, the protein loss can become too great. This is one reason why a surgeon will be reluctant to operate on a poorly nourished patient.

8.7.2 Exercise

anaerobic exercise (Type IIb fibres)

We can distinguish between two types of exercise, aerobic and anaerobic. Long distance running is aerobic, whereas sprinting is anaerobic exercise. During the latter, the muscles (Type IIb fibres) rely mainly on their own stored glycogen, and produce lactate when the effort lasts for more than a few seconds. The lactate is eventually converted back into glucose by the liver. We will describe muscle Type IIb fibres in Chapter 9.

aerobic exercise and glycogen breakdown

Aerobic exercise is more complex. There is not enough glycogen in muscles to provide the energy necessary for a sustained effort, so there is a progressive changeover to the oxidation of fatty acids as the efforts continues. Typically, at the start, fatty acids may supply only 10% of the energy, but after several hours this can rise to 80%. As well as the loss of muscle glycogen, there is also a loss of liver glycogen together with an increase in gluconeogenesis in order to supply the muscles with sufficient glucose. These changes are brought about by increased adrenaline and glucagon and decreased insulin levels. Even though there is a rise in blood fatty acid concentration, there is usually little change in the blood ketone body concentration. After the exercise has finished, however, there is often a rise in ketone bodies because they are no longer being oxidised so quickly, but are still being produced because the fatty acid concentration remains high for some time after the period of exercise.

These changes are caused by the same mechanisms as those involved in the starve/feed cycle, except that adrenaline probably plays a more important role. Adrenaline is actually released in anticipation of exercise, as well as during it.

Π What happens if you are stressed by fear or anger but are not able to respond physically?

You will have experienced the physical symptoms (increased heart rate, sweating, increased breathing) but the adrenaline and nor-adrenaline that are partly responsible for these effects will also cause a release of fatty acids from adipose tissue, and increase glycogen breakdown in muscle, so making more fuel available to muscle. If the muscle do not get used, the fatty acid concentration can get quite high and may cause heart problems. While we are not suggesting you should fight the person who makes you angry, it would probably be beneficial, both physically and psychologically, to do some physical exercise such as fast walking, running, or hitting a punch bag!

The next SAQ, though longer than the others, should enable you to test your understanding of the last three sections.

SAQ 8.6

Complete the following sentences by writing in the space available an appropriate word or words selected from those listed below. Note, one space may represent more than one word, and a word may be used more than once.

When the rate of supply of glucose and other nutrients from the gut falls, the concentration of glucose in the blood supplying the pancreas falls, and this stimulates the secretion of 1 [ ] and reduces the secretion of 2 [ ]. As a result, the liver starts to export 3 [ ] derived from its store of 4 [ ] and from 5 [ ]. It obtains its energy for this from the oxidation of 6 [ ] whose concentration in the blood rises because of an increase in the rate of release from the triglyceride stores in 7 [ ]. This is brought about by the reduction in 8 [ ] concentration, leading to reduced esterification of fatty acids, and increased 9 [ ] and often other hormones, leading to increased 10 [ ]. The gluconeogenesis in the liver depends on the supply of precursors, especially lactate, alanine, glutamine and glycerol. The 11 [ ] is derived from the lipolysis in the adipose tissue, the 12 [ ] from organs such as the red blood cells which oxidise glucose anaerobically and the 13 [ ] skeletal muscle, from the breakdown of 14 [ ]. The increased fatty acid concentration in the blood leads the liver to make and export 15 [ ] which together with fatty acids, become the major fuels for tissues like red skeletal muscle and kidney cortex, thus conserving glucose for other tissues. As a result of these changes, the blood glucose concentration does not fall too low to be able to supply the requirements of the most critical organ, the 16 [ ].

During aerobic exercise, many of the same changes occur, but one difference is the increased importance of the hormone 17 [ ] which stimulates the breakdown of glycogen in skeletal muscle. Insulin and glucogen play the same roles, though. During prolonged stress, such as that due to severe infection, the situation is different from that during prolonged starvation, because the raised metabolic rate, together with raised levels of the hormones 18 [ ], 19 [ ], 20 [ ] and 21 [ ] lead to a rapid loss of 22 [ ] in order to maintain a higher than normal blood glucose concentration.

insulin	glycogen breakdown	pyruvate
glucagon	fat synthesis	proteins
adrenaline	glucose	glutamine
growth hormone	fatty acids	glycogen
cortisol	triglyerides	glycerol
lipolysis	ketone bodies	muscle
gluconeogenesis	amino acids	adipose tissue
protein synthesis	lactate	liver
glycogen synthesis	alanine	kidney
red blood cells	brain	

8.8 Nutrient requirements and energy balance

So far the possibilities for interconversion of the nutrients have been emphasised especially the body's ability to convert carbohydrate and protein into fat, protein into carbohydrate and the oxidation of all three to provide energy, but the inability to convert fat directly into carbohydrate has been considered. Another aspect mentioned in Chapter 4 was the existence of essential amino acids, those ones that cannot be synthesised by the body but must be obtained from the diet. In this section we will look at the extent to which different nutrients can substitute for each other, and the effects of changes in their intakes.

body obeys the laws of thermodynamics

Although thousands of overweight persons would like to deny it, the body obeys the laws of thermodynamics, so that if the food energy absorbed is greater than the energy expenditure, the difference in energy must remain in the body, usually in the form of fat. We define energy balance as the energy intake minus energy expenditure, so gaining body energy is regarded as a positive energy balance. The energy content of foods can be measured by burning them to CO_2, H_2O, N_2 and nitrogen oxides. This, of course, is not quite what the body does; the nitrogen ends of mainly as urea, ammonia and uric acid. It also costs energy to store energy. For example, if excess protein energy is to be stored as fat, the amino acids must be deaminated, catabolised to acetyl CoA and anabolised to fat. Processing costs are least for fats, and highest for proteins. Because of these factors, a kilojoule of carbohydrate is not exactly equivalent to a kilojoule of protein or fat in terms of the net increase in body energy. If you are in energy balance, but then increase your energy intake by 100 kJ of fat, you will store about 96 kJ of this extra energy intake, whereas if it were an extra 100 kJ of protein, only about 67 kJ of this would be stored because of the high energy cost of conversion. The corresponding figure for carbohydrate is about 80 kJ.

Nitrogen balance is analogous to energy balance. A positive nitrogen balance occurs if the intake of nitrogen (mainly in the form of protein) exceeds the excretion of nitrogen (mainly as undigested protein in the faeces, and urea and ammonia in urine). The normal adult will be in nitrogen balance, ie the losses balance the intake. A positive nitrogen balance occurs in pregnancy or childhood, when there is a net increase in body protein. Negative nitrogen balance can result from inadequate dietary protein intake, or inadequate energy intake when some of the dietary protein must be used for energy generation.

The essential amino acids need to be present in adequate amounts. If any one is missing or in too small an amount, insufficient new protein will be synthesised to replace that lost during normal protein turnover, and a negative nitrogen balance results. Most animal proteins contain all the essential amino acids at about the relative proportions required by the human body, but vegetable proteins often lack one or more of the essential amino acids and may also be more difficult to digest. For example, cereal grains are low in lysine, whereas legumes are low in methionine. A mixture of the two, however, is satisfactory.

SAQ 8.7

Complete the table to show the effects of various changes in the intakes of the major nutrients. Choose the most appropriate response(s) from the list below. You may feel you need more information to give a more specific answer. If this is the case, note the alternative answers in brackets.

Assume all other factors remain the same.

Person	Physiological state	Nutritional state	Change	Effect
1) Adult	Normal	In energy and nitrogen balance	Increased protein intake	
2) Adult	Semi-starvation	Net energy and nitrogen loss	Increased carbohydrate intake	
3) Child	Normal	In energy and nitrogen balance	Increased fat intake	
4) Adult	Pregnancy	Net energy and nitrogen gain	Increased protein intake	
5) Adult	Normal	In energy and nitrogen balance with barely adequate lysine and total protein intake	Reduced intake of lysine due to changed source of protein with same total nitrogen intake	

a) Positive energy balance.

b) Negative energy balance.

c) Increased fat storage.

d) Increased protein storage.

e) Positive nitrogen balance.

f) Negative nitrogen balance.

g) Increased body energy loss.

h) Decreased body energy loss.

i) Increased protein loss.

j) Decreased protein loss.

Summary and objectives

In this chapter we have reviewed the functions of the different nutrients and related the use of metabolic fuels to the different tissue functions. We have explored the ways in which nutrient supply is controlled or altered in response to different physiological circumstances, and have discussed what happens when nutrient intake is abnormal.

Now that you have complete this chapter you should be able to:

- relate the nutrient requirements of individual tissues to their functions;
- explain the role of the liver in regulating nutrient supply;
- explain the mechanisms used for the control of nutrient supply;
- describe the changes in nutrient supply that occur after a meal, during fasting or as a result of exercise or stress, and explain how these come about;
- predict the effects of changes in the intakes of the major nutrients in energy balance.

Muscular activity

Muscular activity

Muscle tissue is highly specialised to contract and shorten forcefully and is responsible for the mechanical processes in the body. Muscle moves the trunk and appendages, moves glandular secretions through ducts, pumps blood through vessels and propels food through the digestive system by peristalsis. In addition, the metabolism occurring in the large mass of muscle tissue present in the body produces heat essential to the maintenance of normal body temperature.

In this chapter we will examine the basic structural and functional characteristics of muscle tissue and illustrate the mechanical properties of muscle, its energy requirements and the relevance of these processes to the biochemical and electrical events within muscle cells leading to contraction. We will also examine the central nervous system control and muscular disorders.

9.1 General functional characteristics of muscle

Muscle has four major functional characteristics which are fundamental in the maintainence of homeostasis.

- Contractility is the ability to shorten and thicken, or contract when a stimulus is received.

- Excitability is the ability of muscle tissue to respond to stimulation by nerves and hormones, which makes it possible for the nervous system and, in some muscle types, the endocrine system to regulate muscle activity.

- Extensibility is the ability of muscle tissue to be stretched. Many skeletal muscles are arranged in opposing pairs. Whilst one is contracting, the other is relaxed and undergoing extension.

- Elasticity is the ability of muscles to recoil to their original resting length after they have been stretched.

Π List three important functions that muscle performs through contraction?

The functions we hoped you listed are:

- motion in movements involving the whole body such as those which occur during walking and running, and also in localised movements, for example holding an object, moving an arm or turning the head. All of these movements rely on the functioning of bones, joints and the muscles attached to the bones. Less obvious kinds of motion are the beating of the heart, contraction of the bladder to void urine, the mixing of food in the stomach and peristaltic movements to move food through the intestine;

- maintenance of posture when the contraction of skeletal muscles holds the body in stationary positions as in sitting and standing;
- heat production when skeletal muscles contract. This is an extremely important function and it has been estimated that as much as 85 per cent of all body heat is generated by muscle contractions.

9.2 Types of muscle tissue

The major types of muscle tissue are skeletal, cardiac and smooth. They are categorized by location , microscopic structure and nervous control.

- Skeletal muscle tissue with its associated connective tissues comprise approximately 40 per cent of the body's weight. It is named for its location, is attached to bones and moves parts of the skeleton. It is responsible for locomotion, facial expressions, posture and many other body movements. Skeletal muscle is striated muscle tissue because striations or band-like structures are visible when the tissue is examined under a microscope. Its function to a large degree is under voluntary or conscious control by the somatic nervous system.
- Cardiac muscle tissue is only found in the heart and its contractions provide the major force for propelling blood through the circulatory system. It is striated and involuntary, that is, its contraction is usually not under conscious control.
- Smooth muscle tissue is the most variable type of muscle in the body with respect to distribution and function and is involved with processes related to maintaining the internal environment. Smooth muscle is located in the walls of hollow internal structures, such as the gastro-intestinal tract, blood vessels, bladder and uterus, in the iris and ciliary muscles of the eye, in the dermis of the skin and in other areas. It is referred to as nonstriated because it lacks striations. It is involuntary muscle tissue.

Unlike skeletal muscle, smooth muscle and cardiac muscle are autorhythmic, that is they contract spontaneously at somewhat regular intervals and do not require nervous or hormonal stimulation. Smooth muscle and cardiac muscle are not under direct conscious control but rather are innervated and in part regulated unconsciously or involuntarily by the autonomic nervous system and the endocrine system. Let us firstly deal with some aspects of the cell biology of the different muscle types, concentrating on the best-studied example striated or skeletal muscle.

9.3 Cell biology of skeletal muscle

structure of muscle fibres

Examination of skeletal musculature by microscopy shows the presence of muscle fibres lying parallel to one another. Muscle fibres are cylindrical in shape, multinucleate, between 10 and 100 μ in diameter and may run the full length of the muscle. They are attached to tendons at their extremities. The membrane surrounding the fibre is called the sarcolemma. The sarcolemma surrounds the matrix of the fibre, the sarcoplasm. Within the sarcoplasm of a muscle fibre are many nuclei, mitochondria and enzymes as well as myofibrils and sarcoplasmic reticulum. Each fibre has a single neuromuscular junction and this arrangement and its importance in muscle excitation will be dealt with later.

myofilaments make up myofibrils

The force developed by muscle fibres is generated by intracellular contractile proteins which are arranged into myofilaments. The myofilaments are in bundles, called myofibrils, which run the whole length of the fibre. The myofilaments of a myofibril are arranged into compartments called sarcomeres (Figure 9.1). Sarcomeres are separated from one another by zones of dense material called Z lines. Each myofibril is surrounded by the sarcoplasmic reticulum.

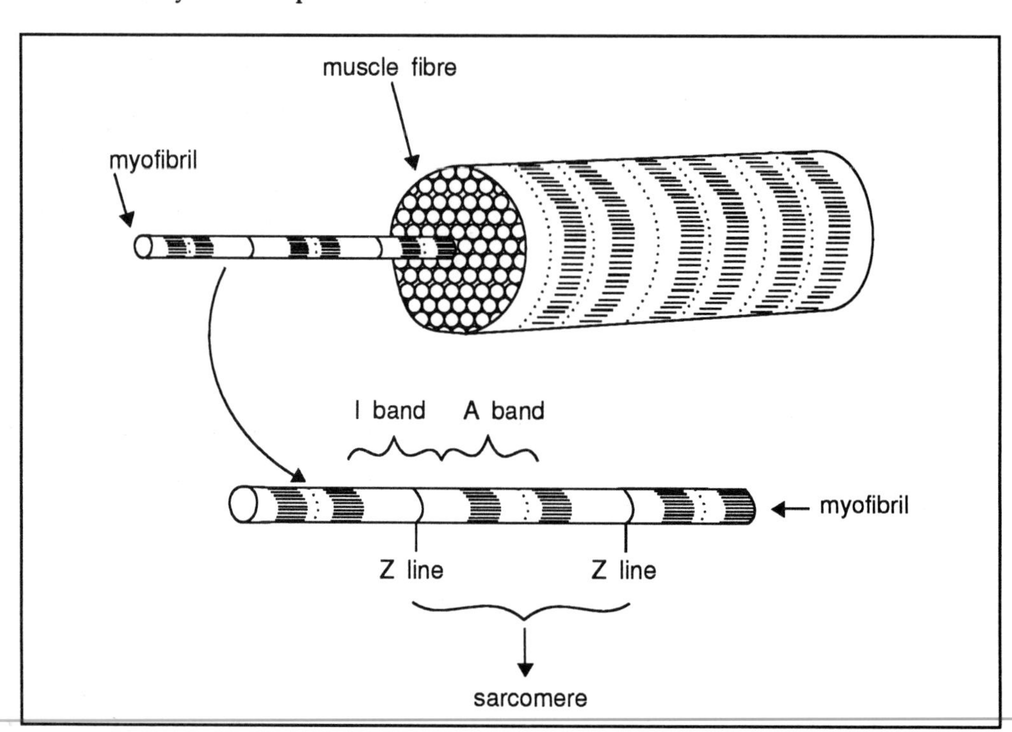

Figure 9.1 Numerous myofibrils in a single skeletal-muscle fibre. Each myofibril shows the banding pattern of the sarcomeres.

9.3.1 Myofibrils

actin and mysoin myofilaments

Each myofibril is about 1 to 3μ in diameter and extends from one end of the muscle fibre to the other. Myofibrils are composed of two kinds of protein myofilaments. Actin myofilaments, or thin myofilaments, are about 8nm in diameter and 1000 nm in length. Myosin myofilaments, or thick myofilaments, are about 12 nm in diameter and 1800 nm in length. The actin and myosin myofilaments are organised in highly ordered units called sarcomeres which are joined end to end to form the myofibrils (Figure 9.2).

Z lines and I and A bands H zone and M line

A sarcomere extends from one Z line to an adjacent Z line. The arrangement of the actin and myosin myofilaments gives the myofibril its banded appearance. Each I band (isotropic band) includes a Z line and the actin myofilaments that extend from either side of the Z line to the ends of the myosin myofilaments. The I band, on either side of the Z line, consists only of actin myofilaments. Each A band (anisotropic band) extends the length of the myosin myofilaments within a sarcomere. The actin and myosin myofilaments overlap to some extent at both ends of the A band. In a cross section of the A band where the actin and myosin myofilaments overlap, each myosin myofilament is surrounded by six actin myofilaments. In the centre of each A band is the H zone where only myosin myofilaments are present as the actin and myosin myofilaments do not overlap in this region. In the middle of the H zone is the M Line which consists of threads that attach to the centre of the myosin myofilaments and hold them in place. (Figure 9.2).

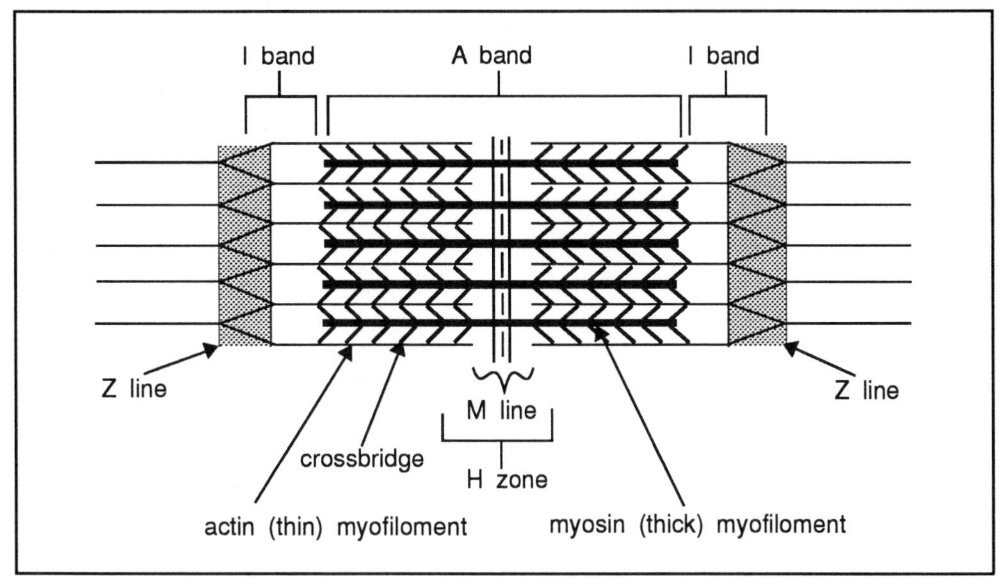

Figure 9.2 Enlarged aspect of a sarcomere showing actin and myosin myofilaments and cross bridges. (Highly stylised).

9.3.2 Actin and myosin myofilaments

F-actin

G-actin

tropomyosin
troponin

Each actin myofilament is composed of two strands of F-actin (fibrous actin) which are coiled to form a double helix that extends the length of the myofilament. Associated with the actin filament are a series of tropomyosin and troponin molecules. Each F-actin strand is composed of a polymer of about 200 molecules of G-actin (globular actin). Each G-actin molecule has an active site to which myosin molecules bind during muscle contraction. Tropomyosin and troponin act as regulator proteins which regulate the interaction between the active sites on the G-actin and myosin. Tropomyosin is an elongated protein that turns along the groove of the F-actin double helix. Each strand of tropomyosin covers seven G-actin active sites. The troponin molecules are situated at the junctions of the tropomyosin monomer molecules. Each troponin is composed of three subunits: one has a high affinity for actin, the second has a high affinity of tropomyosin and the third has a high affinity for calcium ions (Figure 9.3).

globular heads
and cross
bridges

Myosin myofilaments are composed of many myosin molecules each of which has a rod portion and a globular head (Figure 9.4). The rod-like portions are two coiled proteins wound together and form the body of the myosin myofilament whilst the globular heads project laterally from the myofilament. Each myosin myofilament consists of about 100 myosin molecules so that 50 myosin molecules have their heads projecting towards each end of the myofilament. The projecting heads are referred to as cross bridges. The heads of the myosin molecules have binding sites for actin molecules which are required during muscle contraction. The myosin heads also contain ATPase which breaks down ATP to release the energy required for contraction.

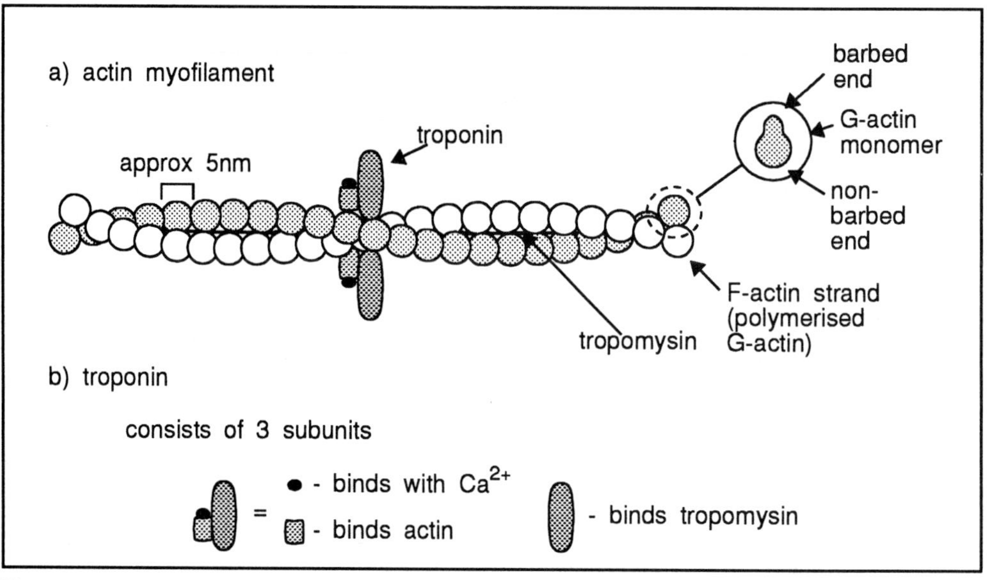

Figure 9.3 a) Structure of a portion of the actin myofilament. Actin myofilaments are composed of individual globular actin (G-actin) molecules, filamentous tropomyosin molecules and troponin molecules, all assembled into a single filament. b) Detail of troponin.

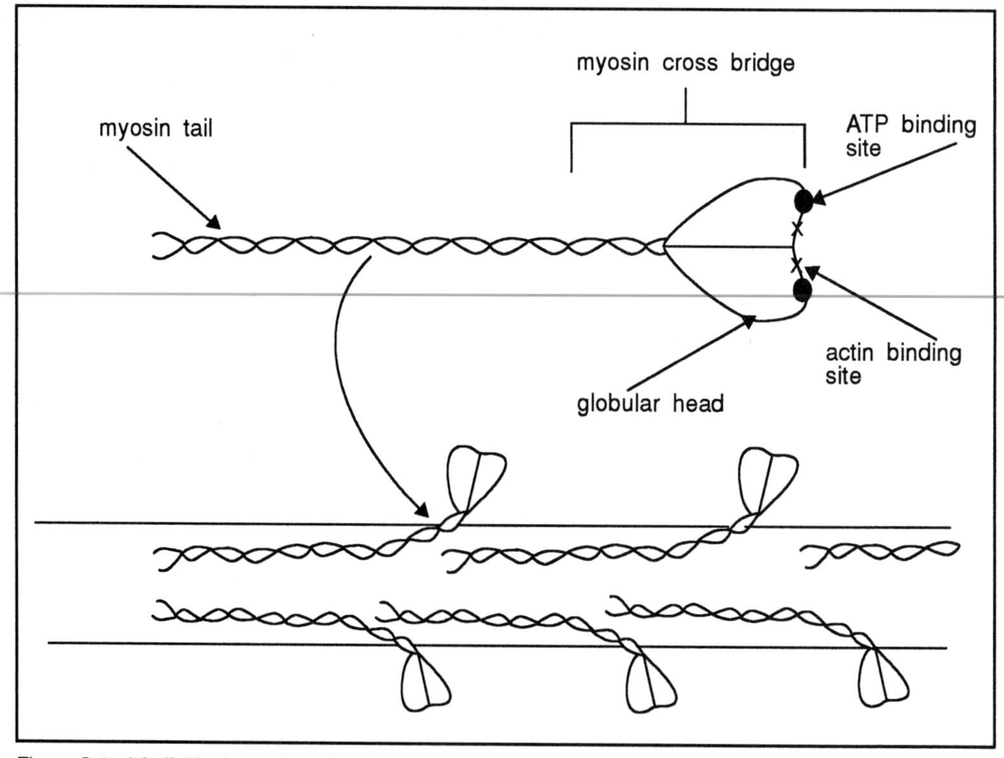

Figure 9.4 a) Individual myosin molecule each of which has a tail portion and a globular head (cross bridge). b) Myosin myofilament composed of myosin molecules. The tails are in a parallel arrangement with the heads (cross bridges) all pointing in the same direction at one end and in the oppostie direction at the other end of the myosin myofilament (see also Figure 9.2).

Transverse tubules and sarcoplasmic reticulum

conduction of muscle action potential through T tubules

The sarcolemma has many tube-like invaginations along its surface called transverse or T tubules, which are regularly arranged and project into the muscle fibre and wrap around sarcomeres in the region where actin and myosin myofilaments overlap (at the junction of the A band and the I band). These T tubules conduct the muscle action potential quickly into the centre of the muscle fibre. Within the sarcoplasm between the T tubules is a highly specialised form of smooth endoplasmic reticulum called the sarcoplasmic reticulum. On either side of each T tubule the lumen of the sarcoplasmic reticulum is enlarged to form a terminal cisterna. A T tubule and two terminal cisterna form a triad. The membrane of the sarcoplasmic reticulum actively transports calcium ions from the sarcoplasm into the lumen of the sarcoplasmic reticulum where they are stored when the muscle is not contracting. The release of calcium ions from the sarcoplasmic reticulum into the sarcoplasm by a muscle action potential initiates muscle contraction (Figure 9.5).

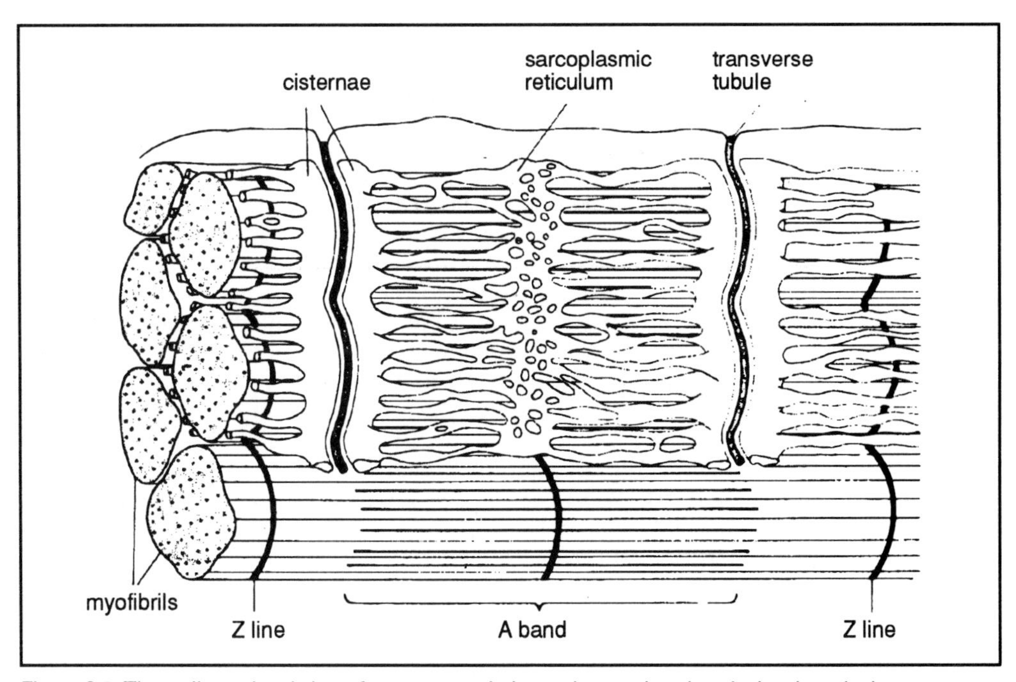

Figure 9.5 Three-dimensional view of transverse tubules and sarcoplasmic reticulum in a single skeletal-muscle fibre (Redrawn from Vander, Sherman and Luciano, (1990), Human Physiology: The Mechanism of Body Function, McGraw Hill, New York).

Muscle contraction - sliding-filament theory

excitation contraction coupling

If a muscle changes its length the sarcomeres also change in length. However, the actin and myosin myofilaments remain the same length and the change in length is achieved as a result of the myofilaments sliding over each other. This sliding-filament theory was developed by Hansen and Huxley in the 1950's. As the sarcomeres shorten, the myofibrils shorten because each myofibril is composed of a series of sarcomeres. Since the myofibrils extend from one end of a muscle fibre to the other, the muscle fibre and consequently the entire muscle also shortens. The production of an action potential in a skeletal muscle fibre leads to contraction of the muscle fibre; a process called excitation contraction coupling. The action potential is generated at the skeletal neuromuscular junction and is then propagated along the sarcolemma of the muscle fibre and penetrates the T tubules. The T tubules carry the action potential rapidly into the interior of the muscle fibre. When the action potential reaches the area of the triads, the membranes of the sarcoplasmic reticulum increase their permeability to Ca^{2+} ions. The concentration of Ca^{2+} ions within the sarcoplasmic reticulum is about 2000 times higher than in the sarcoplasm of the resting muscle; so when the sarcoplasmic reticulum's permeability to Ca^{2+} ions increases, the Ca^{2+} ions diffuse from the sarcoplasmic reticulum into the sarcoplasm surrounding the myofibrils. You will remember that the sarcoplasmic reticulum actively transports Ca^{2+} ions into its lumen.

Ca^{2+} release and troponin-tropomyosin movement

The Ca^{2+} ions released into the sarcoplasm bind to part of the troponin attached to the tropomyosin along the actin myofilaments. This combination causes the troponin-tropomyosin complex to move deeper into the grove between the two F-actin strands and so expose active (myosin binding) sites on the actin myofilaments. The exposed active sites now bind to the heads of the myosin molecules to form cross bridges. An ATP molecule must bind to each myosin head before a cross bridge can be formed. When the heads of the myosin molecules bind to actin, a rapid series of events occur that lead to muscle contraction. The heads move at their hinged region and force the actin myofilament to which the heads of the myosin molecules are attached, to slide over the surface of the myosin myofilament. During the movement of each cross bridge the ATPase in the myosin head breaks down the ATP. Part of the energy released by the breakdown of ATP is utilised for the movement of the cross bridges, and part is released as heat. Once the cross bridge moves, another ATP molecule must bind to the head of the myosin cross bridge before the myosin is released from the actin. After ATP binding, the myosin cross bridge returns to its original position and the cycle of cross bridge formation, movement and release can be repeated. Cross bridge movement while the actin myofilament is attached is called a power stroke whereas return of the myosin head to its original position is called a recovery stroke. Many cycles of power and recovery strokes occur during each muscle contraction.

During muscle contraction, the sarcomeres shorten when the actin myofilaments slide over the myosin myofilaments, although the length of the myofilaments does not change. The H zones and the I bands narrow as the ends of the actin myofilaments within a sarcomere approach each other (Figure 9.6), but the length of the A bands does not change. Relaxation occurs as the result of the active transport of Ca^{2+} ions back into the sarcoplasmic reticulum. As the concentration of Ca^{2+} ions in the sarcoplasm decreases, Ca^{2+} ions diffuse away from the troponin molecules. The troponin-tropomyosin complex then blocks the active sites on the actin molecules. As a consequence of this cross bridges cannot reform once they have been released and relaxation occurs.

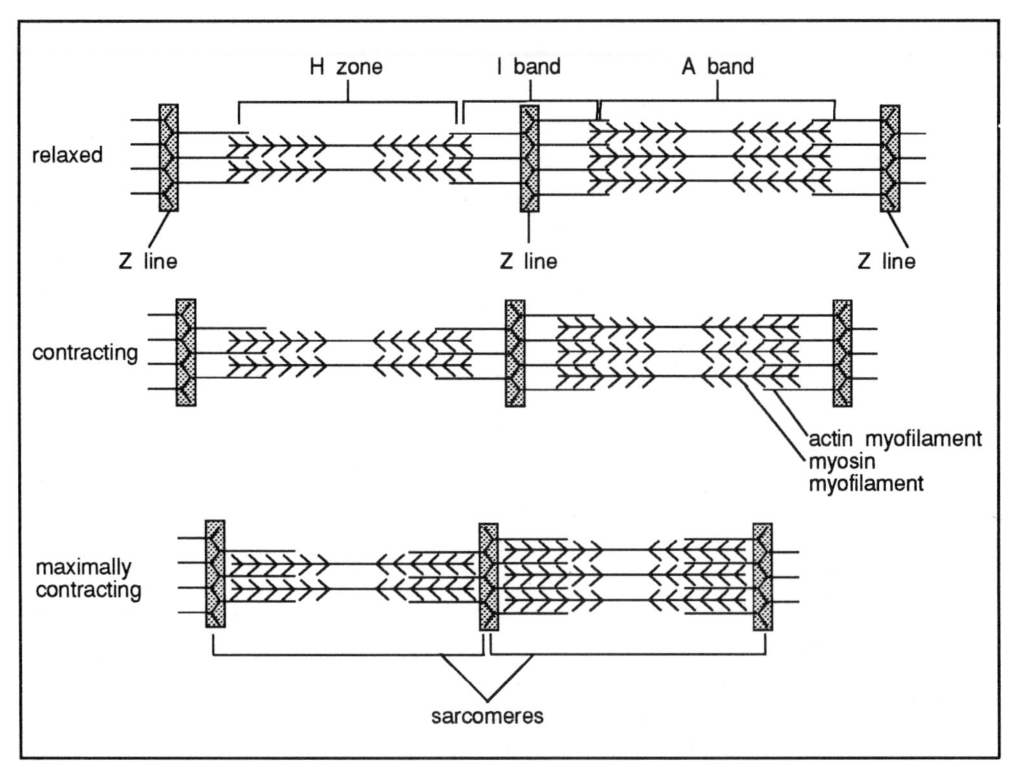

Figure 9.6 Sliding filament theory of muscle contraction. The sarcomere is shown in relaxed, contracting and maximally contracted states. The actin myofilaments move over the myosin myofilaments. Note that the I bands shorten, but the A bands do not. The H zone narrows or even disappears as the actin. Myofilaments meet at the centre of the sarcomere.

SAQ 9.1

Produce a list summarising the sequence of molecular events occurring to produce skeletal muscle contraction.

Π Can you list the processes that require ATP (energy) during excitation contraction coupling?

Energy in the form of ATP is required for a number of processes:

- for the power stroke;
- ATP must be available to bind to the myosin heads to allow cross bridge release;
- for the active transport of Ca^{2+} ions into the sarcoplasmic reticulum; which leads to relaxation;
- the active transport processes that maintain the normal concentration of Na^+ and K^+ ions across the sarcolemma; this ensures normal muscle action potential generation and propagation.

9.5 The neuromuscular junction

initiation of muscle action potentials motor units

We have seen so far that an action potential in the sarcolemma of a skeletal muscle fibre triggers contraction. Muscle action potentials are initiated by stimulation of the nerve fibres to skeletal muscle. The nerve cells whose axons innervate skeletal muscle fibres are known as motor neurons and their cell bodies are located in either the brainstem or the spinal cord. The axons of motor neurons are myelinated and are the largest diameter axons in the body. They are able to propagate action potentials at high velocities, allowing rapid transmission of signals from the central nervous system to skeletal muscle fibres. Within a muscle, the axon of a motor neuron divides into many branches, each branch forming a single junction with an individual muscle fibre. A single motor neuron therefore innervates many muscle fibres, but each muscle fibre has only one nerve junction and is controlled by only one motor neuron. A motor neuron plus the muscle fibres it innervates is called a motor unit. The muscle fibres in a single motor unit are all located in one muscle and do not all lie adjacent to one another. When an action potential is produced in a single motor neuron, all the muscle fibres in its motor unit contract (all or non activity).

motor end-plate and neuromuscular junctions

On entering a skeletal muscle, the axon of a motor neuron branches into axon terminals (telodendria) which terminate close to a portion of the sarcolemma of a muscle fibre. The region of sarcolemma that lies directly under the terminal portion of the axon is known as the motor-end plate. The junction of an axon terminal with the motor end-plate is known as a neuromuscular junction (Figure 9.7).

The axon terminals of a motor neuron contain membrane bound vesicles, the synaptic vesicles that store the chemical transmitter acetylcholine (ACH). When an action potential reaches an axon terminal, it depolarises the terminal membrane, opening voltage sensitive calcium channels thereby allowing calcium ions to diffuse into the axon terminal. Once inside the axon terminal, these calcium ions trigger the release by exocytosis, of ACH from the vesicles into the synaptic cleft. The empty vesicles are refilled with newly synthesised ACH and re-used.

ACH binding and ion channel opening

The ACH diffuses across the synaptic cleft and binds to receptor molecules located within the membrane of the motor end-plate of the muscle fibre. The binding of ACH activates the receptor, which opens ion channels in the end-plate membrane. Both sodium and potassium ions pass through these channels but because of the differences in the electrochemical gradients across the membrane, more sodium moves in than potassium moves out. This produces a local depolarisation of the motor end-plate known as an end-plate potential (EPP). The magnitude of the EPP is normally sufficient to depolarise the sarcolemma adjacent to the end plate membrane to its threshold potential by local current flow, initiating an action potential in the muscle fibre membrane. This action potential is then propagated over the surface of the muscle fibre. As neuromuscular junctions are located in the middle of a muscle fibre the muscle action potential is propagated in both directions towards the ends of the fibre.

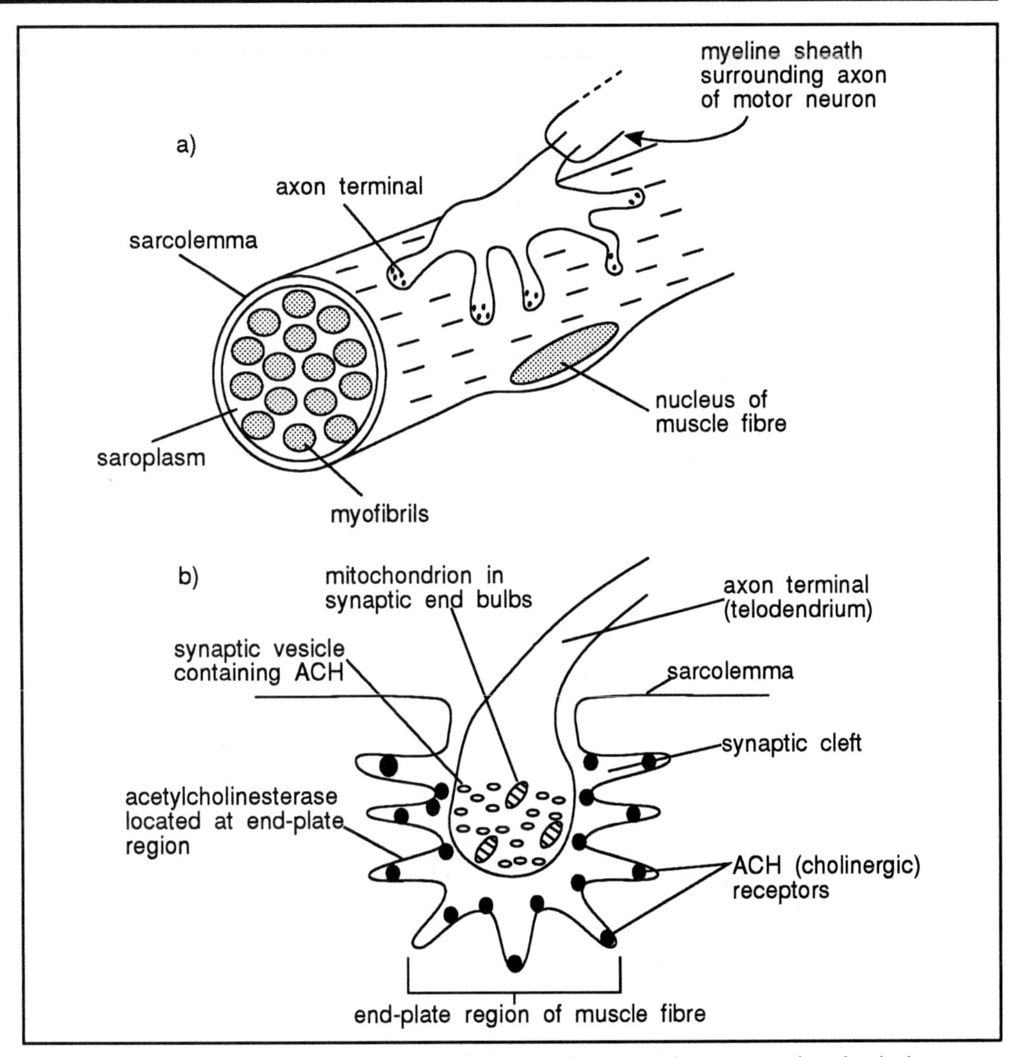

Figure 9.7 a) Neuromuscular junction (motor end-plate). b) Diagrammatic representation of a single axon terminal and end-plate region of muscle membrane.

inactivation of ACH

Acetylcholine released into the synaptic cleft is rapidly broken down to acetate and choline by acetylcholinesterase, an enzyme present at the end-plate which prevents the accumulation of ACH within the synaptic cleft where otherwise it could act as a constant stimulus at the muscle. The release of ACH and its rapid breakdown in the neuromuscular junction ensures discrete transmission, that is one nerve action potential produces only one muscle action potential. The choline molecules are transported back into the axon terminal and then combined with acetate produced within the cell to form new acetylcholine.

SAQ 9.2

Construct a flow diagram to illustrate the sequence of events involved in skeletal neuromuscular transmission.

9.5.1 Drugs acting at the neuromuscular junction

anticholinesterase drugs

Neuromuscular transmission can be increased by anticholinesterase drugs which inhibit acetylcholinesterase and slow down the hydrolysis of ACH in the synaptic cleft. *Neostigmine* and *pyridostigmine* are used in the treatment of myasthenia gravis, an autoimmune disease in which neuromuscular transmission is defective due to loss of functional ACH receptors in skeletal muscle. Anticholinesterase drugs are also used to reverse neuromuscular blockade produced by curariform drugs such as *tubocurarine*. Overdosage of anticholinesterases results in excess ACH and a block of the motor end-plate leading to spastic paralysis.

Neuromuscular blocking drugs are used to relax skeletal muscles during surgical operations and to prevent muscle contractions during electroconvulsant therapy (ECT). Most of the clinically useful neuromuscular blocking drugs compete with ACH for the receptor but do not initiate ion-channel opening. These competitive drugs such as *tubocurarine*, reduce the size of the end-plate potential produced by ACH to a size which is below the threshold for muscle action potential generation and so cause a flaccid paralysis (flaccid = limp). Depolarising blockers such as *botulinum toxin* block neuromuscular transmission by preventing the release of ACH.

9.6 Motor unit

Within a skeletal muscle the individual skeletal muscle fibres are arranged into motor units, each of which consists of a single motor neuron and all of the muscle fibres it innervates. Like individual muscle fibres, motor units respond in an all-or-none fashion. All the muscle fibres of a motor unit contract maximally in response to a threshold stimulus because an action potential in a motor neuron initiates an action potential in all of the muscle fibres it innervates. A complete muscle is made up of many motor units.

Muscles that control precise movements, such as the extrinsic eye muscles, have fewer than 10 muscle fibres to each motor unit. Muscles of the body that are responsible for gross movements, such as the biceps and thigh muscles, may have as many as 500 muscle fibres in each motor unit.

Π How can the total tension in a muscle be varied?

recruitment

Total tension can be varied by adjusting the number of motor units that are activated. The process of increasing the number of active motor units is called recruitment and is determined by the needs of the body at a given time.

muscle tone

Asynchronous firing of motor neurons to skeletal muscle ensures that some motor units are active whilst others are inactive. This prevents fatigue while maintaining contraction by allowing a brief rest for the inactive units. It also helps to maintain a state of partial contraction in a muscle, a phenomenon called muscle tone. Also, recruitment is one factor responsible for producing smooth movements during muscle contraction.

9.7 Nerve and blood supply to skeletal muscle

Skeletal muscles are well supplied with nerves and blood vessels. For skeletal muscle to contract it must be stimulated by impulses from nerve cells. Muscle contraction also requires large quantities of energy and therefore large amounts of nutrients and oxygen. Also, waste products and heat produced by muscle contraction must be removed. Therefore, prolonged muscle action depends on a good blood supply to deliver oxygen and nutrients and remove wastes.

An artery and one or two veins accompany each nerve that penetrates a skeletal muscle. Each muscle fibre is in close contact with one or more capillaries. This arrangement allows for good perfusion of skeletal muscle tissue.

9.8 Smooth muscle

Smooth muscle differs from skeletal muscle in that the characteristic cross-striated pattern of skeletal muscle is not present. It is also sometimes called involuntary muscle because its activity arises either spontaneously or through the activity of the autonomic nervous system.

Π Can you think of various parts of the body where smooth muscle is located?

Smooth muscle is found vessel walls of the vascular system, air passages to the lungs and in various tissues of the intestinal and reproductive systems as well as in the eye and the skin.

Smooth muscle from different sources varies considerably in both its structure and properties, but a number of general features may be distinguished between the different types. The fibres are generally spindle shaped and much smaller than skeletal muscle fibres. Each has a centrally placed nucleus.

irregular arrangement of actin and myosin

Actin and myosin myofilaments are scattered through the cytoplasm, although not arranged in such a highly organised way as those present in skeletal muscle. There are cross bridges present on smooth muscle myosin filaments and the molecular basis of contraction is thought to be similar to that in skeletal muscle. Smooth muscle cells contain only about 10 percent of the actin and myosin of skeletal muscle and have a low ATPase activity, consistent with its low speed of contraction and tension.

9.9 Cardiac muscle

Heart muscle cells have properties somewhere in between those of skeletal and smooth muscle. They are small, striated and branching cells with a single nucleus. Each cell is connected to its neighbour by intercalated discs, within which are gap-junctions. These junctions allow the extremely rapid passage of electrical currants between adjacent cells. The whole heart therefore behaves as a single unit (syncytium) to electrical stimulation, rather than a group of isolated units as does skeletal muscle. Certain heart cells located in the sino-atrial node or pacemaker spontaneously generate action potentials which spread throughout the heart causing cardiac contraction. The actions of the autonomic nervous system and hormones affect this rate of discharge and therefore the heart rate. The sympathetic nervous system and adrenaline will increase heart rate and force of contraction whereas the parasympathetic nervous system will have the opposite effect.

Let us now examine the properties of muscle tissue.

9.10 All or none principle of muscle contraction

In order for a muscle fibre to contract, the fibre has to be stimulated by its motor neuron. An action potential passes down the nerve fibre causing the release of transmitter (acetylcholine) at the neuromuscular junction and the subsequent depolarisation of the muscle membrane. There is a high safety factor in skeletal muscle function in that the transmitter is always released in quantities sufficient to bring about the contraction of the innervated muscle fibre. However, the motor nerve itself has a threshold to stimulation. By this we mean that only above a certain given level of stimulation will the neuron respond with an action potential. This given level is called the threshold or liminal stimulus. The threshold level of stimulus for the nerve is also the minimum stimulus necessary to excite all the muscle fibres it supplies (motor unit), thus causing them to contract. Below this level (subthreshold or subliminal stimulus), the motor unit will not respond. The muscle fibres, therefore, follow the all or none principle which states that below threshold stimulus the muscle fibres will not contract (none), whilst at or above a threshold stimulus, they will contract maximally (all), under otherwise constant conditions. That is to say, individual muscle fibres do not partly contract, it is either an all or none process.

threshold and sub-threshold stimulation

How, then can you explain the way in which a whole muscle is capable of contracting in a smooth and graded fashion? For this explanation you have to remember that a whole muscle is composed of a number of motor units (see Section 9.4). The all or none principle applies to these individual motor units ie all the muscle fibres within a motor unit will contract maximally or not at all depending upon the strength of the stimulus. When nerve impulses are sent to a muscle some but not all of the motor units may receive a threshold (or greater) stimulus and these will contract. The extent of the contraction will depend on how many motor units are recruited. The more motor units excited, the greater the strength of the muscle contraction.

The amount of force (tension) which each muscle fibre is capable of developing depends upon a number of factors including the frequency of the action potentials arriving at the neuromuscular junction. Let us have a look at the different types of contractions which the muscle fibres can produce in response to these excitations.

9.11 Types of muscle contractions

9.11.1 Twitch contraction

latent, contraction and relaxation phases

Given a stimulus at or above threshold a muscle will respond with a brief, rapid rise in tension followed immediately by relaxation. This is known as a twitch contraction and is illustrated in Figure 9.8. The response is characterised by a number of phases. First of all, there is a short time between the application of the stimulus and the beginning of the contraction. This lasts for about 10ms (0.01s) and is called the latent period. It is the time taken for all contractile processes to be developed. Following this period there is an increase in the tension recorded from the muscle which rises to a peak in about 40ms or 0.04s. This is the contraction phase of the response. After reaching a peak, the tension falls back to the baseline as the muscle relaxes. This relaxation phase is approximately 50ms (0.05s) long. These figures are characteristic of a particular frog muscle. Other muscles and other species may have different values for their contractions, but they are usually a matter of fractions of a second or a few seconds only. Characteristically, cardiac and smooth muscles have longer contractions than skeletal muscles.

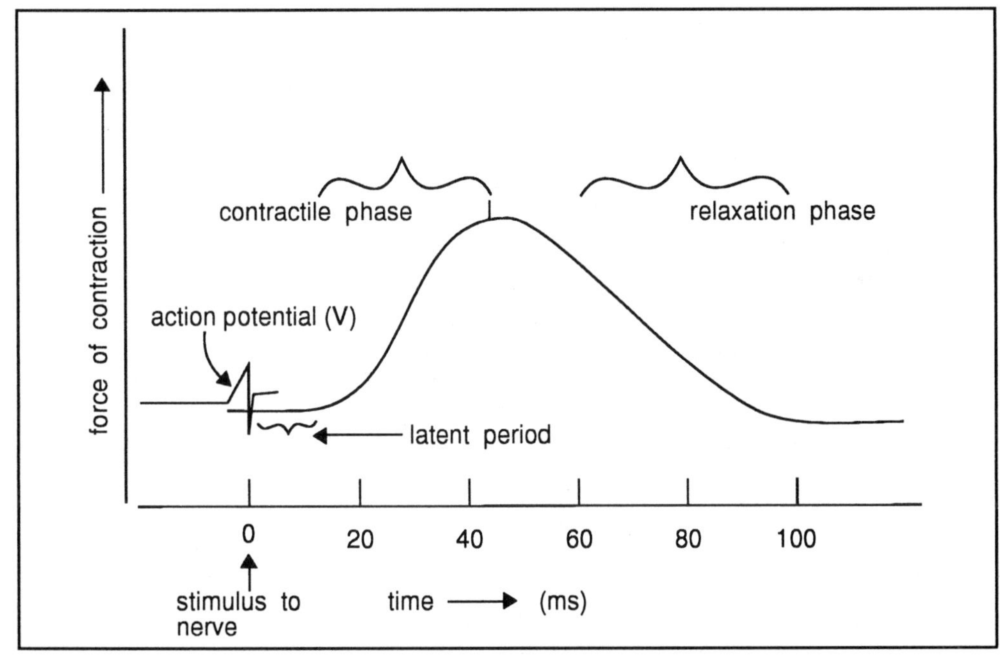

Figure 9.8 Phases in the excitation, contraction and relaxation of a muscle.

refractory period

For a short period following its excitation the muscle is unable to respond to a second stimulus. This so called refractory period allows the muscle to fully recover from the first stimulus. The time required for this is different in the three muscle types. Functionally, this is particulary important for cardiac muscles. They are refractory during almost all of the contraction period in response to the first stimulus. We shall explain this a little later on.

summated contraction

In contrast to cardiac muscle, skeletal muscle recovers quickly from the previous stimulus. In fact, the refractory period is over before the muscle has relaxed, so the muscle is capable of contracting again whilst it is still in the middle of the response to the first stimulus. Consequently, if a second stimulus is given to the muscle after the refractory period has finished, the muscle will contract again to the second stimulus and the response will be added on to what is left of the first response (see Figure 9.9). In this way the muscle gives a larger overall contraction, known as a summated contraction.

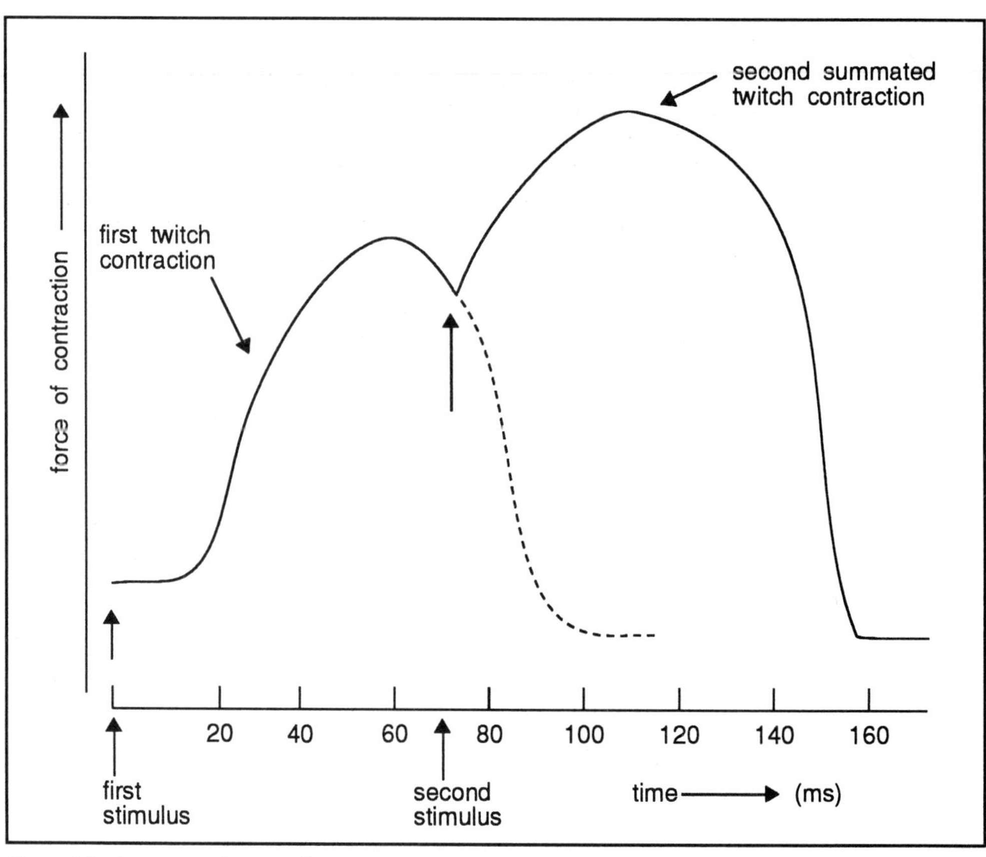

Figure 9.9 A summated contraction.

SAQ 9.3

Using the following data plot a graph of the amplitude of muscle contraction against stimulus voltage applied to the motor nerve supplying the muscle. How can you account for the shape of the graph?

amplitude of contraction (tension in grammes)	stimulus voltage (volts)
0	0.6
2.5	0.8
5.0	1.0
10.0	1.4
14.0	1.6
19.0	1.8
22.0	2.0
29.0	2.6
34.0	3.0
34.0	4.0

9.11.3 Tetanus

unfused and fused tetani

If a continuous train of stimuli at a frequency of about 10 to 20 per second is applied to a skeletal muscle than the muscle can only partly relax in between the resulting contractions. Figure 9.10 illustrates the type of response which will be recorded under these conditions. The trace shows a sustained increase in tension developed by the muscle with incomplete relaxations occurring during the rising phase. This type of response is called incomplete or unfused tetanus. When the stimulus frequency is increased further, the muscle has no time to relax between stimuli so that there is a maximal, smooth, maintained increase in tension called fused tetanus, (see Figure 9.10). This is the absolute maximal tension that a given muscle is able to achieve and it is attained at a particular frequency of stimulation known as the critical or fusion frequency.

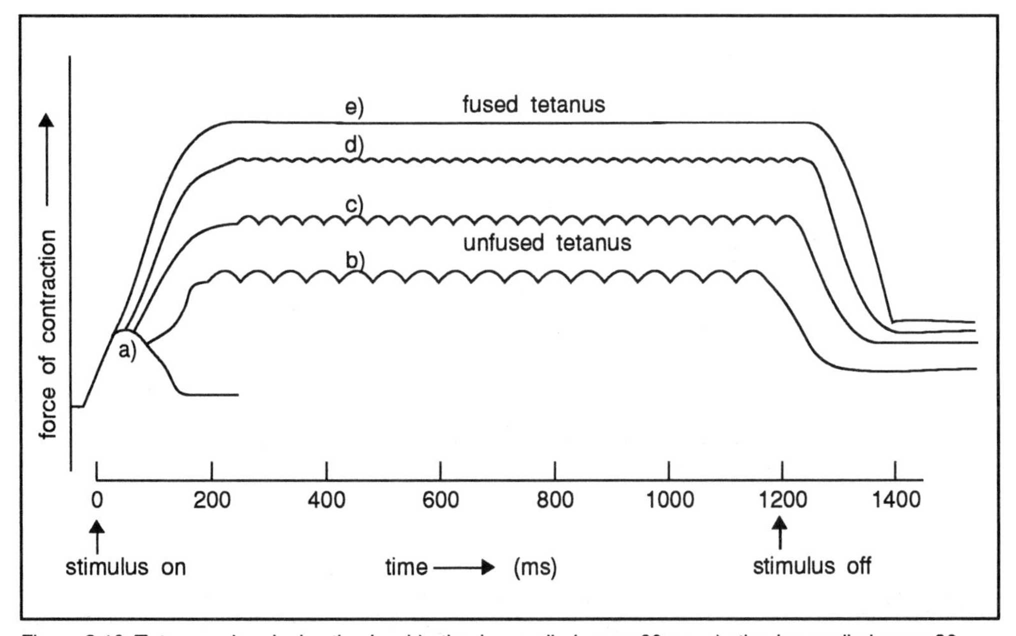

Figure 9.10 Tetanus. a) a single stimulus, b) stimulus applied every 60 ms, c) stimulus applied every 30 ms, d) stimulus applied every 15 ms, e) stimulus applied at higher frequency. See text for further explanation.

Physiologically speaking, most of our skeletal muscle fibres will contract in this manner when stimulated by the nervous system. The response of the whole muscle will depend upon the frequency of action potentials in the motor nerves and on the number of motor units involved. Asynchronous firing of the motor units ensures smooth and graded responses from our muscles appropriate to the task involved. Cardiac muscle is functionally different from skeletal muscle in that it cannot contract tetanically. As the cardiac muscle cells are refractory for almost all of their contraction period, they are almost completely relaxed before they can be stimulated again. Thus, the contractions cannot summate and tetanus cannot occur. Obviously, this is functionally significant, as the heart would be unable to function as a pump if the muscle were capable of going into a tetanic contraction. The heart would not be able to relax and fill.

9.11.4 Treppe

Another type of response can be recorded from a skeletal muscle when the rate of stimulation allows the muscle just enough time to contact and relax with each stimulus but is not sufficiently fast to cause tetanus. In this case the first few responses of the muscle gradually increase in size until the maximum is reached after several stimuli have been applied. This is known as the staircase effect or treppe. It is thought to be brought about by a kind of potentiation of the contractile mechanism (see Figure 9.11).

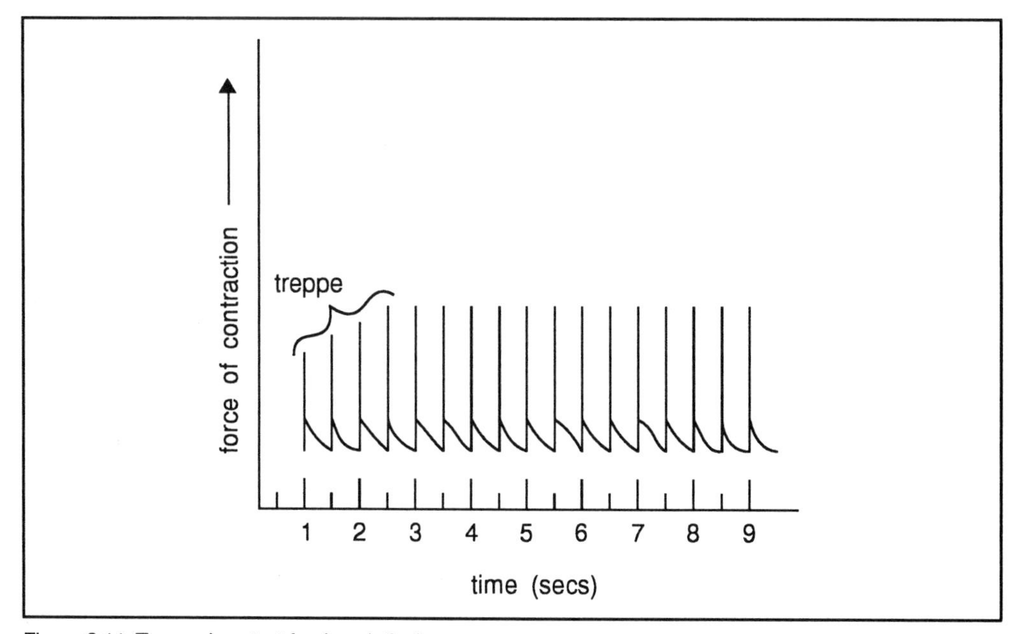

Figure 9.11 Treppe (see text for description).

9.11.5 Isometric and isotonic contractions

A final consideration of the kinds of muscle contractions that can be recorded depends upon whether, or not, the muscle is allowed to shorten when it contracts. When a muscle contracts it develops tension and shortens, thus pulling on whatever it is attached to (eg bone) to produce movement. This shortening contraction is referred to as an isotonic contraction and is the type that produces movement of our skeleton. When the muscle develops tension but does not shorten it is said to be contracting isometrically. These isometric contractions are important in the maintenance of posture and in other situations where static power is required, eg holding a book in your hand with your arm extended. Both types of contractions are used in most activities of the body.

contraction and static power

Π Try to work out whether isotonic or isometric contractions or both are involved in the following activities: running; walking; standing; jumping.

Running would involve both isometric and isotonic contractions. The muscles in the leg on the ground would be contracting isometrically whilst the muscles moving the other leg forwards would contract isotonically to produce the necessary shortening.

Walking would involve the same contractions as running but at a slower rate!

Standing involves only isometric muscle contractions to maintain the posture. No muscle movement and, therefore, shortening is needed so isotonic contractions would not occur.

Jumping, again would require both types of contractions: isotonic to shorten the muscles during take-off and isometric contractions to maintain position on landing.

9.12 Types of skeletal muscle fibres

fibres differ in colour, amount of sarco plasmic reticulum mitochondria capillaries and ATP generation

Skeletal muscle fibres do not all have the same structure and metabolic function. They differ in their colour, the amount of sarcoplasmic reticulum, the number of mitochondria and capillaries and in the way in which they manufacture ATP. The colour is due to the amount of the respiratory pigment, myoglobin, which stores oxygen until it is needed by the mitochondria to make the energy molecule ATP. Myoglobin gives the muscle a red colouration which is deeper the more myoglobin there is present. The capillaries bring blood containing oxygen to the muscle cells, so that the more capillaries there are the more efficient the muscle is at aerobic (oxygen-requiring) metabolism. The sarcoplasmic reticulum is important in the supply of calcium ions to the myofilaments (see Section 9.2) for the contractile processes. A final difference seen in muscles is in the speed with which they are able to contract and reach either maximal isometric tension or maximal isotonic shortening. Based upon these differences in their structure and function, muscle fibres are classified into three main types:

Type Ia fibres

These fibres are also called slow twitch or slow oxidative fibres. They contain a lot of myoglobin and many mitochondria and blood capillaries. This means they are good at manufacturing ATP by oxidative metabolic processes. However, they do not have much sarcoplasmic reticulum and are only able to contract slowly. These fibres are small, red and are very resistant to fatigue. They are found in large numbers in postural muscles, particulary those in the neck.

Type II fibres

These fibres are characterised by fast contraction times as they have a lot of sacroplasmic reticulum and are able to breakdown ATP during contractile processes very rapidly. These fast fibres can be further subdivided into two classes as a result of differences linked to their metabolic activities. They are divided into Type IIa and Tybe IIb.

Type IIb fibres

fast, short duration response by Type IIb fibres

Type IIb are whiter in colour then Type IIa, they contain much less myoglobin then either Type IA or Type IIa fibres. They also have fewer mitochondria and blood capillaries. They are, therefore, not very efficient at the oxidative production of ATP. However, they do have large amounts of enzymes necessary for anaerobic (non-oxygen requiring) metabolic processes to produce ATP. Lactic acid is a common product of this metabolic activity. As the anaerobic generation of ATP cannot be maintained for long, these fibres easily fatique. Type IIb fibre are much larger than Type Ia fibres and are used when a muscle requires a sudden, short lived burst of high power.

Type IIa fibres are almost an intermediate stage between Type Ia and Type IIb fibres.

Type IIa fibres

Like the Type I fibres they have a high content of myoglobin and they contain a large number of mitochondria and capillaries. Thus, they are red fibres which have a powerful oxidative capacity for making ATP. However, they also have a high capacity for glycolytic production of ATP and are relatively resistant to fatigue. Although not found as frequently in human muscle as the other two types of muscle fibres, these type IIa fibres are useful for sustained contractions which require a greater power output than the type I fibres are capable of and which would not be maintained by the easily fatiguable type IIb fibres.

Table 9.1 summarises the properties of these three muscle fibre types.

Property	Type Ia	Type IIa	Type IIb
rate of contraction	slow	fast	fast
myoglobin content	high	high	low
mitochondrial no.	high	high	low
capillary density	high	high	low
sarcoplasmic retic.	little	lot	lot
oxidative metabolism	good	good	poor
anaerobic capacity	poor	good	good
tendency to fatigue	low	moderately low	high
size	small	intermediate	large
power output	low	intermediate	high

Table 9.1 Summary of the properties of the different types of muscle fibres

Most skeletal muscles are a combination of the three types of muscle fibres, their proportion depending upon the function of the given muscle. Postural muscles, for instance, will have high proportion of type I fibres. Arm muscles, on the other hand, which are used intermittently for powerful activities such as lifting and throwing, will have a greater number of type IIb fibres. Muscles which are used for posture and movement, such as the leg muscles, tend to have an equal mixture of the two types.

Even though muscles are a mixture of different fibre types, it has to be remembered that all muscle fibres of a given motor unit will be of the same type. Also, within a given muscle, the type of activity to be performed dictates the fibre type to be employed. For instance, if a task requires only a weak contraction then only type I motor units will be recruited. A stronger contraction would need the contribution of the type IIa fibres and the maximal strength contraction will bring in type IIb fibres. The decision as to which motor units are required for a given task is made by the nervous system.

ratio of fibre types determined genetically

A final point to be made concerning muscle types is that the particular make up of your own muscles is determined genetically. You are born with the muscle types you have and this will decide whether you are good at a particular exercise or sport. At the moment there does not appear to be any convincing evidence that you can alter the muscle types within your muscles by training. So, your muscle type profile might make you either a good sprinter (mostly type IIb) or a good marathon runner (mostly type I) but you could not be a top athlete in both types of sport. Training would only increase your performance of the sport your muscles are designed for, it could not change you from being an accomplished sprinter to being a good marathon runner.

SAQ 9.4

If the blood flow to a skeletal muscle were considerably reduced, which muscle fibres would have their ability to produce ATP for muscle contraction markedly decreased?

SAQ 9.5

Attempt to decide which muscle fibre types would be recruited for the following muscular activities:

squash; football; 100metre sprint; 1000m race; high jump; jogging;

9.13 Muscle homeostasis

Some of the characteristics of muscle function can be considered to play a role in the homeostasis of the muscle itself and in the body as a whole. In the following section we shall look at the contributions of oxygen debt, muscle fatigue and heat production.

9.13.1 Oxygen debt

lactic acid and homeostasis

During muscular exercise, muscle blood flow is dramatically increased (by dilation of the blood vessels) in order to provide the extra oxygen required for the aerobic breakdown of pyruvic acid to make ATP. Initially, sufficient oxygen can be provided to fulfil all the needs of the contracting muscle, but as the exercise becomes more intense or prolonged, the demand for oxygen outstrips the supply. Under these conditions the ATP is manufactured by anaerobic mechanisms (no oxygen required) with the pyruvic acid being converted into lactic acid. Much of the lactic acid is taken to the liver and changed back to glucose or glycogen, but some accumulates in the muscle tissue. The resulting build up of the acid causes discomfort and a reduction in the contractile ability of the muscle. Furthermore, once exercise has ceased, the lactic acid has to be metabolised into carbon dioxide and water. This requires oxygen, as do the replenishment of the ATP, glycogen and phosphocreatine supplies of the muscle. In addition, the oxygen stores of the body have to be restored eg in the haemoglobin, myoglobin, lungs and body fluids. Thus, once exercise has stopped, there is still an increased demand for oxygen to service all the above events. Consequently, the respiratory rate remains high to provide this additional oxygen, otherwise known as the oxygen debt. Therefore, in effect, the build up of lactic acid in the muscle acts as a homeostatic mechanism which results in the cessation of the exercise until the oxygen and energy stores of the body and muscles are restored.

9.13.2 Muscle fatigue

When a muscle is continuously stimulated at a maximal contraction, eventually, the tension produced by the ensuing muscle contractions begins to decline until, finally, the muscle ceases to contract. This failure of the muscle to maintain the previous contractile force is known as muscle fatigue. It is thought to be related to a reduction in the availability of either ATP or glycogen, but the mechanisms are not fully understood.

homeostatic function of muscle fatigue

The homeostatic function of muscle fatigue is thought to lie in the fact that such a fatigue reaction prevents the muscle from going into a rigor. This would occur if contractile activity were to continue and deplete the ATP supplies to the point where the muscle fibres were unable to relax. This would be very damaging to the muscle tissue. So, it is a protective mechanism which ensures the homeostatic balance of the energy stores in the muscle tissue.

9.13.3 Heat production

85% of energy released as heat

Muscle contraction involves the conversion of chemical energy to mechanical energy with the production of heat as a by-product. As much as 85% of the energy is released as heat and the body uses this as a means of maintaining body temperature (thermoregulation). When the body is cold, excessive heat has been lost through the skin and lungs and the central nervous system initiates the process of shivering as one of the homeostatic mechanisms to restore normal body temperature. Shivering is the involuntary contractions of muscles throughout the body and can result in a 100% increase in the amount of heat which is generated. This extra heat serves to counterbalance the excessive loss of heat and, thus, goes a long way towards returning the body temperature to normal values.

9.14 Disorders of muscle functions

When muscle does not function in the normal manner the fault may lie in the muscle itself or in the nerve supplying the muscle or in the parts of the central nervous system which control the output to the muscle. In this section we shall examine, briefly, at a few of the more common muscular disorders.

9.14.1 Atrophy

enervation and disuse atrophy

Atrophy of a muscle means a loss of muscle mass whereby the contractile tissue is replaced by some non-contractile tissue such as connective tissue. This condition may occur after the nerve to a muscle is severed (enervation atrophy) or following prolonged disuse of muscle such as might happen whilst immobilised in plaster cast (disuse atrophy). If the nerve can be restructured then the atrophy of the muscle can be halted, although regeneration of the nerve may take some considerable time. Electrical stimulation of the muscle, in the mean time, may prevent further atrophy until reinnervation can take place. In the case of disuse atrophy, physiotherapy usually restores the muscle mass and function. However, without such measures the muscle atrophy would become irreversible and the function of the muscle would be permanently reduced.

9.14.2 Ageing

After the age of about 30 there is a progressive replacement of muscle tissue with non-contractile fat tissue. This change is associated with a loss of muscular strength and alterations to skeletal muscle reflex function.

9.14.3 Muscular dystrophies

Duchenne dystrophy

This group of muscle diseases is characterised by degeneration of the muscle tissue associated with a progressive loss of function. The muscle contractile tissue is replaced with fatty tissue and muscle fibres degenerate and decrease in size leading to atrophy of the muscles. The dystrophies vary in their age of onset and clinical features, but all are genetically inherited. The commonest form of the disease is Duchenne muscular dystrophy. It is usually diagnosed in young males (age about 6-7 yrs) who rarely survive beyond their late teens or early twenties. Another type of the disease occurs in later life and is seen in both males and females. Recently a protein called dystrophin has been discovered to be present in normal muscle but is not found in the muscles of patients suffering from Duchenne dystrophy. The role of this protein is, as yet, unknown but may be associated with the leakage of calcium ions from the sarcoplasmic reticulum and consequent degeneration of muscle fibres. In addition, the gene responsible for this dystrophy has been identified and its nucleotide sequence worked out. This may lead eventually to a possibility to treat or prevent the disease.

9.14.4 Myasthenia gravis

acetylcholine receptors destroyed

This disease is characterised by weakness in the muscles and it is due to a malfunction of the junctions between the nerves and the muscles. It is an autoimmune disease in which the receptors for the neurotransmitter acetylcholine are destroyed so that the muscle loses the ability to respond to the messages being sent from its controlling nerves. An increasing number of motor units become affected until the muscle may cease to function altogether. Treatment is usually with drugs which increase the levels of the transmitter available to bind to the receptor. Also, steroids are used as immunosuppressive drugs to lower the levels of the receptor antibodies.

9.14.5 Fibrosis

pneumoconiosis

Fibrosis is the process by which the muscle contractile tissue is replaced by connective tissue as a result of trauma or disease. This can occur in all types of muscle eg cardiac muscle will fibrose following a heart attack; skeletal muscle after an injury and smooth muscle in the lungs is often damaged by industrial diseases such as pneumoconiosis (coal miners disease). The loss of contractile tissue, obviously, decreases the functional performance of the muscle in question.

9.14.6 Fibrositis

This is the condition in which pain and inflammation of the muscle occurs. Again, this often follows injury or prolonged or inappropriate use of the muscle.

9.14.7 Abnormal contractions

- Cramp: a sudden maximal tetanic contraction of a muscle which reaches a much higher than normal level of recorded tension. For this reason the spasm is very painful. The cause is not fully understood.

- Tremor: rhythmic, involuntary contractions of opposing muscle groups eg as seen in Parkinson's disease.

- Fasciculation: involuntary, brief twitches of a muscle which are visible under the skin. These may be seen in such demyelinating diseases as multiple sclerosis.

- Fibrillations: similar to the above but are not visible to the naked eye, they can be recorded by electromyography.

- Tic: a spasmodic twitching of a muscle usually under voluntary control eg. eyelid or face muscles. The muscle movements often have a psychological origin.

9.15 Control of movement

Movement of the mammalian body varies from the gross movements associated with running or walking, through to fine manipulative movements, for example playing the piano. Such movements may be produced either in response to some external stimuli or alternatively, they may be produced in response to some central command from the brain. This means therefore, that control of movement occurs at several different levels within the central nervous system (ie the brain and spinal cord). Important control areas include: spinal cord, motor centres of the cerebral cortex, the cerebellum and the basal ganglia. In reality they almost certainly work in co-operating with each other. We covered many aspects of this when we discussed the nervous system in Chapter 2. Here we will remind you of a few of the main points.

spinal cord

cerebral cortex

cerebellum and basal ganglia

9.15.1 The spinal cord

spinal reflexes

Movement mediated by the spinal cord is typically that which is made without the need for conscious thought. Such movements are termed spinal reflexes. An everyday example of such a response would be touching a very hot object with the hand. The response to this is removal of the hand from this (painful) stimulus. This occurs without the need to consciously think about removing the hand.

Attempt SAQ 9.6 to test your re-call of a spinal reflex arc. If you cannot answer this question we would suggest you re-read the appropriate section in Chapter 2.

SAQ 9.6

Draw a diagram illustrating the organisation of a spinal reflex which contains two interneurons.

The number of interneurons contained within a spinal reflex varies considerably. In the knee-jerk reflex there are no interneurons. The flexor-withdrawal reflex (ie removal of a limb from a painful stimulus) contains three or four interneurons. In the scratch reflex (ie scratching of the skin in response to an irritant substance) there may be twenty or more interneurons. Spinal reflexes operate over a very short period of time between stimulus and response. This is known as the reflex time.

reflex time

Π List the factors which will contribute to the total reflex time.

reflex time and number of synapses

The reflex time will take into account the time taken for the action potential to be transmitted along the sensory neuron from the receptor (eg pain receptor in the case of the flexor withdrawal reflex and the time for transmission of the action potential across the sensory - motor neuron synapse and then along the motor neuron. Finally, the action potential must cross the synapse between the motor neuron and the muscle (Neuromuscular junction) and alter the activity of muscle. The total reflex time for a typical monesynaptic reflex is about 20ms.

Π What will be the effect of adding more interneurons to a spinal reflex on the total reflex time.

The effect of adding interneurons to the reflex time is to increase it. This happens because of the addition of extra synapse where delay occurs due to the release of neurotransmitters which convey the action potential across the synapses. Each synapse adds an extra 1 msec or so to the reflex time.

postural and protective reflexes

It is possible to distinguish between two different types of spinal reflex. These are postural reflexes and protective reflexes. Postural reflexes, as the name suggests allow an animal to maintain its posture whilst protective reflexes protect from injury. Protective reflexes take precedent over postural reflexes.

9.15.2 Sensory receptors in muscles: muscle spindles

There are sensory receptors within muscles which pass information regarding the state of the muscle, to the central nervous system. These sensory receptors which lie in the body of the muscle are called muscle spindles.

Muscle spindles are for example involved in the knee-jerk reflex where they cause reflex contraction of the quadriceps muscle in response to its stretching. The structure of the typical muscle spindle is shown in Figure 9.12.

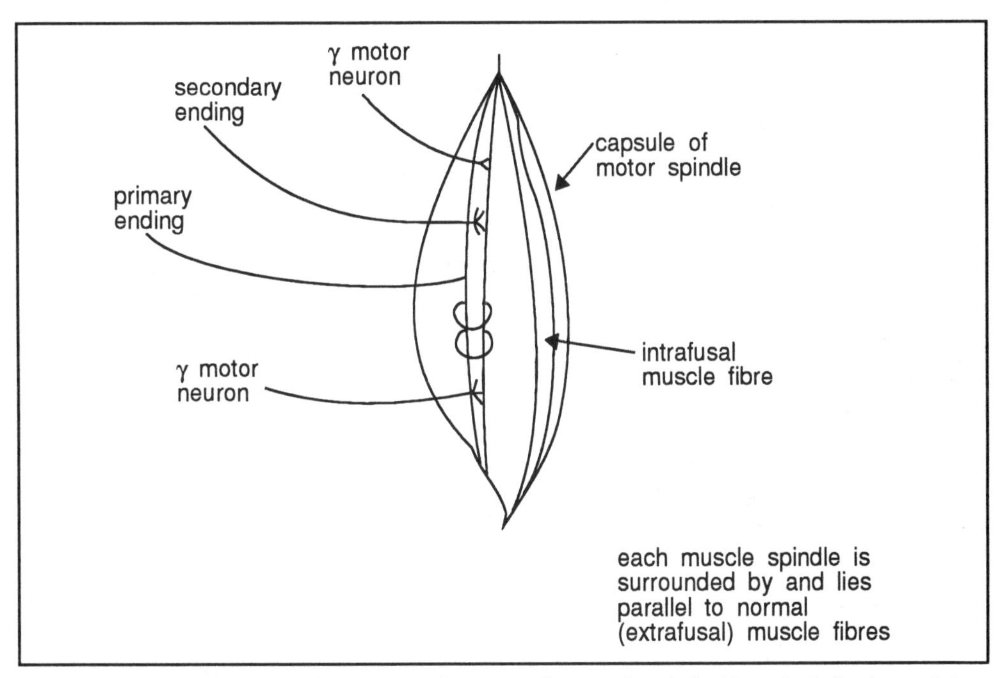

Figure 9.12 Diagram illustrating the structure of a mammalian muscle spindle. The spindle lies in parallel with the muscle fibres in which it lies (see text for details).

intrafusal and extrafusal fibres

Muscle spindles are composed of special muscle fibres which are generally thinner than the normal fibres which form the typical muscle. The fibres making up the muscle spindle are called intrafusal fibres to distinguish them from the extrafusal fibres which make up the majority of the muscle. Intrafusal fibres of the muscle spindle (unlike the extrafusal fibres) have a sensory (afferent) innervation. This afferent innervation is of two types. The first comes from the centre of the muscle spindle and is called the primary or annulospiral ending. The second come from the region next to the centre and is called the secondary or flower-spray ending. In addition to this afferent innervation there is also a motor (efferent) innervation. The so called motor neurons innervate the extrafusal fibres of the muscle. The intrafusal fibres of the muscle spindle also receive a motor innervation at the extremities of their fibres (these are the only contractile regions). This motor innervation is called the γ (gamma) motor neuron innervation. The role of this γ-motor neuron innervation is to allow the muscle spindle to work over a wide range of muscle lengths without any loss in sensitivity. There is a resting discharge of action potentials from the sensory nerves. If the muscle is stretched, this discharge increases, whilst if the muscle shortens ie contracts the resting discharge decreases. Activation of the sensory neurons forms part of the spinal reflexes.

9.16 Higher levels of motor control

The spinal reflex constitutes a fairly basic mechanism of controlling movement. Higher levels of control are organised within the brain. It may be worth while reviewing the gross structure of the human brain and the higher levels of muscle control (see Chapter 2). Here we remind you of some important features.

control of voluntary movement

The CNS is particularly important in the control of voluntary movement. It is possible to distinguish several levels of control (in Chapter 2 we described levels 1-5). The highest level involves part of the cerebral cortex particulary those involved with memory, emotion and regions of the cortex associated with movement. This region forms the plans or intention to move. The next level (cerebellum, brain stem) converts movement plans into a number of smaller sub-plans. These plans can then be transmitted (via so called motor neurons) to cause movement. Let us begin by looking at the higher levels of control.

9.16.1 Highest levels of control: Cortex

Probably the highest level of control originates in the so called supplementary motor area of the cerebral cortex. This is where the 'decision' to move is made. The question that needs to be asked is what initiates this decision. The answer to this is unclear although it is known that memory and motivation are important. If we now look at other regions of the CNS which are involved in the control of movement we shall see that much more is understood of their role.

memory and motivation

9.16.2 Brain stem nuclei - the role of the reticular formations

The brain stem, which, as you will recall consists of the pons, medulla and midbrain, has groups of neurons (called nuclei) scattered throughout it. These nuclei collectively constitute the reticular formation. Within the reticular formation are areas involved in motor control. In the medulla are the 'vital centres' which control such reflexes for breathing (respiratory centre), heart rate and blood pressure.

reticular formation

basal ganglia and posture control

The basal ganglia, are a group of sub-cortical (ie beneath the cortex) nuclei which lie deep within the brain. The basal ganglia play an important role in the control of posture, in particular selecting and maintaining desired movements whilst at the same time suppressing other unwanted movements. Much of what is known about the role of the basal ganglia comes from observations in humans with damage to this region of the brain. The most common disorder of the basal ganglia is Parkinson's disease. The most common features of this disease are:

- a disinclination to move, or at least a greater motivation required. This is termed akinesia;
- involuntary movements ie not under voluntary control - usually tremor;
- rigidity, the underlying cause of Parkinson's disease is a loss of neurons within the basal ganglia, particulary those which use dopamine (Dopa) as their neurotransmitter.

Π How might Parkinson's disease be treated

One method of treating Parkinson's disease is via replenishment of dopamine. This is achieved by the administration of the pre-cursor L-Dopa, although this is not without problems. the major one being the ability to get sufficient amounts of it into the brain. Much of the ingested Dopa is metabolised in the liver before it reaches the brain.

9.16.4 The cerebellum

cerebellum and balance and learnt skills ataxia, adidokinesia and hypotonia

The cerebellum is located at the base of the brain. A major role of the cerebellum is concerned with the maintenance of balance. A second major role of the cerebellum is to act as a comparator, ie comparing actual movements with desired movements and initiating corrective action to correct differences between the two. A final role of the cerebellum includes the co-ordination of fine movements and the learning of skilled movements. Damage to the cerebellum usually presents itself as a failure in the integrated action of various muscle groups. This results in phenomena such as ataxia ie unsteadiness on ones feet, adidokinesia ie the inability to perform rapidly alternating movements (eg sitting and standing) and hypotonia ie a reduction in muscle tone making them appear 'floppy'.

9.16.5 The motor cortex

The parts of the cortex which are involved in the control of movement are the primary motor cortex and pre-motor area. In general regions requiring fine movements have a large area of the cortex devoted to them. From the motor cortex and pre-motor area the neurons initially exit which either directly or indirectly, connect with the motor neurons innervating muscle. However, it is not the prime initiator of movements. This region, together with the cerebellum, basal ganglia and brain stem regions are simply relay stations in the control of movement, although there is some capacity for them to act independently. The general organisation of movement control is shown in Figure 9.13.

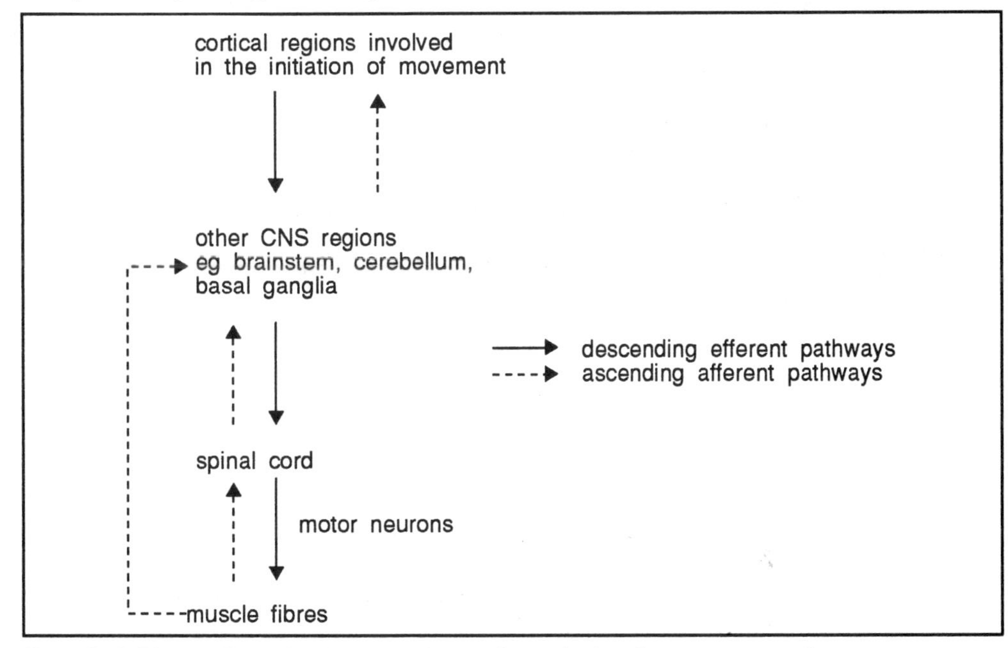

Figure 9.13 Diagram illustrating the connections and organisation of movement control

Summary and objectives

In this chapter we have examined muscle activity. We have considered the general functional characteristics of muscle together with the cell biology and molecular biology of such tissues. We considered how nerve impulses are transmitted via neuromuscular junctions to produce action potentials in muscles and how this is propagated within motor units. We also explored the various types of muscle contraction and the diversity found in muscle fibre types. In the final parts of the chapter we discussed some common disorders of muscle functions and examined the higher levels of muscle control.

Now that you have completed this chapter you should be able to:

- describe the banded structure of skeletal muscle and the arrangement of molecules within this structure;
- list the sequence of molecular events which lead to the contraction of skeletal muscle;
- construct a flow diagram which illustrates the sequence of events involved in skeletal neuromuscular transmission;
- interprete data relating amplitude of muscle contraction to applied stimulus voltage;
- predict the types of muscle fibres that would be used in for a variety of muscular activities;
- identify muscle fibre types, from described properties;
- draw spinal reflex arc's;
- explain the basis of a variety of muscular disfunction characteristics of a variety of disorders.

Human reproduction

Human reproduction

10.1 Introduction

gonads

tootoo

ovarys

In this chapter we will be examining the physiology of the system which is involved in the continuation of the human species. The primary reproductive organs in our bodies are called the gonads, the testes in the male and the ovaries in the female. One function of the gonads is to produce the special population of germ cells, called the gametes. The ovary releases the ovum (plural ova), whilst the testes produce spermatozoa (singular spermatozoan). During fertilisation an ovum and a spermatozoan fuse to form the zygote which is the beginning of the new organism. The second important function of the gonads is to secrete steroid hormones, the sex hormones, which are vital in the control of the reproductive function of the body and the development of the secondary characteristics (these are the external physical differences seen between the male and the female). We shall begin our study of reproduction by looking at the anatomy and physiology of the male and female systems, to include the hormonal control of gametogenesis. We will then examine the process of fertilisation and development of the foetus; pregnancy; placental function and parturition. We shall complete the chapter with a consideration of prenatal diagnostic tests, *in vitro* fertilisation and contraception.

10.2 The male reproductive system: anatomy and physiology

10.2.1 Anatomy

The male reproductive system consists of the following paired components: testis, epididymis, vas deferens, seminal vesicle, ejaculatory duct and bulbourethral gland in addition to a single prostate gland and the penis. The relationship of these various structures is illustrated in Figure 10.1.

scrotum

structure of testes

The testes are suspended outside the body in an outpouching of the abdominal wall called the scrotum, which is divided, internally, by a septum into two sacks, each holding one of the testes. The purpose of this position outside the body is to ensure that the testes are kept at a temperature which is slightly less than the internal body temperature. This is essential for normal development of the spermatozoa. Each of the testes is divided into a number of lobes containing highly coiled seminiferous tubules in which the process of spermatogenesis (the development of the sperms) takes place (Figure 10.2). A cross section of one of these tubules is shown in Figure 10.3. You can see that it is bounded by a basement membrane and has a fluid filled central lumen. The rest of the tubule is composed of the germ cells in various stages of development (the most immature cells being at the edge of the tubule) enveloped by Sertoli cells which extend from the basement membrane to the central lumen.

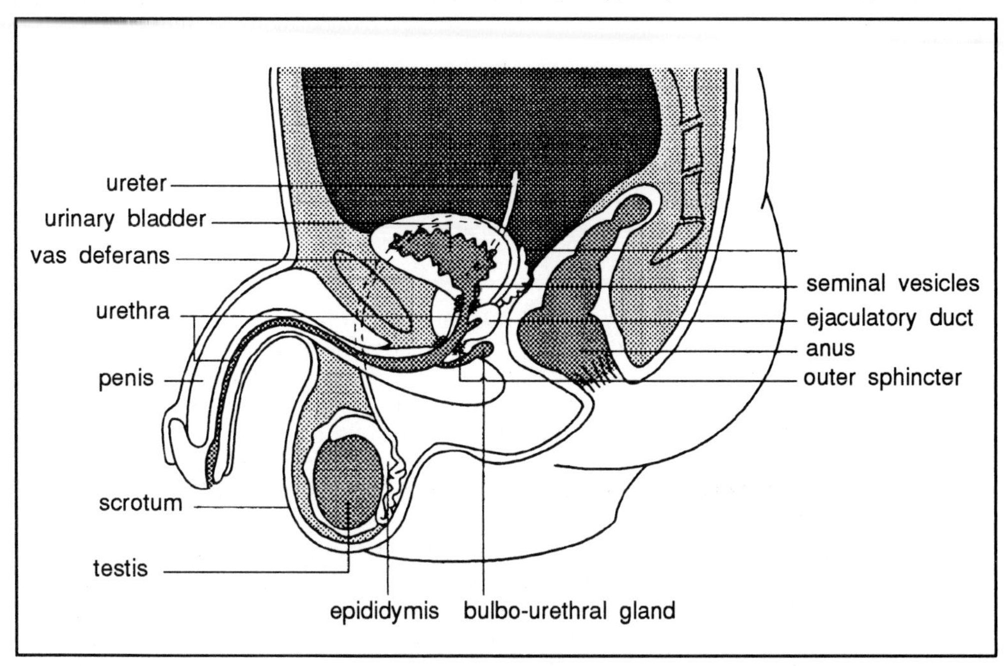

Figure 10.1 Diagram to illustrate the anatomy of the male reproductive tract.

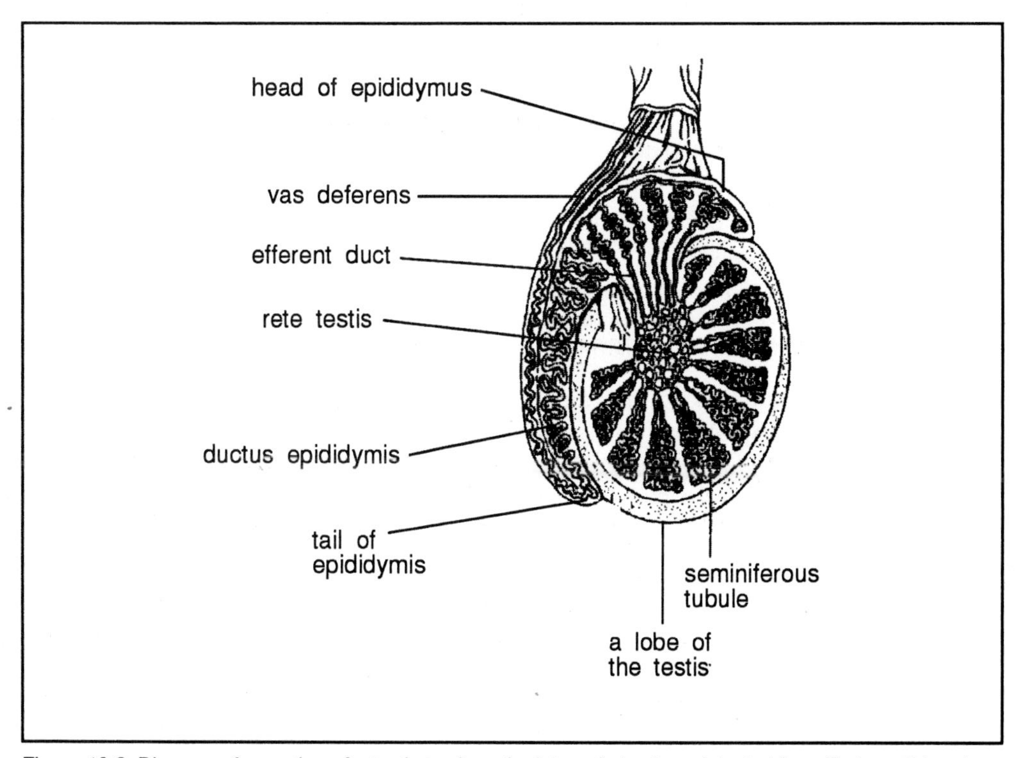

Figure 10.2 Diagram of a section of a testis to show the internal structure. Adapted from Tortora, GJ and Anagnostakos, N.P. (1990), Principles of Anatomy and Physiology, Biology Science Textbook Inc, USA, p881.

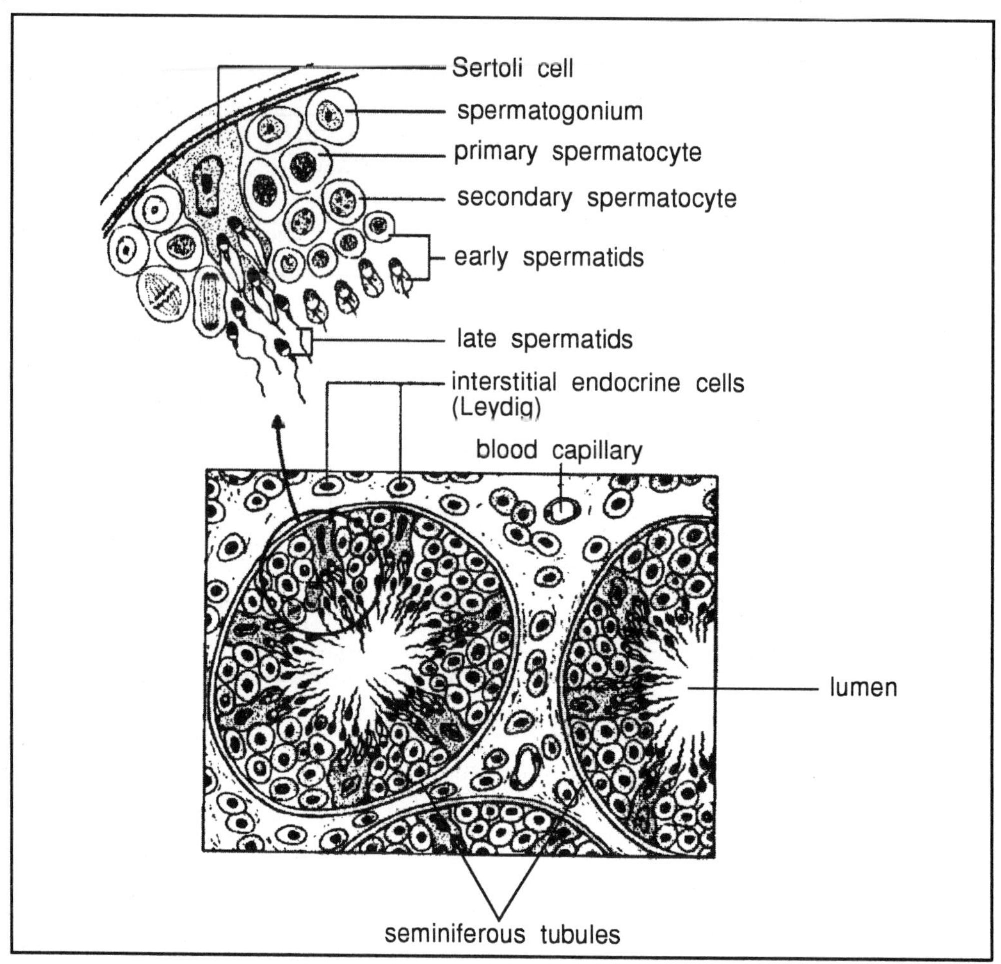

Figure 10.3 Section of a seminiferous tubule to show the positions of the spermatozoa in various stages of their development. Note the close relationship between the developing sperms and the Sertoli cells. The tight junctions act as a blood/testes barrier to protect the sperms from foreign chemicals and to prevent an immune response to their presence.

Lying in small connective tissue spaces in between the seminiferous tubules are the Leydig cells which are endocrine cells producing the male sex hormone, testosterone.

Once fully matured, the sperms are transported through the coiled seminiferous tubules to a network of ducts in the testis called the rete testis and, thence, out of the testis via the efferent ducts to a structure called the epididymis. This is a long coiled duct loosely attached to the outside of the testis which leads in to a straight, thick walled tube known as the vas deferens. Together, these two tubes are important structures for the final maturation and storage of the sperms prior to their release. The vas deferens is bound together with the blood vessels and nerves supplying the testis in the spermatic cord which enters the abdomen through the inguinal canal. Once inside the abdomen the vas deferens is routed towards the bladder and become known as the ejaculatory ducts when they merge with the ducts arising from the large glands, the seminal vesicles. These glands produce viscous fluid which is alkaline and rich in fructose. The alkalinity of the secretions helps to neutralise the acidic vaginal fluids, whilst the fructose is used as an energy source by the sperms. Also found in the secretions of the seminal vesicles are prostaglandins (so called because they were originally thought to be produced by the prostate gland) whose function, it has been suggested, may be to induce contractions in the female genital tract to assist the sperms in their journey through it. The ejaculatory ducts continue into the tissue of the prostate gland and join the urethra from the bladder. The prostate gland lies below the bladder and surrounds the upper part of the urethra. It secretes a fluid which constitutes about 33% of the volume of semen and which is important in the motility and viability of the sperms.

Lying below the prostate are the bulbourethral glands which drain into the urethra just as it leaves the prostate. These glands produce a mucous rich fluid which helps to lubricate the penis during penetration and increases the motility of the sperms. The urethra is the terminal duct of the male reproductive tract as it passes through the prostate and terminates in the tip of the penis. It serves to transport both sperms and urine to the exterior of the body.

ejaculation and the release of sperms

As the sperms pass through the reproductive tract they are mixed with the various secretions of the different parts of the tract to form the semen which is released from the urethra during ejaculation from the penis. The penis is used to introduce the spermatozoa into the vagina. It is cylindrical in shape consisting of three vascular cords extending the entire length of the organ. Normally these cords contain little blood as the blood vessels supplying them are constricted. However, during sexual excitation, the arteries are dilated allowing a lot of blood into the penis. This, in turn, compresses the veins which empty the vascular cords, therefore, the blood is trapped in the penis. This has the effect of making the penis become erect and rigid so that it can then penetrate the vagina. The mechanical stimulation of the penis elicits a spinal sympathetic reflex in which the semen is released into the vagina through the urethral opening in the tip of the penis. This process is known as ejaculation and is accompanied by many physiological changes in the body. Simultaneous closure of the sphincter at the base of the bladder prevents entry of the sperm into the bladder and the release of urine during ejaculation. The penis then returns to its flaccid state as the arteries once again constrict and the pressure on the veins is relieved. There is a latent period which has to elapse before a further erection is possible.

10.2.2 Spermatogenesis

The process of spermatogenesis in the seminiferous tubules involves a number of different stages of development. The undifferentiated germ cells (spermatogonia) which line the seminiferous tubules close to the basement membrane (see Figure 10.3) divide mitotically to produce a cluster of virtually identical cells (mitosis is the cell division in which genetically identical sister cells are produced). All but a few of these undergo further differentiation to become primary spermatocytes (the remaining few serve as a reservoir of germ cells to prevent depletion of the supply). The primary spermatocytes are the first cells to divide by meiosis, each producing two secondary spermatocytes which then undergo the second meiotic division to form four haploid spermatids. (Meiosis is the cell division in which the chromosome number is halved to produce what are called the haploid cells. When these cells fuse, (at fertilisation), the normal diploid chromosome number is restored). Thus, each primary spermatocyte, containing 46 chromosomes, gives rise to four spermatids, each of which contains 23 chromosomes.

primary spermatocytes

secondary spermatocytes

spermatids

We can represent this process as follows:

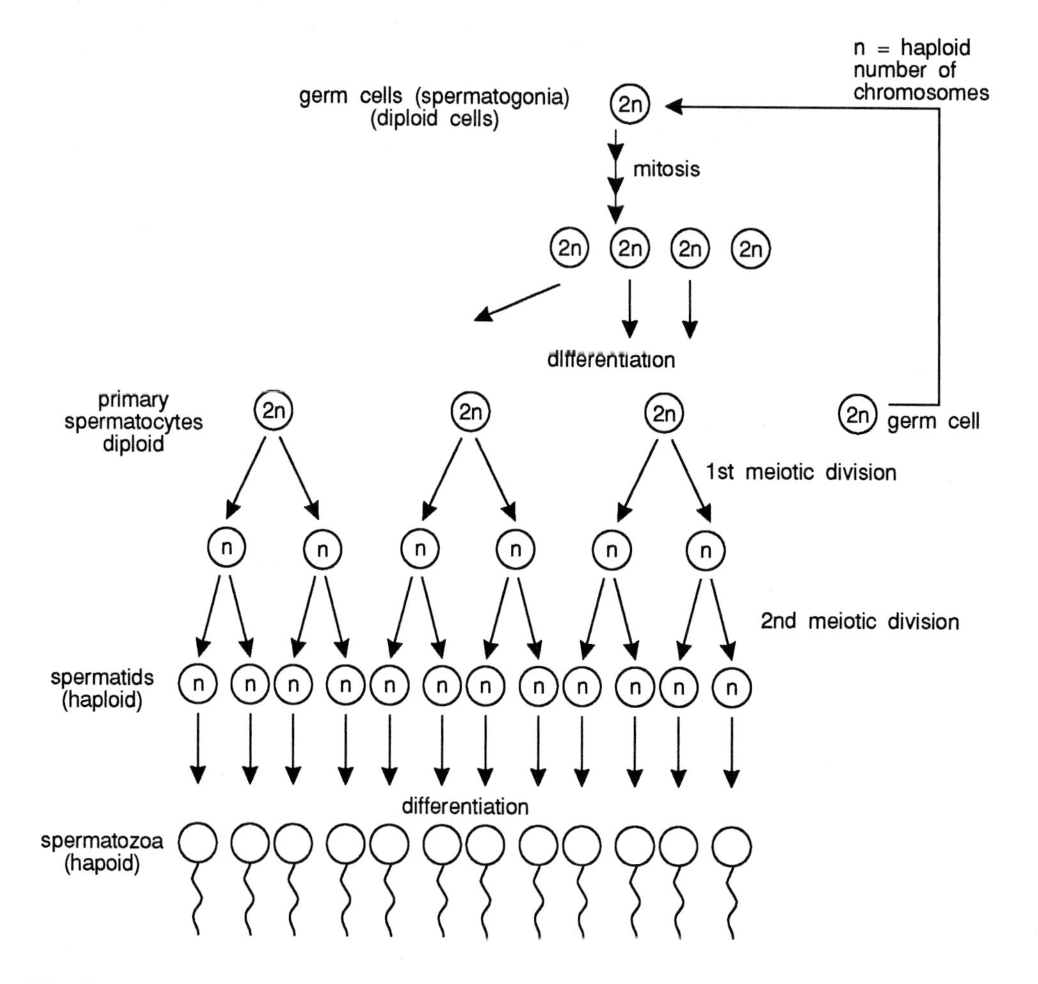

The final stage of spermatogenesis consists of the gradual maturation of the haploid spermatids into spermatozoa. This takes place towards the central lumen of the seminiferous tubules.

Sertoli cells and cell nourishment

The whole sequence of events from primary spermatocytes to spermatozoa takes approximately nine to ten weeks. During this time the cells are nourished by special nurse cells called Sertoli cells with which the developing sperms are intimately associated, as illustrated in Figure 10.3. Just below the basement membrane these cells are joined together by tight junctions, thereby dividing the seminiferous tubule into two compartments, one above the junction and one below. In this way the Sertoli cells form a blood-testes barrier which functions to prevent the passage of many compounds, such as proteins, from the blood into the lumen of the tubules. This barrier is important in preventing an immune response to the developing sperms whose surface antigens could be recognised as foreign by the immune system. The compartmentalisation also serves to separate the initial mitotic divisions of the spermatogonia, which occur in the upper or basal section, from the later meiotic divisions taking place in the more central sections.

Sertoli cells, androgen binding proteins, inhibin secretion and phagocytosis

The Sertoli cells also function to support and protect the developing sperms. They provide nourishment, control movements of the spermatogenic cells and the release of the sperms into the lumen of the seminiferous tubule. They secrete most of the fluid of the tubular lumen which has a specific ionic composition and contains androgen binding protein (ABP) an important protein for binding with testosterone. The Sertoli cells also secrete the hormone, inhibin, which is important in the regulation of sperm production and, finally, they remove, by phagocytosis, any defective or degenerating sperms.

mature sperm structure and ovum penetration

The spermatozoa enter the lumen of the seminiferous tubule and migrate to the epididymis where they complete their maturation over the course of 10 to 14 days. They are stored in the vas deferens where they can maintain their fertility for several weeks. The structure of the mature sperm is illustrated in Figure 10.4. It is composed of a head, a midpiece and a highly motile tail. The head contains the nucleus which is packed with DNA bearing the sperm's genetic instructions. The tip of the nucleus is covered by the acrosome which contains enzymes and contractile proteins which aid the sperm's penetration of the ovum. The midpiece contains mitochondria to generate the energy required for the movement of the sperms. This movement is carried out by the tail which is a group of contractile filaments forming a flagellum capable of producing whiplike propulsion of the sperm along the female reproductive tract.

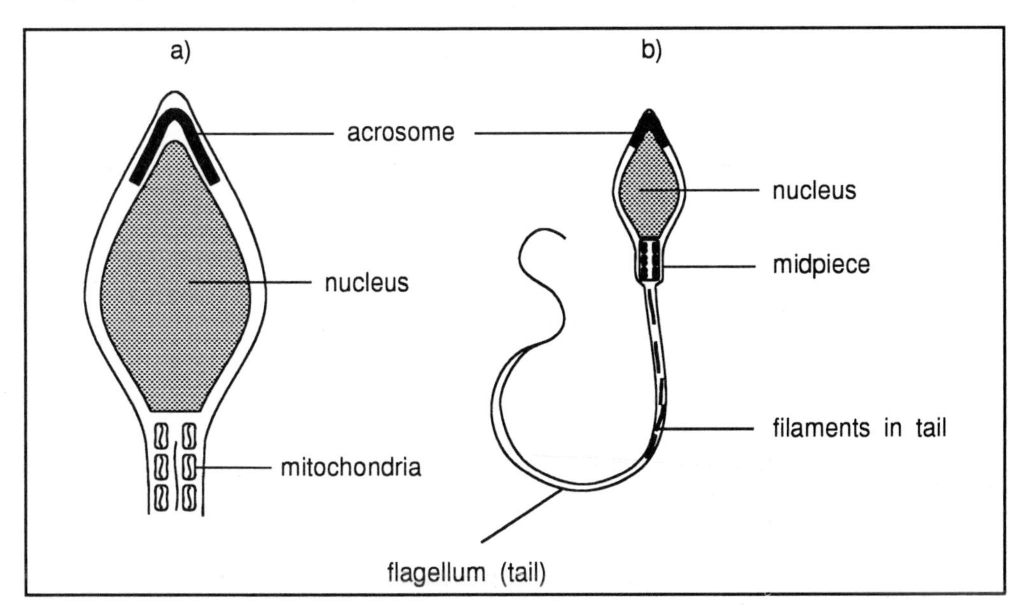

Figure 10.4 Diagram to illustrate the structure of a mature spermatozoan. a) Close-up of the head. b) The whole spermatozoan.

The normal human male produces about 30 million sperms a day, this manufacture being under complex hormonal control.

Π Draw a summary table of the various stages of spermatogenesis. (Do this without looking back at our flow diagram. Then when you have completed your table, check the stages with our flow diagram.)

Π Make a list of the functions of the Sertoli cells.

The main properties of Sertoli cells are as follows:

- nourishment of developing sperms;
- secretion of luminal fluid containing androgen binding protein (ABP);
- acting as a blood/testis barrier to foreign substances;
- secretion of the hormone inhibin;
- control of entry of sperms into lumen of seminiferous tubules;
- phagocytosis of defective sperms.

10.2.3 The hormonal control of male reproductive functions

gonadotrophins secretion from hypothalamus

follicle stimulating and lutenising hormone from anterior pituitary

Normal male reproductive function is controlled by the effects of a series of hormones. The hypothalamic neurohormone, gonadotrophin releasing hormone, is secreted, episodically, from neurosecretory cells in a certain area of the hypothalamus and passes to the anterior pituitary where it stimulates the production of two further hormones. These are follicle stimulating hormone (FSH) and luteinising hormone (LH). Both these hormones act on the testes, the result being the maturation of the sperms and the secretion of testosterone. The testosterone, itself, has a number of different functions which include the stimulation of spermatogenesis, the development of the secondary sexual characteristics and sexual behaviour in addition to the growth of the male reproductive tract and other body tissues.

testosterone secretion by Leydig cells

The pituitary hormones, FSH and LH, influence different parts of the testes. Luteinising hormone stimulates the Leydig cells to secrete testosterone, whilst FSH is thought to act through the Sertoli cells to stimulates spermatogenesis. The actual mechanism involved in the latter process is uncertain, although it has been established that LH is essential for the process to occur; LH, therefore, has an indirect effect on spermatogenesis. One way in which the two hormones may interact is that LH promotes the secretion of testosterone which then exerts a local effect on the Sertoli cells. In conjunction with FSH, this causes the Sertoli cells to secrete a protein hormone called androgen binding protein (ABP). This particular hormone appears to avidly bind testosterone, thereby increasing the concentration of testosterone in the seminiferous tubules which then functions to encourage maturation of the sperms.

You are aware that all four hormones, GnRH, FSH, LH and testosterone, are essential for the normal processes of spermatogenesis. It is, therefore, not surprising that the plasma levels of these hormones are closely controlled. As with many hormone systems, such control is brought about by negative feedback - a process in which an increase in the plasma levels of the hormones will result in a decrease in the output of the controlling hypothalamo-pituitary hormones (see Chapter 3).

An increase in the plasma level of testoserone above the normal values will lead to an inhibition of the output of both gonadotrophin releasing hormone and LH. Further production of testosterone is decreased and plasma levels will return to normal.

inhibin secretion by Sertoli cells

The output of FSH is negatively controlled by another hormone secreted from the Sertoli cells, called inhibin. An increase in the plasma concentration of inhibin acts on the pituitary to inhibit the output of FSH with the result that there is less stimulation of the Sertoli cells and consequently less facilitation of spermatogenesis.

SAQ 10.1

From the above description of the hormonal control of male reproductive function try to construct a flow diagram to show the output and feedback pathways of all the contributing hormones.

SAQ 10.2

If LH were injected into a mature male mammal, which of the following results should be expected?

1) Increased production of oestrogens.

2) Increased spermatogenesis.

3) Increased production of testosterone.

4) Decreased testicular activity.

5) Increased output of GnRH.

10.3 The female reproductive system: anatomy and physiology

10.3.1 Anatomy

ovaries

ova

The female reproductive system consists of the gonads called the ovaries (singular ovary) which produce gametes (ova) in a regular monthly cycle and the reproductive tract consisting of the oviducts, the uterus and the vagina. Figure 10.5a illustrates these components.

The ovaries are almond shaped glands which lie in the pelvic abdomen on either side of the uterus. They are held in position by a number of ligaments attaching them to both the uterus and the pelvic wall. As well as producing and releasing the ova, the ovaries also secrete the hormones oestrogen and progesterone (see Section 10.3.4).

fallopian tubes

Lying close to the ovaries are the funnel shaped openings of the oviducts (fallopian tubes). These openings are fringed with fingerlike projections which help to guide the ova into the lumen of the oviducts and, thence, into the cavity of the uterus (Figure 10.5b).

myometrium and endometrium components of the uterus

cervical canal

The uterus is a hollow, thick walled, pear shaped structure lying on the floor of the pelvic abdomen above the bladder and in front of the rectum. The thick muscular wall is called the myometrium whilst the inner lining is the endometrium. It is this lining which breaks down during menstruation, the monthly event occurring when no implantation of a fertilised ovum has taken place. The lower part of the uterus is the cervix. An opening in this, the cervical canal, opens out into the vagina which connects the uterus with the exterior of the body.

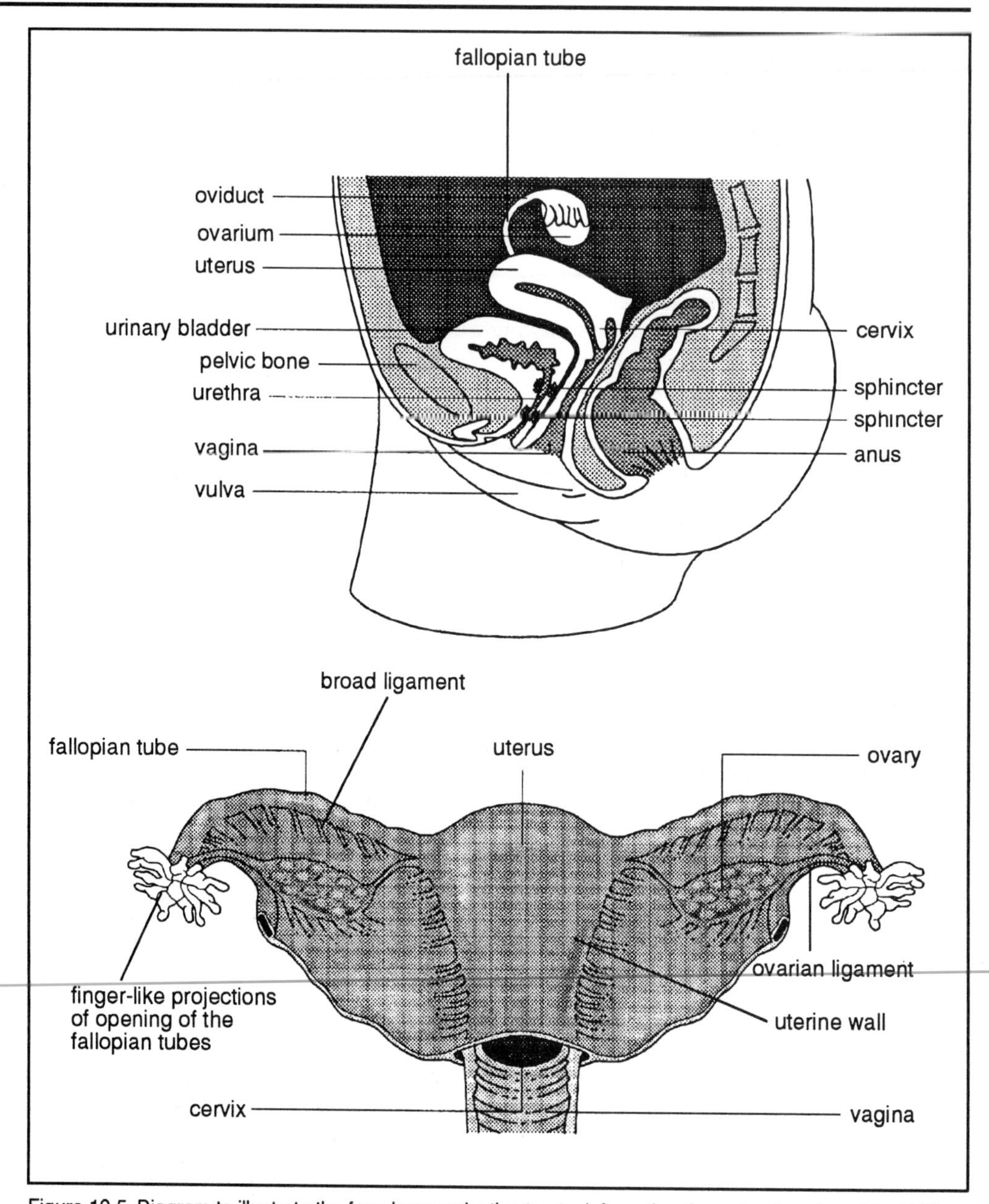

Figure 10.5 Diagram to illustrate the female reproductive tract. a) A section through a female pelvis. b) The relationships between the different parts of the female reproductive tract.

vulva

The vagina is a muscular tube in which the sperms from the male are deposited during sexual intercourse. Surrounding the opening of the vagina are the female external genitalia, known collectively as the vulva. A number of different structures are included, namely the mons pubis, the labia majora and labia minor, the clitoris, the vestibule and the vestibular glands.

menstrual or oestrus cycle

Unlike the continuous production of spermatozoa, the maturation and release of the female gametes occurs in a monthly cycle, called the menstrual cycle in humans (known as the oestrus cycle in other animals). On average this occurs every 28 days, although it may range from approximately 24 to 35 days. The whole cycle of events is under the influence of the hormones produced by the ovaries.

10.3.2 Ovarian function

Here we will divide discussion of ovarian function into the generation of oocytes (oogenesis) and the hormonal control of ovarian function.

10.3.3 Oogenesis

Oogenesis is the process analogous to spermatogenesis in the male in that it involves the formation of the haploid ova, the female gametes, in the ovaries. The events begin in the foetus because, unlike the male, the female is born with all the germ cells she will ever have. (This may explain the increased incidence of children with birth defects being born to older women as the ova to be fertilised may have undergone some aging process). In early foetal life the undifferentiated germ cells (oogonia, cf spermatogonia in the male) undergo numerous mitotic divisions for a limited length of time. Beyond this time no further oogonia are generated. The oogonia already formed develop into primary oocytes which begin to divide meiotically. However, they do not complete this division in foetal life, but remain at this stage until puberty brings about renewed activity of the ovaries.

oogonia form primary oocytes

Each primary oocyte is surrounded by a layer of cells called the granulosa cells, thus forming a structure known as the primary follicle. Further development of these follicles only occurs under the influence of hypothalamo-pituitary hormones which begin to be secreted at puberty. Each month a certain number of primary follicles are stimulated to progress further. The oocyte grows and the number of granulosa cells increases, the oocyte becoming separated from those cells by the zona pellucida. The outer layers of granulosa cells differentiate further to form the theca. On reaching a certain diameter the follicle develops a fluid filled antrum as a result of secretions of the granulosa cells. At this stage the oocyte has reached its maximum size and any further increase in size of the follicle is due to the expansion of the antrum. Under specific hormonal signals only one of these primary follicles continues to develop to maturity. It is usually the largest follicle and this is called the Graafian follicle. The remaining follicles that had previously been growing now degenerate. (The commonest cause of multiple births is when more than one follicle matures beyond this stage). Figure 10.6 illustrates these events.

primary follicles and the production of granulosa cells

zona pellucida

theca

antrum

Graafian follicle

The Graafian follicle now resumes meiosis and two cells of unequal size are formed which are both haploid (have half the normal chromosome number). One cell receives most of the cytoplasm to become the secondary oocyte, whilst the smaller cell is known as the first polar body. The follicle becomes so large that it balloons out of the surface of the ovary. At ovulation the follicle and adjoining ovarian surface rupture thereby releasing the secondary oocyte along with its polar body and some of the surrounding supportive cells. The oocyte enters the fallopian tube where meiosis continues only if fertilisation occurs. Following such division, the secondary oocyte produces two haploid cells of unequal size. The larger one develops into the mature ovum whereas the smaller cell forms a second polar body. Thus, since all the polar bodies disintegrate, each oogonium gives rise to only one secondary oocyte and, therefore, to one mature ovum. This contrasts with the male gametogenesis in which the spermatogoniun gives rise to four spermatozoa.

secondary oocyte and first polar body

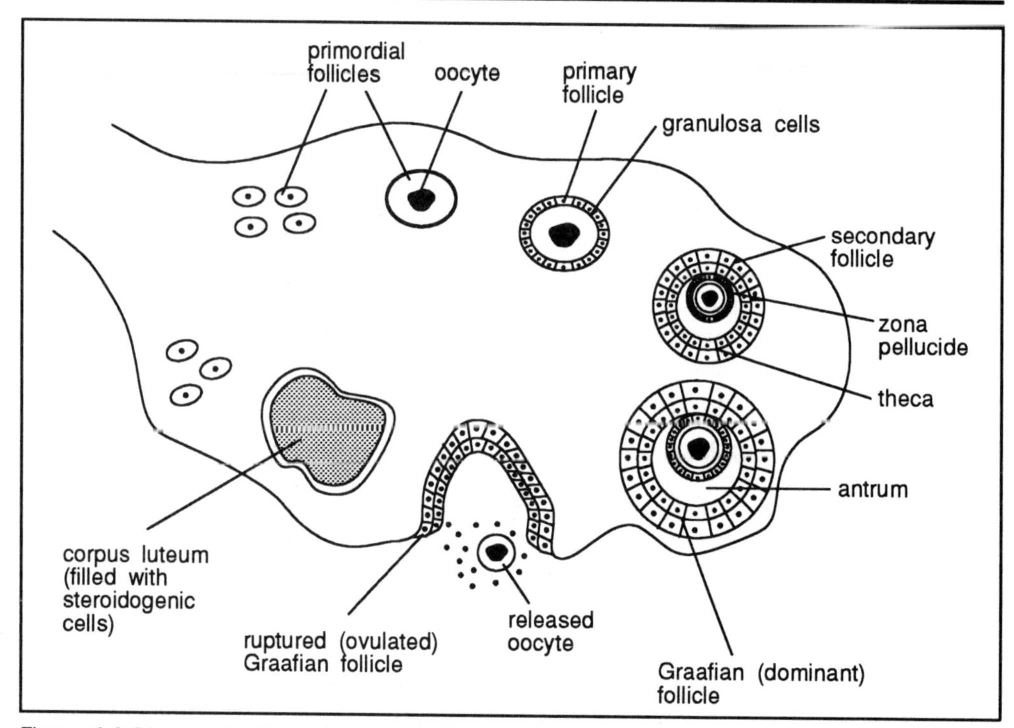

Figure 10.6 Diagram showing various stages of oogenesis in the ovary (stylised).

corpus luteum

Following ovulation the remnants of the follicle in the ovary change rapidly to become the corpus luteum. If no fertilisation of the ovum in the fallopian tube occurs, then the corpus luteum continues to grow for about 10 days after which it degenerates. It is this event which triggers menstruation, the shedding of the uterine lining. In the event of fertilisation the corpus luteum continues its development beyond this point and provides hormonal support for the developing zygote until such time as the placenta is capable of taking over this function.

This cycle of ovarian events can be divided into two phases, the follicular phase prior to ovulation and the luteal phase following it up to the demise of the corpus luteum. The whole cycle is under hormonal control and it is this facet of the events which we shall examine in the next section.

SAQ 10.3

Draw up a flow diagram to illustrate the ovarian events which occur during the two phases of the menstrual cycle.

10.3.4 Hormonal control of ovarian function

hypothalamic production of GnRH and anterior pituitary and production of FSH and LH

As with the hormonal control of testicular function, the same hypothalamo-pituitary hormones control ovarian function in the female. Namely, gonadotrophin releasing hormone (GnRH) is released from the hypothalamus and causes FSH and LH to be secreted from the anterior pituitary. These hormones, in turn, influence the events in the ovaries which themselves produce oestrogens and progesterone, the female sex hormones. The plasma levels of these controlling hormones follow a set pattern throughout the menstrual cycle and lead to the timing of the changes which occur in the ovaries. Figure 10.7 summarises the blood levels of the different hormones during

the ovaries. Figure 10.7 summarises the blood levels of the different hormones during the menstrual cycle and the ovarian and endometrial changes with which they are associated.

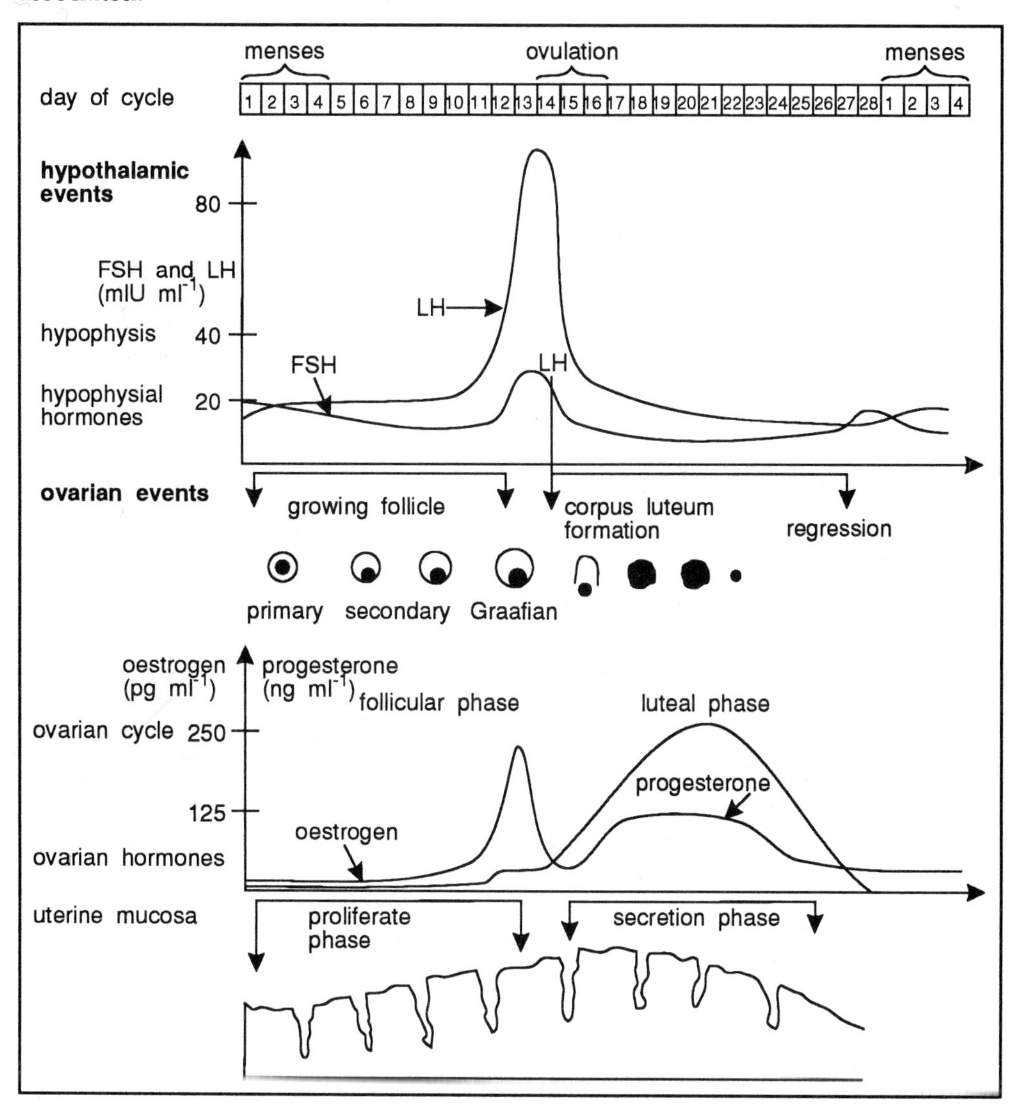

Figure 10.7 Diagram relating the plasma hormone levels and ovarian events during the menstrual cycle (see text for details).

follicular phase

From the figure it is evident that during the early follicular phase (from day 1 of menstruation) the level slowly rise, FSH levels are at a fairly high level. This causes some of the primary follicles to start developing further, with FSH and LH having separate effects on these follicles.

The FSH has three affects on the granulosa cells (which only have FSH receptors):

- it stimulates them to proliferate;
- it activates an enzyme system which enables the cells to manufacture oestrogens from androgens;

- it causes the development of LH receptors in the cells.

The LH influences the thecal cells causing them to produce androgens which, then, diffuse into the granulosa cells to be converted into oestrogens. The FSH together with the oestrogens continue to stimulate the follicles to develop further. It is important to note, however, that both LH and FSH are required for oestrogen production and, therefore, for this follicular growth.

preparation of uterus for implantation

The rising oestrogen plasma levels stimulates the repair of the uterine lining and growth of the tissues in order to prepare the uterus for a possible implantation of a fertilised ovum. However, in addition, the oestrogen also exerts a negative feedback onto the output of FSH and LH from the pituitary and, possibly, on the secretion of GnRH from the hypothalamus. Thus, the levels of LH and FSH, in particular, start to fall. By the beginning of the second week of the menstrual cycle one follicle has become the dominant one and this, alone, continues to develop further. It is thought that this may be the follicle which possesses the most FSH receptors and is, therefore, in a position to respond maximally to the falling levels of FSH in the blood. The output of FSH is further reduced by the negative feedback effect of inhibin which now begins to be secreted from the granulosa cells. At day 10/11 see the falling levels of FSH accompanied by the rising levels of LH resulting in follicle degeneration leaving the single Graafian follicle to continue development. There is, now some production of progesterone from this follicle but only in small quantities.

The next phase in the menstrual cycle is ovulation which is caused by a sudden change in the feedback control of the pituitary hormones, in particular LH. It would appear that the rising oestrogen levels, possibly aided by the small rise in progesterone, reverse their hitherto negative feedback control and now exert a positive feedback effect on the output of LH and FSH from the pituitary and GnRH from the hypothalamus. The result is a huge surge in the secretion of LH (see Figure 10.7) accompanied by a smaller rise in output of FSH. The surge in LH induces ovulation which also initiates the production of enzymes and prostaglandins which break down the ovarian wall. This occurs around day 14 of the cycle.

ovulation

luteal phase

The cycle now enters the luteal phase which is characterised by the development of the corpus luteum from the remnants of the ruptured follicle. This occurs under the influence of LH. The cells of the corpus luteum secrete progesterone as well as oestrogens so levels of these hormones are seen to rise during this phase. The result of this is that there is a negative feedback influence on the output of LH, FSH and GnRH whose levels progressively fall during the luteal phase. This means that no further development of follicles can take place.

Without fertilisation the corpus luteum survives for about 10 to 14 days after which it 'self destructs'. Once it begins to degenerate its output of oestrogens and progesterone falls. This means the uterus is deprived of its hormonal support and the endometrium begins to disintegrate. Eventually the tissue debris and blood are passed out of the vagina. This is day 1 of the cycle and the start of menstruation. The falling levels of oestrogens and progesterone also release the pituitary and hypothalamus from their inhibitory influence, therefore the levels of GnRH, FSH and LH start to rise and stimulate follicular development once again. The cycle begins anew.

You should note that the body temperature also changes during the cycle. From menses until ovulation, the mean body temperature is about 36.6°C. At ovulation there is often a transitory fall to about 36.4°C, but quickly rises to about 36.8°C. This temperature is maintained through the luteal phase. At the end of this phase, it falls again to 36.6°C.

SAQ 10.4

Having read through this account of the hormonal control of ovarian function, construct a flow diagram to summarise the pathways involved.

10.4 Fertilisation and early development

10.4.1 Fertilisation

The fusion of the ovum and a sperm occurs in the fallopian tube following the ejaculation of the spermatozoa from the penis into the vagina during sexual intercourse. This fusion is fertilisation. A number of factors contribute towards the arrival of the sperm at the surface of the ovum prior to their fusion. Firstly, the consistency of the vaginal mucous is altered by oestrogens released from the ovary; secondly, the vaginal cilia help to waft the sperms up the female reproductive tract towards the uterus and fallopian tubes and, thirdly, the whiplike actions of the sperm tails also aid this movement. There may also be a role for the uterine smooth muscle contractions which are part of the female sexual response to intercourse. In addition to aiding the transport of the sperms, the vagina also gives the sperms the capacity (capacitation) to fertilise the ovum. The sperms have to be in the vagina for at least 10 hours before they are capable of fusing. During this period, a process called capacitation takes place. Capacitation involves alteration of the surface of the acrosome on the head of the sperms. The second stage, sperm activation, ensues in which enzymes are released from the altered acrosome membrane. It is these enzymes that allow the sperms to begin to penetrate the ovum through the layer of granulosa cells known as the corona radiata (see Figure 10.8). Once through this layer, the sperms bind with receptors in the zona pellucida. This binding, normally, is followed by further penetration of only one of the sperms which, with the help of more acrosomal enzymes, moves through the zona pellucida to fuse with the plasma membrane of the ovum. The prevention of entry of other sperms (polyspermy) is brought about by the release of enzymes from the ovum which harden the zona pellucida and inactivate the sperm binding sites and prevent the binding of any late arrivals. The successful sperm continues its journey by entering the cytoplasm of the ovum, losing its tail in the process, while the nucleus in the head becomes surrounded by a membrane to form the male pronucleus. The ovum, then, completes its second meiotic division and forms the female pronucleus. These two pronuclei fuse to form a segmentation nucleus in the centre of the ovum. This fertilised ovum is then known as a zygote. It contains 23 chromosomes from the female pronucleus and 23 from the male pronucleus, thus restoring the diploid number of chromosomes. The zygote is now ready for further development and it continues its passage down the fallopian tubes to the uterus wafted by the beating of cilia.

capacitation, sperm activation and the prevention of polyspermy

Occasionally, a fertilised ovum remains in the fallopian tube where there is insufficient room for its continued development. Unless spontaneous abortion occurs, the zygote has to be removed by surgery (which usually involves removal of the tube as well) otherwise its continued growth would result in fatal maternal haemorrhage. If undiagnosed these tubal (ectopic) pregnancies are one of the commonest causes of maternal death. Even more unusually the fertilised ovum may be pushed out into the abdominal cavity where implantation can occur and pregnancy can continue to term. Delivery is then a surgical procedure.

tubal (ectopic) pregnancies

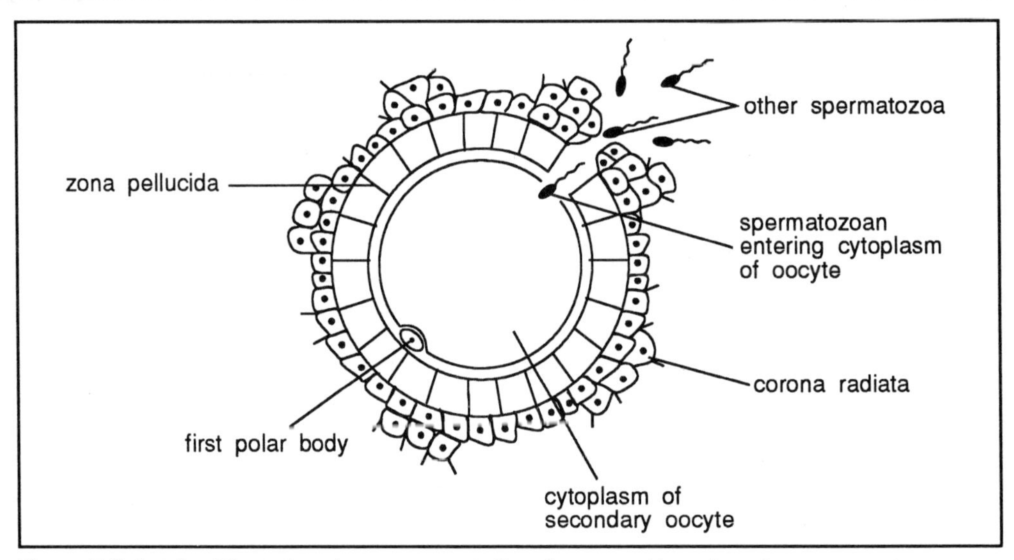

Figure 10.8 A stylised illustration of a sperm penetration of an ovum. Normally, only one spermatozoan successfully enters the cytoplasm of the oocyte. Entry of other sperms is prevented (see text).

dizygotic, monozygotic and Siamese twins

Fraternal (dizygotic) twins result from the simultaneous release followed by fertilisation of two ova by different sperms. These individuals, therefore, are genetically different from one another and need not, necessarily be of the same sex. On the other hand, identical twins (monozygotic) are derived from a single fertilised ovum that splits into two at an early stage of its development. These twins, therefore, contain the same genetic material and are always of the same sex. Where the division of the single fertilised ovum is not entirely complete, the resulting twins are born joined together in varying degrees. Such babies are called Siamese twins and their survival depends on the extent of the conjoining and the possibility, or otherwise, of surgical separation.

In the next section we shall look at the continuing stages of the development of the zygote as it continues its leisurely 3-4 day passage down the fallopian tube towards the uterus where implantation should occur.

Π Put the following statements into the correct order to give a description of the fertilisation of an ovum.

1) One sperm enters the zona pellucida to fuse with the plasma membrane and enter the cytoplasm of the ovum.

2) Ejaculated spermatozoa are helped to travel up the female reproductive tract following sexual intercourse.

3) Polyspermy is prevented by the release of enzymes from the ovum following fusion of the plasma membrane with the one successful sperm.

4) On encountering the ovum, the activated sperms pass through the corona radiata and bind with receptors on the zona pellucida.

5) Whilst in the female reproductive tract the sperms are, first, capacitated and then activated in order to enable them to penetrate the ovum.

6) On entering the cytoplasm of the ovum the sperm loses its tail and the nucleus becomes the male pronucleus.

7) The zygote contains the segmentation nucleus with the normal diploid number of chromosomes; it is the successfully fertilised ovum.

8) The nucleus of the ovum becomes the female pronucleus which fuses with the male pronucleus to form the segmentation nucleus.

You should have placed the sentences in the following order 2, 5, 4, 1, 3, 6, 8, 7.

10.4.2 Early development and implantation

formation of the morular and then the blastocyst

After fertilisation, whilst still in the fallopian tube, the zygote undergoes a number of mitotic cell divisions (cleavage) to form a cluster of cells which remain surrounded by the zona pellucida. This cell mass is known as the morula. It slowly progresses down the fallopian tube, helped by cillial beating, into the uterine cavity where it floats around in the uterine fluid. By this time the structure of the morula has changed from a dense ball of cells to one which is hollow and now referred to as the blastocyst. The blastocyst consists of an outer layer of cells, called the trophoblast, an inner cluster of cells and a central fluid filled cavity, the blastocoel (see Figure 10.9). The trophoblast secretes the hormone chorionic gonadotrophin which helps to maintain the corpus luteum and in turn, prepares the uterine lining for implantation of the blastocyst. It is the presence of this hormone in the mother's urine (it is excreted unchanged like many gonadal hormones) which is the basis of early pregnancy tests. The blastocyst floats freely in the uterine fluid for up to 4 days before it becomes attached to the endometrium (the uterine lining).

chorionic gonadotrophin and early pregnancy testing

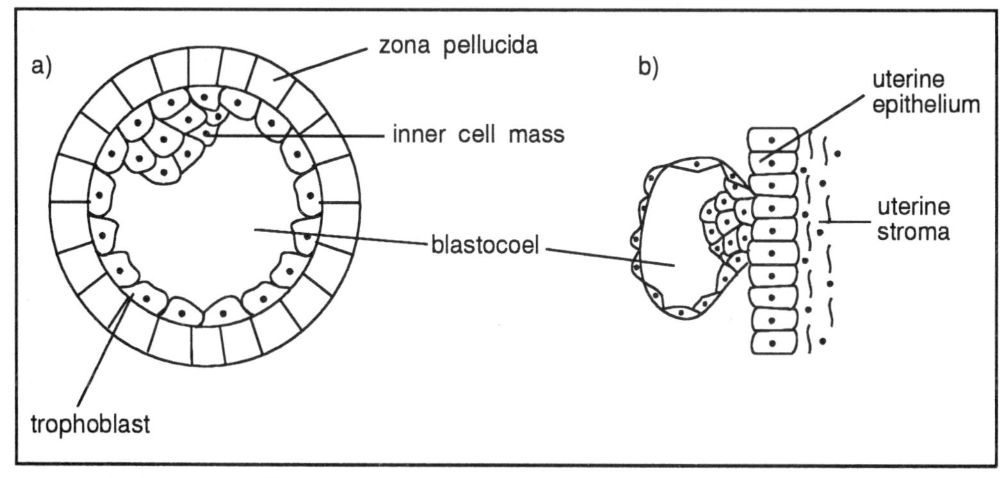

Figure 10.9 a) Diagram to illustrate the structure of a blastocyst. b) A blastocyst beginning to implant into the endometrium, (note the zona pellucida has been lost). Adapted from Tortora GJ and Anagostakos, N.P. (1990) Principles of Anatomy and Physiology Biology Science Textbook Inc, USA p931.

attachment of blastocyst to the endometrium

The period of time over which these events occur corresponds to days 14 to 21 of the menstrual cycle during which the uterine lining is being prepared by the actions of the two hormones oestrogen and progesterone secreted from the corpus luteum. The attachment of the blastocyst to the endometrium takes place 7 to 8 days after fertilisation and the process is known as implantation. Basically it consists of the blastocyst burrowing into the endometrium. The trophoblast cells overlying the inner cell mass are particularly sticky and it is this area which becomes attached to the endometrium (see Figure 10.9). The initial contact is followed by rapid growth of the trophoblast so that the cells penetrate between the endometrial cells. The endometrial cells also undergo many changes and provide the necessary nutrients for the further development of the inner cluster of cells which is destined to become the embryo. The blastocyst eventually becomes totally embedded in the endometrium (see Figure 10.10).

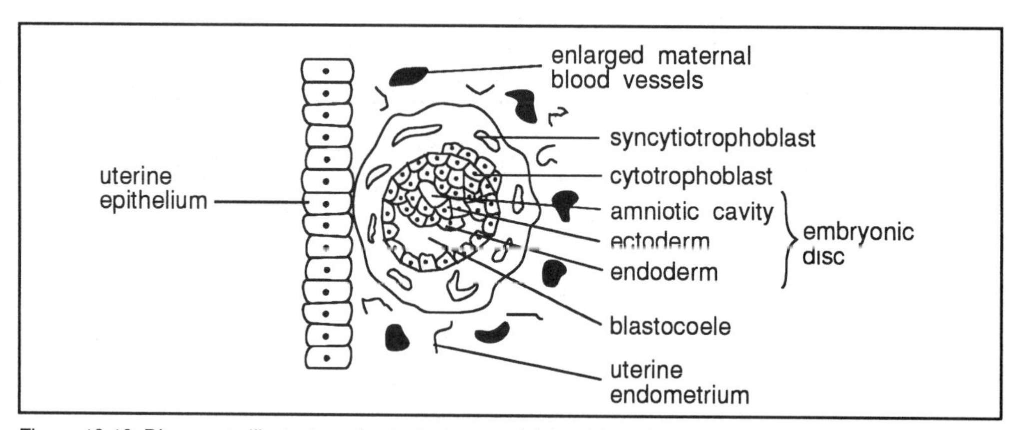

Figure 10.10 Diagram to illustrate an implanted ovum. Adapted from Starr and Taggart (1990), The Unity and Diversity of Life, Wadsworth Publishing Co. Inc, USA.

development of the placenta from chorion and endometrium

After about two to three weeks this simple nutritive system is inadequate for the growing needs of the embryo. Continuing development of the area of contact between the blastocyst and the endometrium gives rise to the placenta which takes over all future nutrition of the growing embryo and becomes the organ of exchange between embryo and mother until the end of the pregnancy. This process of placentation is well under way by the third week after fertilisation and very well established after five weeks. The foetal part of the placenta is derived from the outer layer of the trophoblast called the chorion, whilst the maternal part develops from the underlying endometrium. Fingerlike extensions of the chorion project into the endometrium, each containing a network of capillaries. These chorionic villi secrete enzymes which alter the composition of the endometrial cells overlying them so that each villus is surrounded by a pool (sinus) of maternal blood. The uterine blood vessels supply the sinuses whilst the umbilical vessels carry blood to and from the chorionic villi. The umbilical blood vessels are carried in the umbilical cord which connects the placenta to the embryo (see Figure 10.11). The placenta acts to pass nutrients to the embryo and to take waste products away, some of which move by diffusion, whilst others are carried by mediated transport mechanisms. It must be emphasised, however, that this is an exchange of substances as there is never (except as a result of a trauma of some kind) any mixing of the maternal and foetal blood. The final development is that of the amniotic cavity between the inner cell mass and the trophoblast. The membrane surrounding the cavity is the amnion which is also derived from the inner cell mass and eventually becomes fused with the inner chorion so that only one membrane surrounds the embryo. The cavity is filled with amniotic fluid which is identical with the embryonic extracellular fluid and acts as a buffer against temperature changes and mechanical disturbances.

The embryo floats in the amniotic cavity attached to the placenta by the umbilical cord and continues its development to a foetus (the name used after two months of growth) and, thence, into a viable infant, if all goes well.

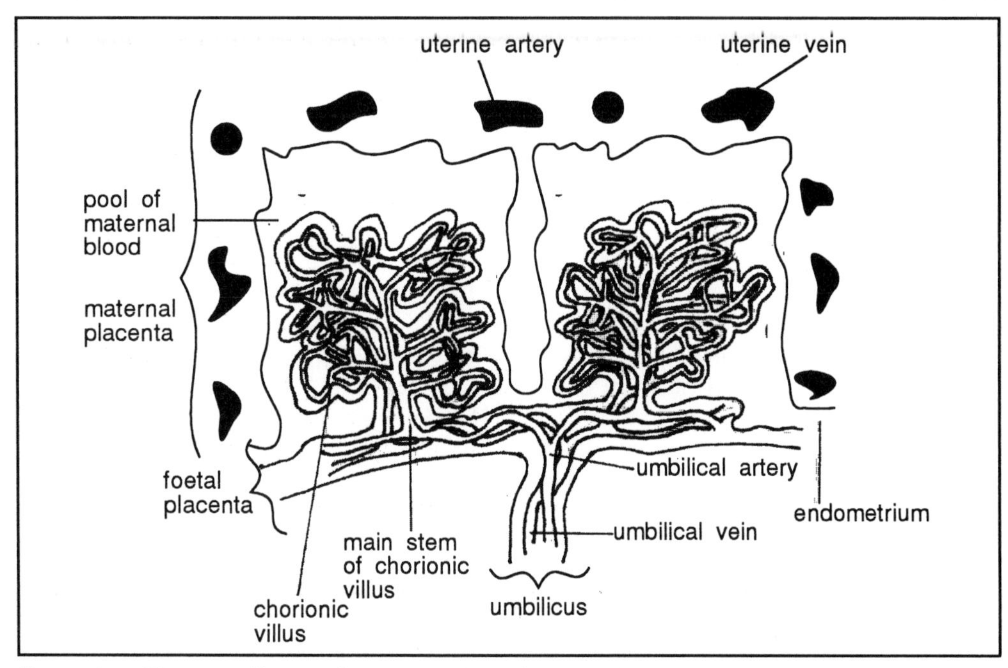

Figure 10.11 Diagram to illustrate the structure of the placenta. Adapted from Tortoria G.S. and Anagnostakos, N.P. (1990), Principles of Anatomy and Physiology, Biological Science Textbook Inc, USA p936.

amniocentesis and chorionic villus sampling

Foetal diagnosis can be achieved by sampling the amniotic fluid (amniocentesis) and by chorionic villus sampling. Until recently, the latter procedure was thought to offer advantages in that it could be performed at an earlier stage and, therefore, provide information prior to the 16 weeks needed before amniocentesis can be carried out.

However, there is now some evidence being put forward that chorionic villus sampling has led to a greater number of infants being born with some defect, possibly due to the sampling procedure itself. As a result of this some hospitals have stopped offering this procedure for foetal assessment.

In the next section we shall study, briefly, the organ development in the foetus.

10.5 Embryonic development

10.5.1 Beginnings of the organ systems

ecto-, meso- and endoderm

gastrulation

Following implantation, the inner cell mass of the blastocyte begins to differentiate into three separate germ layers: the ectoderm, the mesoderm and the endoderm from which all tissues and organs of the body will develop. The processes by which this occurs are known as gastrulation.

Gastrulation begins with the development of the amnion and amniotic cavity from the inner cell mass as described in the previous section. The layer of cells which lies closest to the amniotic cavity develops into the ectoderm, whilst those cells which border the blastocoel form the endoderm. These ectodermal and endodermal cells form the embryonic disc (see Figure 10.10). The mesoderm develops between the endoderm and ectoderm and later gives rise to the heart, muscles, bone and other connective tissue and organs. The endoderm, on the other hand, becomes the epithelial lining to many of the internal organs such as the respiratory tract and the gastrointestinal tract. Finally, the ectoderm will form the skin and nervous system. This rapid process of gastrulation begins in the second week after fertilisation and continues to the fourth week. The next four weeks involve the slower development of the main organs with the nerve cord, the heart and the respiratory organs forming, but not functioning. Fingers, arms, toes and legs also become visible. During the eight or nine weeks of organogenesis the developing embryo is very susceptible to damage by external factors. It is the time when the mother should be very careful to avoid any possible causes of harm to the embryo such as drinking excess alcohol, taking drugs and developing infections all of which may interrupt the normal sequence of events that are occurring.

After the first two months of development the new individual is referred to as a foetus. A period of foetal development occurs in which all the major organs are developed, bone is deposited in place of the cartilaginous embryonic skeleton and foetal movement becomes large enough to be felt by the mother. Development appears to be complete by the seventh month of gestation although infants born at this time do not often survive unless good quality medical facilities are available. The main reason for this is the lack of a particular chemical in the lungs (surfactant) which is essential for normal expansion of the lungs to facilitate breathing. Approximately 280 ± 14 days following fertilisation the fully developed foetus is ready for its entry into the world.

In the following sections we shall look at the way in which hormones control pregnancy and the physiological changes which they bring about.

SAQ 10.5

State which of the following statements are true and which are false. Give reasons for your answers.

1) The blastocyte consist of an outer layer of cells called the trophoblast, an inner cell mass and a central cavity.

2) The morula becomes embedded in the endometrium during implantation.

3) The placenta acts as an exchange mechanism whereby nutrients and waste are moved between the maternal and foetal circulations.

4) Maternal and foetal blood is mixed while passing through the placenta.

5) Gastrulation is the development of the stomach in the foetus.

10.6 Hormonal control of pregnancy

oestrogen, progesterone and chorionic gonadotrophin

Use Figure 10.12 to help you to follow the hormonal events during pregnancy. Both oestrogen and progesterone are produced throughout pregnancy to stimulate the growth of the uterine muscle, maintain the endometrium and promote development of the mammary glands in preparation for lactation. Initially, for the first 3-4 months, these hormones are secreted by the corpus luteum which, in turn, is maintained by the output of another hormone, human chorionic gonadotrophin, from the trophoblast. The human chorionic gonadotrophin (HCG) stimulates the continued production of the steroids by the corpus luteum until the placenta is sufficiently developed to take over the function. The level of HCG in the maternal plasma rises rapidly during these early months of the pregnancy (thereby providing the basis for pregnancy testing) and reaches a maximum at 60-80 days after the end of the last menstrual period.

Following this peak the levels fall to a low value which is maintained for the rest of the pregnancy. It has been shown that, after the first 3 months, the ovary is not essential for the maintenance of pregnancy, yet its removal during the first 3 months causes immediate abortion. This effect is entirely related to the important role of the corpus luteum.

oestrol as a diagnostic indicator

After the third month, the placental output of oestrogen and progesterone gradually increases, taking over from the corpus luteum during the fourth month. So from this point until the end of the pregnancy the placental hormones take over the management of the mothers body in preparation for the birth of the foetus (parturition). In fact, the placenta is only able to manufacture progesterone directly as it lacks the necessary enzymes for making oestrogen. These enzymes are found in the foetal adrenal gland and, therefore, the production of oestrogen is carried out by the placenta and foetal adrenal glands working together with various intermediates being transported between them and the foetal circulation. This is clinically very useful as measuring the levels of oestriol (the major oestrogen produced) in the maternal blood gives a good indication of foetal well being. If the level falls late in pregnancy it is likely that placental function is declining and an induced completion of the pregnancy is medically indicated in order to avoid compromise of the foetus.

As well as maintaining the endometrium and stimulating the growth of the uterine muscle and the mammary glands, another important function of the oestrogen and progesterone is to exert a negative feedback on the output of GnRH, LH and FSH from the hypothalamopituitary axis, thereby preventing further follicular development and ovulation.

human placental lactogen and breast development

Another important hormone produced by the placenta is human placental lactogen (HPL). Its production begins at about the same time as that of HCG but follows a different pattern. The output increases with placental mass so that as the placenta grows the amount of HPL produced increases until it reaches a maximum at about 32 weeks. This level is maintained until parturition. The function of HPL is thought to be involved in the growth of breast tissue in preparation for lactation and, also, in the alteration of maternal metabolism to support the additional needs of the foetus.

relaxin and parturition

Relaxin is secreted by the placenta as well as by the corpus luteum and this has the physiological role of relaxing the various ligaments in the pelvis to ease the passage of the foetus through the birth canal. The output and functions of these placental hormones is summarised in Figure 10.12.

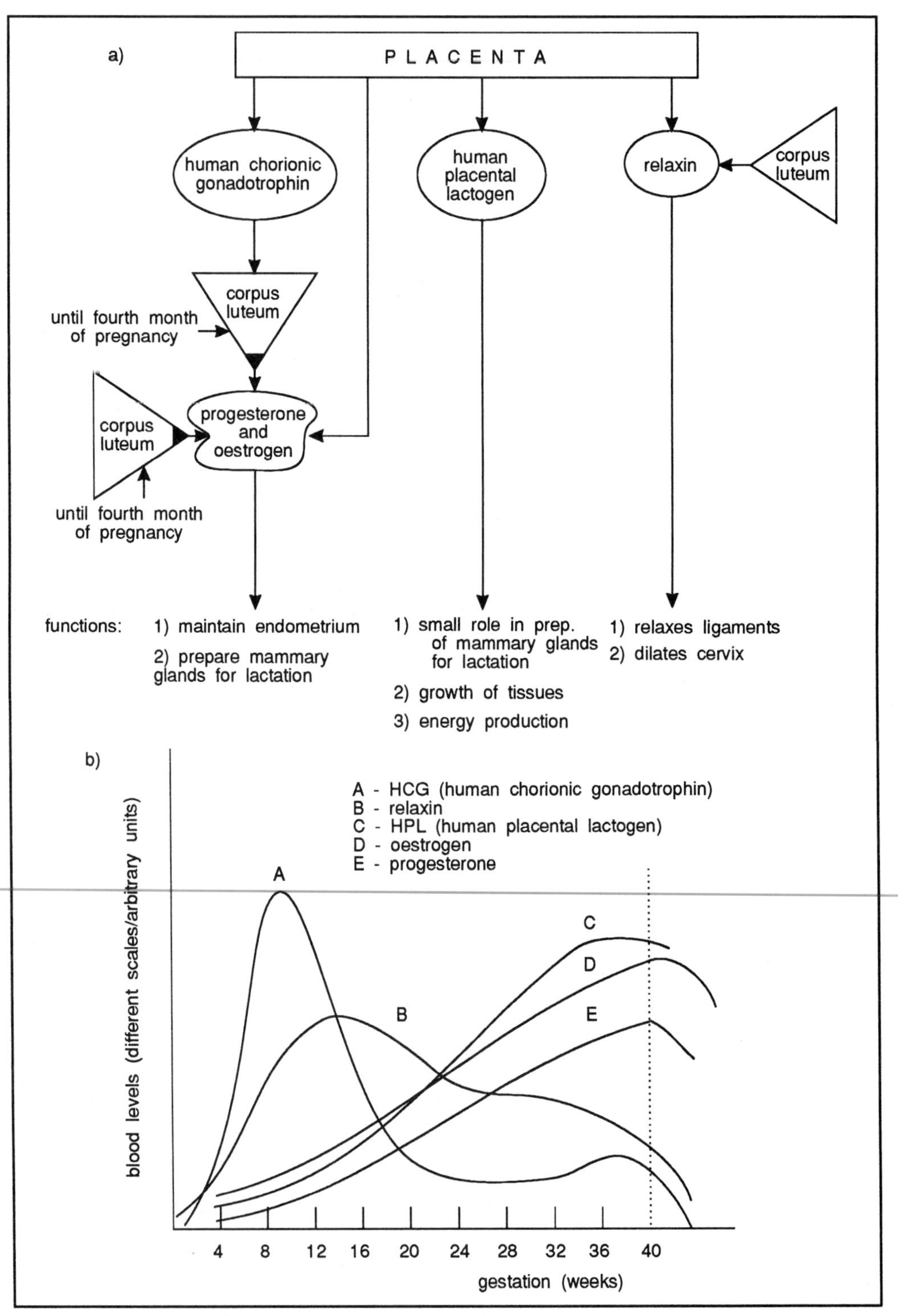

Figure 10.12 The hormones of pregnancy. a) Source and functions of the hormones. b) Output of the hormones during the period of gestation, (note, the relative amounts of the blood levels are not to scale). Modified from Tortora, GS and Anagnostakos, N.P. (1990), Principles of Anatomy and Physiology, Biological Science Textbook Inc, USA, p140.

10.7 Physiological changes in pregnancy

The period of time that the zygote, embryo and foetus is developing inside the uterus is known as gestation. In the human this lasts for about 280 days and a number of anatomical and physiological changes occur throughout this period.

changes in the abdomen

The most profound anatomical changes are caused by the expanding uterus which fills the pelvic abdomen by the end of the third month. After this it can be palpated above the symphysis pubis. It continues to grow higher and higher in the abdomal cavity until, at term, it fills, virtually the whole of the cavity and almost reaches the bottom of the sternum. The maternal intestines, liver and stomach are pushed aside to accommodate it, the diaphragm is elevated and the thoracic cavity is widened. The ureters and bladder in the pelvis become compressed and this may result in altered function of the maternal urinogenital system in the later stages of pregnancy.

The physiological changes are very widespread and involve most systems in the body. There is, obviously an increase in weight which is due, not only to the increasing size of the uterus and the foetus, but also to the amniotic fluid, the enlarged breasts and to the increased water and food stores found in the mother's body.

lung function changes

Lung function is altered in the mother to allow a 10 to 20% increase in oxygen consumption. The expiratory reserve volume and functional residual capacity are both decreased to a marked degree. This means that more of the total lung capacity is actually used for normal breathing than in the non-pregnant state. In addition to this the airway resistance in the bronchi is reduced so that air can flow more freely and tidal volume is increased by as much as 30 to 40%. The overall effect is that more air enters the lungs each minute (minute volume) and, therefore, more oxygen is provided to cover the needs of both the mother and the foetus. At the same time more carbon dioxide can be removed. However, breathlessness (dypsnoea) may occur. Sometimes the cause is not apparent, but usually it is due to the high position of the uterus in the abdomen which makes the act of inspiring (breathing in) physically difficult.

dypsnoea

changes to cardiovascular system

blood volume, sodium, water retention, cardiac output increased, blood pressure

The cardiovascular system also undergoes some profound changes during pregnancy. Blood volume increases because of an increase in both the volume of the plasma by retention of sodium and water and in the total number of erythrocytes (red cells). The heart increases the amount of blood it pumps around the body. The heart rate and stroke volume both increase with the result that, by the beginning of the third trimester, the cardiac output increases by about 20 to 30%. This allows for the extra blood that is required to supply the placenta and to service the increased metabolic demands of the tissues of both the mother and the foetus. Blood pressure usually remains the same, if not slightly reduced. This is attained by a decrease in the total peripheral resistance brought about by vasodilation occurring in the blood vessels supplying the uterus, skin, breasts, gastrointestinal tract and the kidneys. Combined with the increase in cardiac output, the fall in peripheral resistance keeps arterial blood pressure constant. Sometimes, the blood pressure does rise and, if uncontrolled, this can lead to retention of fluid and may progress to the serious condition of toxaemia of pregnancy from which epilepsy and maternal death may occur.

The gastrointestinal tract is also affected by pregnancy. There is an increase in appetite to ensure that the mother takes in sufficient food to supply energy for the needs of the foetus as well as herself. The motility of the tract is usually decreased and this often gives rise to constipation and delayed gastric emptying. Nausea, vomiting and heartburn are common problems associated with pregnancy.

bladder compression

Compression of the bladder in the pelvis causes such common urinary symptoms as frequency, urgency and stress incontinence, all of which become more severe as the pregnancy progresses and the uterus gets larger. In addition the amount of plasma which the kidney filters is increased because both the blood flow to the kidney (renal plasma flow) and the glomerular filtration rate are increased.

pigmentation and striae development in skin

Many changes are also seen in the skin and in the reproductive tract itself. Often there is increased pigmentation of the skin, particularly around the areolae of the breasts, around the eyes and cheekbones (chloasma) and down the centre of the lower abdomen (linea nigra). Stretch marks (striae) may develop around the abdomen as it increases in girth and there is an alteration in the balance of hair growth and loss. In the reproductive tract, as well as the obvious changes in the size of the uterus, the vulva increases in size and vascularity and the vagina becomes more pliable and vascularised. These are all in preparation for parturition.

In addition to these specific changes there are also alterations in the output of many hormones in the mother and in the general metabolic processes leading to the provision of energy supplies to the tissues of both mother and foetus.

The breasts enlarge and develop glandular tissue and ducts leading from these to the nipple. These tissues, after birth, initially produce colostrum and then milk, which is passed to the infant during suckling.

All these physiological changes in the various systems of the mothers body ensure that a perfect environment is produced in which the developing foetus can survive and grow to maturity, whilst at the same time being non-detrimental to its host.

Π Fill in the gaps in the following passage using the information in the preceding two sections.

During the first three months of pregnancy the [ ] [ ] secretes oestrogen and [ ] which function to stimulate the growth of the endometrium and the mammary glands. After the fourth month the [ ] takes over this role and continues to produce these hormones in ever increasing quantities until the end of the pregnancy. In order to supply the placenta with sufficient blood to perform this function the heart [ ] its cardiac [ ] by as much as 30%. Blood pressure remains the same because the [ ] [ ] [ ] falls as a result of vasodilation in many blood vessels. The increased amount of oxygen necessary is provided by an increase in the [ ] [ ] (the amount of air entering the lungs each minute). Air can enter the lungs more easily because there is a lowering of the [ ] [ ] and more of the lung capacity is used as indicated by a decrease in both the functional [ ] [ ] and the [ ] reserve volume. Other physiological changes include a [ ] in gut [ ] and an increase in the [ ] filtration rate and [ ] [ ] [ ] in the kidney.

The words you should have used are: corpus luteum; progesterone, placenta, increases, output, total peripheral resistance, minute volume, airway resistance, residual capacity, expiratory, decrease, motility, renal (kidney), renal plasma flow.

10.8.4 Parturition

The processes by which the foetus and the placenta are delivered are known as parturition or more commonly, 'labour'. It is brought about by strong rhythmical contractions of the uterus and usually occurs after 280 days or 40 weeks of gestation. In fact weak, uncoordinated uterine contractions begin at around 30 weeks and these gradually increase in strength and frequency up to term. Towards the end of the pregnancy a softening of the cervix occurs. This is initiated by the hormone relaxin produced by the corpus luteum. During the last month of pregnancy the foetus moves down in the abdominal cavity until it comes into contact with the cervix in readiness for birth. This process is often referred to as quickening.

role of relaxin

The actual stimulus for labour is, as yet, not entirely known. Many factors appear to contribute to the start of events. These include:

- a decrease in the ratio of progesterone to oestrogen in the maternal plasma so that the increased levels of oestrogens are sufficient to overcome the inhibitory effects of progesterone on uterine contractions;
- oxytocin is secreted from the posterior pituitary and stimulates the contractions of the uterine muscle. The release is reflexly stimulated by the excitation of receptors in the uterus. The number of receptors also increase markedly during pregnancy, possibly as a result of oestrogen influence and uterine stretch nearer term;
- a prostaglandin is secreted by the uterus at the end of pregnancy and acts as a powerful stimulator of the uterine muscle. The output of this may be stimulated by oxytocin;
- the stretch of the uterine muscle itself may result in an increase in the strength and coordination of the contractions. (This is an inherent property of smooth muscle of which the uterus is composed).

uterine contractions, rupture of amniotic sac and dilation of the cervix

All these factors work together to bring about a sudden increase in the strength and coordination of the uterine contractions which begin at the top of the uterus and pass downwards. At, or just prior to this time, the amniotic sac ruptures and amniotic fluid is released into the vagina (a show). For the first stage of labour the contractions result in the dilation of the cervix to a maximum of 10 cm. Once full dilation is achieved, the second stage of labour ensues. During this phase the uterine contractions force the foetus to pass down through the cervix into the vagina, helped by the pushing efforts of the mother. Following delivery, the umbilical and placental blood vessels constrict thereby stopping blood flow from the placenta to the foetus. The foetus is, therefore no longer provided with oxygen and its CO_2 levels rise. This stimulates the respiratory medullary centres to initiate inspiration and the newborn infant takes its first breath. The placenta becomes separated from the uterine wall and along with the umbilical cord, is expelled from the vagina by another series of strong uterine contractions (the afterbirth).

Normally, in the majority of births, these events follow an automatic and trouble free pattern. However, in various instances, some medical intervention is required in order to deliver the infant safely and preserve the life of the mother. Such occasions include feet-first (breech) delivery, failure of labour to progress, distress of the foetus and exhaustion of the mother, to name but a few. Usually, there is a happy conclusion to the initial meeting of the sperm and ovum at fertilisation.

SAQ 10.6

Construct a flow diagram to show the factors influencing the process of parturition.

10.9 Prenatal diagnostic tests

There are several test available to detect genetic disorders in the foetus and to assess the wellbeing of the foetus *in utero*. We have already mentioned one or two of these in preceding sections. In this section we shall look at some of the more commonly used investigative techniques.

10.9.1 Amniocentesis

Downs Syndrome, muscular dystrophies, sickle cell anaemia

This involves the removal of a sample of 10-20 ml of the amniotic fluid by way of a hypodermic needle introduced through the maternal pelvic abdomen. Palpation and ultrasound (see later) are used to determine the position of the foetus and placenta before insertion of the needle. The sample of fluid is then subjected to microscopic, biochemical and chromosomal examination. The test is usually carried out at weeks 14 to 16 in women who are either over 35 years or known to be carriers (or their spouse) of certain genetic disorders. Amniocentesis is sometimes performed where there is a medical condition that requires early delivery. Typical genetic disorders which may be detected by this method include Trisomy 21 (Downs Syndrome), some muscular dystrophies and sickle cell anaemia. As the chromosome makeup (karyotype) of the foetus is examined in amniocentesis its gender is determined. One possible complication of this test is that it may induce a miscarriage.

10.9.2 Chorionic villus sampling

incidence of sampling and foetal disorders

As mentioned in previous sections, this technique can be performed earlier than amniocentesis, usually at 8-10 weeks of gestation. Ultrasound is again used, this time to guide a probe through the vagina and cervix into the uterus where a 30 mg sample of tissue from the chorionic villus is taken and subjected to chromosomal analysis. The procedure, therefore, avoids penetrating the abdomen or uterine wall or amniotic cavity. However, recent medical evidence has shown that these tests have been associated with a greater incidence of foetal abnormalities than might have been expected. It would appear that there is some danger of removing foetal tissue as well as placental tissue and this, in some instances, leads to malformations in the developing infant. Some medical establishments have stopped using this technique until more data is available to show that the risk of such eventualities is negligible.

10.9.3 Ultrasound examination of the foetus

Ultrasonography involves the passage of a probe, which emits high frequency sound waves over the abdomen of the pregnant women. The sound waves bounce off the foetus and are transduced to a visual image on a screen. Using a full bladder as a landmark, the size and development of the foetus may be monitored and an estimate of its gestational age obtained. It is a useful technique for determining the position of the foetus and also for ascertaining the presence of multiple foetuses. In Great Britain ultrasound is used routinely to confirm pregnancy and estimate probable delivery dates. It is felt to be a harmless, noninvasive technique. However, in America it is only used for specific medical indications.

10.9.4 Alphafoetoprotein (AFP)

Alphafoetoprotein (AFP) is a substance produced by foetal liver cells as well as liver, ovarian and testicular cells in adults. It is measured from a blood sample taken from the mother at 15-16 weeks gestation and is usually used to indicate the possible occurrence of neural tube defects in the foetus. Positive results are followed up by further investigations, such as ultrasonography or amniocentesis.

10.10 *In vitro* fertilisation

The first recorded birth of a test tube baby ie one who was born as a result of *in vitro* fertilisation, was on July 12th, 1978 near Manchester in the UK. For those couples who find the woman cannot conceive, this method (IVF) offers hope of a normal pregnancy. There are many reasons why conception may be difficult or impossible and the cause can lie with either the man or the woman, eg the sperm count of the male may be too low to be affective; the woman may not ovulate or her reproductive tract may present hostile conditions for the ejaculated sperm. Using *in vitro* fertilisation, the woman is given hormone treatment to control her menstrual cycle and to ensure the production of several secondary oocytes, rather than the usual single one. These oocytes are collected by laparoscopy (an operation in which a laparoscope is inserted into the abdomen via a small incision near the umbilicus) and are placed in a medium that simulates the fluids of the female reproductive tract. This medium also contains the partner's sperm and fertilisation can take place. Once fertilisation has occurred, the fertilised ovum is transferred to another solution which allows cleavage to occur. Two to four days later (when the 8 to 16 cell stage has been reached) the ovum is introduced into the females uterus so that implantation and further growth can take place. In 5 to 10% of cases the transplant will be successful and pregnancy will continue to completion. It is also possible to freeze the ova or sperms to enable future attempts to be made without having to collect further gametes.

ova collected by laparoscopy

embryo transfer

A further type of IVF is embryo transfer in which the males sperm is used to inseminate artificially, a fertile female donor. Once fertilisation has been successfully achieved, the morula or blastocyte is transferred to the uterus of the infertile female partner who carries the foetus to term. In this technique the menstrual cycles of the two women have to be synchronised to ensure the blastocyte is transferred at the time when the uterus is prepared for implantation. Embryo transfer is usually reserved for women who are infertile or who have irretrievable blocked fallopian tubes or who are likely to pass on genetically inherited disorders to their own offspring.

Gamete intrafallopian transfer (GIFT) is yet another version of IVF where secondary oocytes are aspirated from the female partner, mixed with a solution of the male's sperm and then immediately transferred to her fallopian tubes.

Ova and spermatozoa can be donated and frozen for use by infertile couples in any of the above procedures. The supply of these has recently been threatened by the introduction of the Human Fertilisation and Embryology Act 1992 under which all donors have to give personal details to be kept on computer files. Previously, it has been an anonymous procedure which ensured that only the physical characteristics, but not the identity, of the donor were known. Although the law states that only licensed personnel may have access to the information, the loss of anonymity has discouraged a lot of potential donors.

10.11 Contraception

oral contraceptives and the alteration of feedback control of GnRH

Opposite to the problem of infertility are those of limiting the ability of the sperms to successfully fertilise an ovum. This is the realm of contraception or birth control and many different methods are available. In this section we shall briefly look at a few of these techniques beginning with the oral contraceptive pill. This is a hormonal control of fertility which involves manipulation of the menstrual cycle and alteration of the preparedness of the female reproductive tract. The contraceptive pill consists of varying concentrations of oestrogens, and progesterone, the most commonly used containing either a high concentration of progesterone combined with a low concentration of oestrogen or progesterone alone. These contraceptive pills function by altering the feedback control of gonadotrophin releasing hormone (GnRH) so that its secretion from the hypothalamus is inhibited. This, in turn, results in a decrease in the amounts of FSH and LH secreted from the pituitary so that ovulation is usually prevented. In the event that ovulation does occur, the cervical mucous is rendered hostile to the sperms and the uterus is not receptive for implantation. This combination of events ensures that fertilisation is an unlikely occurrence following sexual intercourse. The success rate of the contraceptive pill is about 99.8% but it is associated with a number of side affects, some of which have only come to light as a result of statistical studies following many years of use of oral contraceptives by fertile women. Today its use is preferable in women under 40 who do not smoke and who have no history of blood clotting disorders. It is thought to confer some protection against ovarian and uterine cancer, but may be associated with some risk of breast cancer. There is thought to be a possible risk of infertility in women who have been taking the pill for many years from an early age. Nevertheless, its use is associated with a lower fatality rate than pregnancy itself.

Oestrogen and progesterone can also be taken orally once a month, either separately or together, as a pill on about day 22-25 of each cycle. The storage and subsequent slow release of the oestrogen prevents ovulation. In addition to this, injection and implants of steroids have also been tried eg capsules of progesterone under the skin or intrauterine or intravaginal implants or injections, all of which can be effective for 1 year or more. The latter method has been used in the third world as a potentially controllable way of providing contraception on a large scale.

intrauterine devices (IUD)

Another method of contraception is provided by the use of intrauterine devices (IUD). These small devices, which are usually made of copper, plastic or stainless steel, are inserted into the cavity of the uterus via the vagina. They are thought to function by preventing implantation of the blastocyte in the endometrium but the exact mechanism is unknown. This likely abortifacient mechanism of action makes this particular method morally unusable by certain religious groups. An alternative method of birth control is the rhythm method whereby sexual intercourse is avoided during the days immediately surrounding the ovulation time in the woman's menstrual cycle. The difficulty with this technique is knowing when ovulation is likely to occur as most women do not have a strictly regular cycle such that the event can be accurately predicted. Usually body temperature is routinely monitored and ovulation is taken as the time when basal temperature rises. Also, ovulation is associated with other changes such as the production of abundant, clear, elastic mucous, softening and elevation of the cervix and sometimes pain. If intercourse is avoided when these conditions exist then the chances of fertilisation are markedly reduced.

rhythm method of contraception

cervical cap, diaphragm, spermicides, condom

Fertilisation can also be prevented by the use of barriers, either physical or chemical or both, which are designed to stop the initial meeting of the sperms and ova in the female genital tract. The barrier can be a cervical cap or diaphragm both of which are designed to fit over the cervix and have to be fitted, initially, by a physician to get a correct fit. They may be left *in situ* for 24 to 48 hours after intercourse and are most effective when used in conjunction with a spermicide. The latter is a chemical which disrupts the plasma membranes of the sperms, thus killing them. The spermicide agent can be in the form of a foam, a jelly, a suppository or a douche. The barrier method has a failure rate of about 10 to 20%, failure usually resulting from incorrect fitting or displacement of the cap or diaphragm during intercourse. A third barrier method is the condom which is a nonporous, elastic covering for the erect penis used to prevent the release of the ejaculated sperm into the vagina. This is a particularly important method of helping to prevent the spread of sexually transmitted diseases including HIV (human immunodefficiency virus) which results in AIDS (acquired immunodefficiency syndrome).

coitus interruptus

Coitus interruptus (withdrawal) is a further contraceptive method which refers to the withdrawal of the penis from the vagina before ejaculation occurs. However, the failure rate of this method is quite high due to the inherent unreliability of the male's ability to succeed in his intentions or to the release of sperm containing fluid from the urethra prior to ejaculation.

sterilisation by tubal ligation or vasectomy

A further method to be considered in this section is that of sterilisation. This is a surgical technique which involves tying off the fallopian tubes in females (tubal ligation) or the severance of the vas deferens in males (vasectomy). A possible disadvantage of this method is that it is generally held to be irreversible, so a change in the circumstances of the subjects could lead to future disappointment.

induced abortion

Should any of these methods fail or not be put into use one further way of preventing pregnancy from continuing to term is by an induced abortion. This is the removal of the foetus by various physical or hormonal methods. The time limit for this procedure has recently been set at 20 weeks as it is now possible for a foetus of 23 weeks to survive. Most induced abortions would be performed well before this time, but it should never be seen as a routine method of contraception.

A 'morning after' pill has also been developed to prevent pregnancy following the act of sexual intercourse. This method involves the use of a very large dose of oestrogen or progesterone taken within 72 hours of coitus. The progesterone stimulates endometrial growth, but this is followed by the breakdown of the endometrium when the high levels of the hormone are not maintained. Thus, implantation is prevented, irrespective of whether fertilisation has taken place or not. Similar effects are brought about by the large dose of oestrogens. However, the effectiveness and acceptability of this method remains to be fully evaluated.

GnRH antagonists

A novel method of contraception being currently investigated is that of using antagonistic peptide analogues of gonadotrophin releasing hormone (GnRH). These peptides, which are similar in structure to GnRH, can be synthesised and when administered bind to GnRH receptors in the pituitary but do not stimulate the subsequent release of LH and FSH. Thus, they effectively block the activity of any circulating GnRH. The timing of the administration of these peptides have yet to be studied but, potentially, this is a very exciting development in the field of contraception.

Π Which method of administration of such peptides would you not be able to use and why not?

These could not be administered orally as they would be degraded in the digestive tract before they were absorbed.

SAQ 10.7

Which of the following treatments, theoretically, could have contraceptive affects?

1) One injection of LH.
2) One injection of FSH.
3) Oral adminstration of a GnRH antagonist (a peptide).
4) Daily injection of a GnRH antagonist.
5) Daily oral administration of progesterone and oestrogen.
6) Implantation of progesterone which would be slowly released.

Summary and objectives

This has been a long chapter in which we have described the processes of human reproduction. This has included a description of the organs involved. A key point in understanding reproduction is to understand the roles of hormones in regulating reproduction processes.

Now that you have completed this chapter you should be able to:

- describe the basic anatomy of the male and female reproductive tract;
- describe the processes involved in spermatogenesis and oogenesis;
- explain hormonal control of male reproductive function;
- explain hormonal control of female reproductive function;
- describe the mechanism of fertilisation, the early development of the zygote and the implantation of the fertilised ovum into the uterus;
- give a brief outline of embryonic development and explain the hormonal support of pregnancy;
- list the physiological changes associated with pregnancy;
- describe parturition;
- explain how prenatal diagnostic tests and contraceptives work;
- explain the processes and potential of *in vitro* fertilisation.

Responses to SAQs

Responses to Chapter 1 SAQs

1.1

You probably found one or two of these difficult because of the rather concise descriptions of their functions. This was deliberate and was to emphasise the overlapping roles of many of the systems.

System	Representative organs	Functions
Integumentary	Skin, hair, nails	Protects body, helps to regulate body temperature, receives environmental stimuli
Skeletal	1	20
Nervous	9	13
Muscular	10	19
Endocrine	8	14
Cardiovascular	7	12
Lymphatic	6	16
Respiratory	5	18
Digestive	4	15
Urinary	11	3
Reproductive	2	17

1.2

1) Adipose tissue: connective tissue. A rather unusual examples, in that the cells are often touching, but it certainly does not fit any of the other types.

2) Tissue that lines body cavities: epithelium. Also covers body surfaces and some organs.

3) Blood: connective tissue. Again, rather specialised, but the cells are widely spaced and there is much intercellular material.

4) Tissue whose cells actively secrete substances into the blood: glandular epithelium, the only actively secreting cells.

5) Tissue with a rich blood supply, widely scattered cells and much intercellular material: connective tissue.

6) Tissue containing cells that produce fibres in the intercellular material: connective tissue. (Note that muscle cell fibres are intracellular).

1.3

1) True. The epithelium lining the intestine is involved in active absorption of nutrients.

2) False. Exocrine glands secrete into ducts or directly onto surfaces. They may exist as single cells. Endocrine glands secrete into the blood.

3) False. Smooth muscle cell is usually involuntary, for example you do not voluntarily control the contraction of blood vessels.

4) True.

5) False. Skeletal muscle is voluntary.

6) False. They stimulate separate cells individually, so that a variable number of cells contracts according to need.

7) False. Cardiac muscle has to contract the heart, so is arranged as a network of fibres.

8) False. Axons conduct away from the cell bodies, dendrites conduct electrical impulses towards the cell bodies.

9) True. Examples are blood, bone.

10) True. These neurotransmitters are released from nerve endings when the electrical impulse reaches them.

Responses to Chapter 2 SAQs

2.1

The Nernst equation is $E_m = \frac{RT}{zF} \ln \frac{[C_o]}{[C_i]}$

In this situation: $[C_o] = 100$ mM; $[C_1] = 150$ mM.

Therefore, $E_m = \frac{310 \times 8.3146}{96486.6} \ln \frac{100}{150} = -0.0108 \text{ V} = -10.8 \text{ mV}$

It demonstrates that the K^+ concentration gradient between the interior and exterior of the neuron is important in determining the resting membrane potential. Lower the concentration gradient of the ions on each side of the membrane and you reduce the resting membrane potential.

2.2

Several experiments may be done. Probably the easiest in to replace the extracellular Na^+ ions, with another positively charged substance but one which is unable to enter the neurone and then measure the resting membrane potential. Such a substance is choline. Since it is unable to enter the neuron, the resting membrane potential is only dependent on K^+ ions and therefore becomes equal to the potassium equilibrium potential.

2.3

Inhibition of the Na^+-K^+-ATPase pump will result in the running down and abolition of the ionic gradients across the neuronal membrane. Thus there will be a loss of the resting membrane potential. It is analogous to a battery running out.

2.4

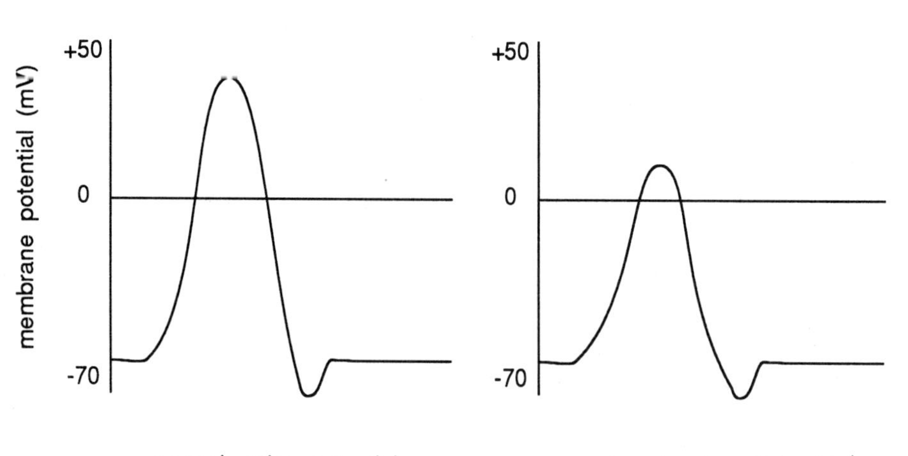

When extracellular Na^+ is reduced, the height of the action potential is reduced, less charge may flow into the neuron.

2.5 Your response should have included at least four of the properties listed below.

Action potential	**Local potential**
All or nothing response	Potential may be graded
Always depolarising	May be depolarising or hyperpolarising
Has a threshold	Has no threshold
Has a refractory period	Has no refractory period
Conducted without decrement	Conducted decrementally
Cannot be summed	Can be summed

2.6

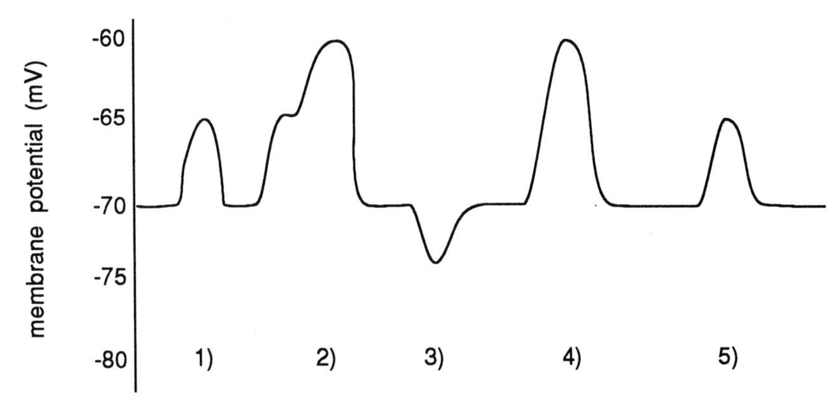

All that we are doing is generating an EPSP of +5mV (with inputs A and B) and an IPSP of -5mV (with input C). In 2) above we have temporally summated two EPSPs. In 4) we have spatially summated two EPSPs. In 5) we have spatially summated two EPSPs and one IPSP to produce a single EPSP.

2.7 The four ways in which the magnitude of the generator potential can be varied include:

- changing the intensity of the stimulus;
- altering the rate of change of the stimulus intensity;
- by the length of time the stimulus is applied (adaptation);
- by applying several different stimuli (summation of individual generator potentials).

You may also have included administration of pharmacological agents which block the transduction of the stimulus into a change in membrane potential (eg Lidocaine which block ion channels in pain fibres).

2.8 The fastest reflexes are the monosynaptic reflexes. This is because in any neural system, the principal sites of delay are the synapses. Therefore, increasing the number of synapses within a reflex increases the time delay between stimulation and response.

2.9

1) Brain stem nuclei.

2) Ascending pathways.

3) Level 2.

2.10 Thus was not a simple question to answer mainly because many parts of the brain play a role either directly or indirectly in the functions described. Nevertheless here we were looking for the principal parts involved.

1) The cerebellum is regarded as the centre controlling the maintenance of posture.

2) The medulla oblongata contains the centre which predominantly controls heart beat rate (via the parasympathetic system (see Figure 2.14).

3) The hypothalamus has a major role to play in the control of hormone secretion.

4) The medulla oblongata contains the centre which controls peristalsis. Nerve impulses are sent by the vagus nerve (see Figure 2.14).

5) Initation and planning is undertaken by the cortex of cerebrum.

Responses to Chapter 3 SAQs

3.1

Nervous system: communication is rapid, of short duration, acts at precise locations and over short distances. It co-ordinates activities involving rapid adjustments eg movement.

Endocrine system: communication is slower in onset, of longer duration and acts over longer distances often involving widely scattered cells. It co-ordinates long term adjustments eg metabolic processes.

3.2

1) False. Neurotransmitters are secreted into the synaptic cleft between the pre- and postsynaptic neurones, not into the bloodstream.

2) True.

3) False. A neurohormone is secreted from a nerve ending not from an endocrine gland.

4) True.

3.3

1) False. It is produced in the anterior lobe.

2) True. See Figure 3.5.

3) False. TSH release is inhibited by thyroxine, not by testosterone.

4) True (see Figure 3.4).

3.4

Your table should contain the following points:

Peptides Hormones	Neurotransmitters
hormone stored in vesicles.	transmitter stored in vesicles.
release caused by stimulus to the gland cell.	release caused by stimulus to the nerve.
influx of calcium involved.	influx of calcium involved.
release of hormone by exocytosis.	release of transmitters by exocytosis.
synapsin I not involved.	synapsin I is involved.

3.5

The correct order is 5, 6, 2, 3, 4, 1.

3.6

1) True.

2) False. Steroid hormones are synthesised as required; they are not stored in the cell.

3) True.

4) False. Steroid and thyroid hormones do not require calcium ions for their release.

5) True.

3.7 No 3 is the correct answer. If a cell with receptors for a given hormone is exposed to a high plasma level of that hormone for a prolonged time, there will be a reduction in the number of receptors on the cell.

3.8 The correct order is 2, 6, 4, 5, 7, 1, 3 or 2, 6, 4, 5, 7, 3, 1.

The reason we have given two sequences is that events 1 and 3 can occur simultaneously. They are independent of each other.

3.9 No, the response would not be completely abolished because the intracellular calcium levels could still be increased by being released from intracellular stores. This would be mediated by inositol triphosphate.

3.10

1) False. IP_3 acts to increase the cytosolic concentration of calcium ions by releasing them from intracellular stores.
2) True.
3) False. Phosphodiesterase inactivates cAMP only.
4) False. The different transduction mechanisms are mediated by different G-proteins.
5) True.

3.11

A	cAMP	Maximum responses are achieved when adenylate cyclase is activated by forskolin and when theophylline is added. Both these would result in an increase in the concentration of cAMP.
B	Calcium ions	The ionophore, where given, results in a maximum response ie when the concentration of calcium ions is increased.
C	cAMP+ calcium ions	Only 25% of the maximum response is achieved when agents altering either cAMP or calcium ion levels are given individually. When the theophylline and the ionophore are given together, a full response is seen, indicating that both second messenger systems act synergistically to give the full response in this case.

3.12 Your flow diagram should contain the following information.

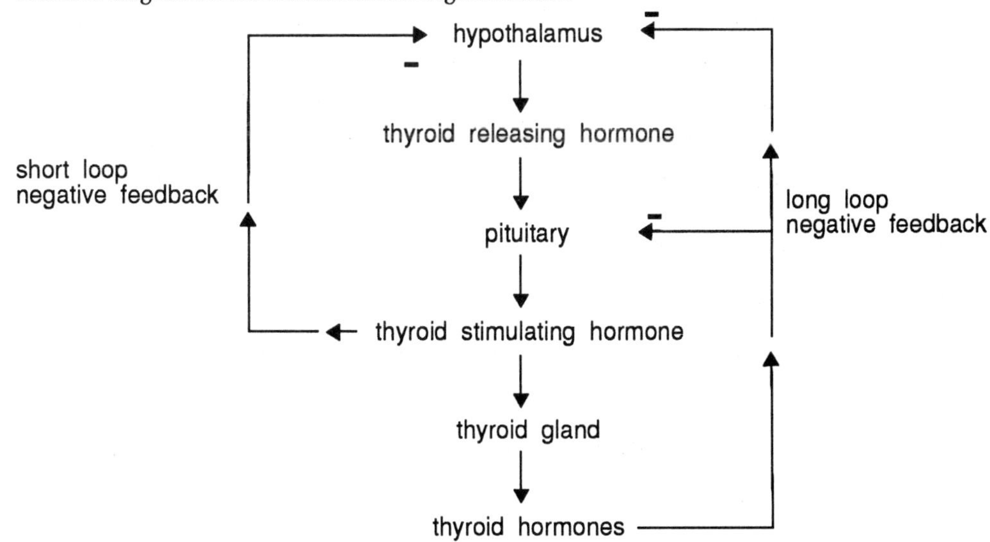

Responses to Chapter 4 SAQs

4.1

Salivary gland 5, 6, 7.

The saliva secreted by the salivary gland does contain significant amounts of an enzyme, but no digestion occurs in the gland, and very little in the mouth because food is not there for long enough. The saliva pH is buffered and helps keep the pH of the mouth near neutral.

Oesophagus 1, 7.

All parts of the gastrointestinal tract secrete mucus.

Stomach 2, 4, 5, 6, 7.

Although some digestion occurs in the stomach it is insignificant compared with that occurring in the small intestine.

Small intestine 3, 5, 7, 9, 10.

Large amounts of water and electrolytes are absorbed by the small intestine, as well as carbohydrates, fat and amino acids.

Large intestine 7, 10, (3).

How much chemical breakdown of food is carried out by the large intestine depends very much on the amount and type of fibre in the diet.

Liver 8.

The liver is the site of synthesis of bile.

Pancreas 5, 6.

The pancreas secretes a range of enzymes, but also bicarbonate to increase the pH of the acid chyme arriving from the stomach.

4.2

Enzyme	Substrate	Product
1) Salivary amylase	starch, glycogen and other α1-4 linked glucose polymers	branched oligosaccharides
2) Pepsin	proteins	polypeptides
3) Trypsin	Protein and polypeptides	small peptide
4) Chymotrypsin	Protein and polypeptides	peptides
5) Carboxypeptidase	Protein and polypeptide	peptides
6) Ribonuclease	RNA	nucleotides
7) Pancreatic amylase	starch and other α1-4 linked glucose polymers	maltose & dextrose
8) Lipase & colipase	triglyceride	monoglycerides, fatty acids & glycerol
9) Aminopeptidase	polypeptides	smaller peptides and amino acids
10) Dipeptidase	dipeptides	amino acids
11) Sucrase	sucrose	glucose and fructose
12) Lactase	lactose	galactose and glucose

Explanatory notes

1) Starch or other digestible (α1-4 linked) polysaccharides such as glycogen. Cellulose and some other polysaccharides cannot be digested because of the substrate specificity of the amylase.

2) Proteins. Pepsin is the first proteolytic enzyme to be secreted, and brings about some digestion of protein in the stomach.

3) Small polypeptides. Trypsin is one of a group of proteolytic enzymes secreted by the pancreas, whose products are small polypeptides rather than single amino acids.

4) As for 3.

5) As for 3.

6) The substrates for ribonuclease are nucleic acids.

7) Starch (see 1).

8) Fat (triglyceride). To be fully effective, bile salts are necessary as well.

9) Smaller peptides and amino acids. This enzymes removes amino acids from the NH_2 end of polypeptides.

10) Amino acids. A dipeptide is only composed of two amino acids.

11) Glucose and fructose are the constituent sugars of the disaccharide sucrose.

12) Lactase is the enzyme which specifically hydrolyses lactose to galactose and glucose. Although present in nearly all infants, a very significant number of adults lack this enzyme, are therefore unable to digest lactose and are consequently intolerant of lactose containing foods such as milk.

4.3

1) False. Absorption refers to the process of nutrient molecules moving across the epithelial cells of the intestine into the blood or lymph.

2) True. Some people confine the word to the chemical breakdown, while others also include mechanical breakdown.

3) False. Although glucose and galactose are absorbed by active transport, fructose is absorbed by facilitated diffusion.

4) False. Much protein is only digested in the lumen as far as di- and tripeptides, which are absorbed and digested to amino acids inside the absorptive cell.

5) True.

6) True. They are needed to form mixed micelles which effectively solubilise lipids and make them more accessible to the digestive enzymes.

7) False. Most electrolytes are absorbed by active transport as they may be moving up their concentration gradients.

8) True.

9) False. The bile acids, after release into the small intestine, are mostly reabsorbed, pass into the portal vein, and are taken up by the liver for reuse. This is the circulation.

4.4

1) villi	These modifications are necessary for efficient absorption, the greater the area in contact with chyme, the faster absorption can occur.
2) or 3) blood	
3) or 2) lymph	
4) microvilli	
5) brush border	
6) surface area	
7) fibre	Bacteria in the large intestine can ferment some of the undigested material, mainly carbohydrates, forming short chain fatty acids.
8) mucosal	The side facing the lumen.
9) basal or basolateral	The side facing away from the lumen.
10) Na^+	Although the actual entry or glucose and Na^+ into the cell does not directly involve ATP hydrolysis, ATP is required to maintain the low intracellular Na^+ concentration, which is the driving force for Na^+/glucose entry to the cell.
11) Na^+	
12) basolateral (Basal)	
13) Na^+	
14) mucosal	
15) down	

4.5

1) False. A secretagogue is a substance that stimulates the secretion by a cell of another substance, for example, CCK is a secretagogue acting on pancreatic cells, causing release of the zymogen granules.

2) True. The pancreatic zymogen granules contain inactive proenzymes such as trypsinogen and chymotrypsinogen.

3) False. Saliva secretion is inhibited by sympathetic nerves.

4) False. It is true that gastric acid secretion is stimulated by the vagus nerve and by gastrin, but gastrin is released into the blood.

5) True. It is also inhibited by GIP.

6) False. The signal is a low pH, ie an acid chyme, thus signalling the need to reduce acid output.

7) True. It also decreases stomach motility and reduces gastric emptying.

8) False. Gastrin increases, the others decrease, gastric emptying.

9) True. These include VIP, GRP, GIP, neurotensin, and substance P.

10) False. The hypothalamus does appear to be a centre for the control of feeding but the signals from the gut appear to be important for this.

4.6

1) a) ↓ because impulses from the vagus stimulate acid secretion;
b) ↑ because of the acid in the pickles;
c) ↑ because of caffeine in coffee stimulates acid secretion.

2) a) and b) → Most likely there would be no significant effect because most people with this condition suffer from a lack of intrinsic factor which must be present for vitamin B_{12} absorption rather than a lack of vitamin B_{12} in their diets. c) This is an effective treatment since it bypasses the absorption process.

3) a) → Little effect since they are losing water via the intestinal mucosa due to the osmotic effect of the excreted chloride and lack of uptake of sodium.
b) → This will have little effect as the intestine is still unable to absorb the sodium ions because the cholera toxin inhibits the Na^+ carrier.
c) ↓ The inclusion of glucose means that Na^+ can now be absorbed using the glucose-Na^+ cotransporter, so the concentration of Na^+ in the lumen will be decreased, and more water will be absorbed.

Responses to Chapter 5 SAQs

5.1

1) We hope that you included the following four categories. Nutrients, gases, wastes, hormones. You may have included other items such as antibodies. It is, however, on the four items listed above that we wish to focus on in this chaper.

2) Temperature regulation, defense, pH balance, prevention of fluid loss.

5.2

1) Oedema. The nephrotic syndrome is another name for kidney disease. The condition described in the question is one of the symptoms of this disease, but the nephratic syndrome has many other symptoms (eg anaemia, uraemia). Haemophilia is a term which is used to describe several different condition in which blood fails to coagulate properly even after only minor trauma.

2) Albumin. If you got this wrong, re-read 5.1.2-5.1.4.

3) Adrenalin. The other components listed are secreted by platelets (see 5.2.2).

4) A proteolytic enzyme (see 5.2.4).

5) A blood clot (see 5.2.4). Both heparin and warfarin are anticoagulants and are likely to reduce the occurence of thrombosis. Fibroblasts are cells that involved in the depostion of protein fibres in scar tissue. Thrombosis can hardly be attributed to the production of these cells.

5.3

1) d).

2) b).

3) c).

4) a).

5) e).

5.4

1 - aorta
3 - superior vena cava
8 - left atrium
2 - pulmonary artery
12 - bicuspid (mitral) valve
13 - inferior vena cava
5 - pulmonary veins
11 - aortic valve
4 - right atrium
6 - left ventricle
10 - pulmonary valve
9 - tricuspid valve
7 - right ventricle

If you have got these all right, try and add the appropriate words to the unnumbered labels in Figure 5.2. You will be able to check your answers with Figure 5.1.

5.5

1) True.

2) False. The output would be 6400 ml min^{-1}. The cardiac output is the volume of blood flowing through either the systemic or the pulmonary circulation per minute, not the sum of these two circulations.

3) False. Starling's law indicated that the stength of contraction is, within limits, directly related to the length of the cardiac muscle fibres.

4) False. The threshold potential remains the same. Parasympathetic stimulation simply slows the rate at which this potential is achieved (see Figure 5.6).

5) True. The more distended the ventricles, the greater the force of contraction.

6) False. Some cations such as Ca^{2+} increases the rate and strength of contraction. Other (for example Na^+ and K^+) decrease the rate and strength of contraction.

5.6 The correct order is 1, 5, 7, 2, 3, 6, 4, 8. If you got this wrong, we suggest you write out the statements in this order then you will see the logic of this order. You should note that the baroreceptors will also send an input to the cardiac centre which will stimulate the cardioaccelerating centre and inhibit the cardioinhibitory centre. This will lead to an increase in cardiac output and therefore this will also increase the arteral blood pressure.

5.7 Exercising muscles produce a lot of heat and this is dissipated by increasing blood flow to the skin thereby enabling heat to be lost to the surrounding. The digestion of food is not a priority during exercise so blood is redistributed from the abdominal organs to the muscles whose metabolic needs increase dramatically during exercise.

Responses to Chapter 6 SAQs

6.1 Three factors which ensure that the process of gas exchange is carried out efficiently are: a large surface area; close proximity to the circulatory system; gases are moist.

6.2 Damage to the intrapleural space abolishes its subatmospheric pressure. Therefore, the transpulmonary pressure which prevents the lung from collapsing is abolished. Therefore the lung collapses. In addition the chest wall will spring outwards.

6.3 The flow chart should look something like this:

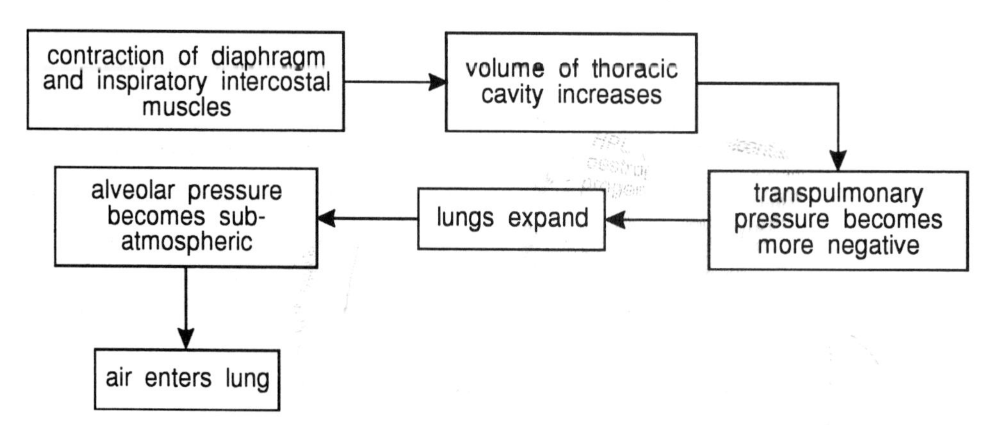

6.4 The change in intrapleural pressure would have to be greater ie become more negative. This is because the compliance of the lung is decreased by the absence of surfactant ie the lungs are more difficult to inflate.

6.5 The lung volumes are simply read from the spirometer trace. The values are as follows:

1) Tidal volume = 800 ml.

2) Inspiratory reserve volume = 2950 ml.

3) Expiratory reserve volume = 1700 ml.

4) Vital capacity = 5450 ml.

6.6

The minute volume is simply volume is defined as tidal volume multiplied by frequency of breathing.

ie 600 ml x 15 breaths min^{-1} = 9000 ml min^{-1}

If dead space = 150 ml, then of each breath only (600 - 150) ml is available for gas exchange, ie volume of air used for gas exchange is 450 ml x 15 breaths min^{-1} = 6750 ml min^{-1}.

6.7

If the person is hypoventilating then there will be a build up of carbon dioxide in the blood. Since carbon dioxide can dissolve to form carbonic acid which can generate hydrogen ions, then the amount of such ions will increase, therefore blood pH will decrease.

Responses to Chapter 7 SAQs

7.1

1) Skin.

2) Kidneys.

3) Lungs.

4) Liver.

7.2

Your answer should be 7200 ml. The blood flow is 1.2 l min^{-1} and 50 per cent is plasma, therefore, 600 ml min^{-1} plasma is presented to the kidneys. One fifth is filtered = 120 ml min^{-1}. Therefore the volume of plasma filtered per hour = 120 x 60 = 7200 ml.

7.3

1) a) ADH secretion is stimulated.

b) Osmotic pressure, due to the high solute concentration of the interstitial fluid compared with that of the tubular contents.

2) a) False, in the kidney the blood flow changes very little with increased arterial pressure due to autoregulation (provided the pressure is within physiological limits).

b) False, glomerular capillary pressure and glomerular filtration rate are largely independent of changes in the arterial blood pressure in healthy kidneys due to autoregulation of blood flow.

c) False, the rate of urine production is determined by tubular function not by glomerular filtration rate.

d) True, the loss of one-sixth of the plasma volume as glomerular filtrate results in the increased concentration of red cells (raised haematocrit) and plasma proteins.

e) True, this results from the activity of the countercurrent multiplier system (loop of Henle and vasa recta).

f) True, this is due to active transport of sodium chloride out of the ascending loop of Henle.

g) False, the permeability is controlled by ADH.

7.4

1) From the Henderson-Hasselbalch equation

$$pH = pKa + \log \frac{[HCO_3^-]}{[H_2CO_3]}$$

Therefore $\log \frac{[HCO_3^-]}{[H_2CO_3]} = pH - pKa = 5.4\text{-}6.4 = -1$

Thus $\frac{[HCO_3^-]}{[H_2CO_3]} = 0.1 = \frac{1}{10}$

2) Similarly $\log \frac{[HCO_3^-]}{[H_2CO_3]} = pH\text{-}pKa = 8.4\text{-}6.4 = 2$

Thus $\frac{[HCO_3^-]}{[H_2CO_3]} = 100$

7.5

1) True, urine smells of ammonia under these circumstances.

2) False, sodium (Na^+) is reabsorbed in exchange for hydrogen.

3) True.

4) False, the hydrogen reacts with alkaline sodium phosphate (Na_2HPO_4) releasing a sodium ion. The Na^+ is reabsorbed into the blood and acid sodium dehydrogen phosphate (NaH_2PO_4) is excreted in the urine.

5) False, although carbonic acid is produced in the renal tubular cells, this dissociates into bicarbonate and hydrogen ions. The hydrogen ions are secreted into the tubular fluid whilst the bicarbonate ion is reabsorbed in the blood. Thus we do not find high concentrations of carbonic acid in urine (see Table 7.5).

Responses to Chapter 8 SAQs

8.1

Glucose 1, 2, 3, 9

The major role of glucose is for energy supply, but it can be used as part of some structural compounds and, via many metabolic steps, can be converted into triglycerides and steroid hormones.

Fatty acids 1, 2, 3, 10

Fatty acid oxidation is a major source of energy for many cells. Fatty acids can be stored by combination with glycerol to form triglycerides, and also form part of phospholipids, the major constituent of cell membranes.

Cholesterol 3, 10

Cholesterol is a precursor of the steroid hormones, but is also a major constituent of cell membranes. It cannot be used as an energy source and can only be removed by secretion in the bile after conversion to bile salts.

Amino acids 1, 2, 5, 8, 9, 10

Proteins are made up from amino acids, and this is the most obvious function. However, amino acids can be metabolised to form triglycerides and to provide energy. Some are involved in cell signalling as neurotransmitters.

Fibre 2, 11

Although fibre is well known for its role in increasing gut motility (ie reducing constipation) some components can be fermented in the large intestine to produce short chain fatty acids which can provide energy.

Vitamins 6, 7, 8

Vitamins are not used as an energy source, but by acting in these ways, play a regulatory role in metabolism.

Minerals 4, 6, 8

Minerals play a regulatory role but also can be structural. For example, calcium plays an important part in cell signalling but is also a major component of bone, and can act directly on enzymes to alter their activities.

If you found this SAQ difficult, it might be helpful for you to read the BIOTOL texts relating to metabolism especially 'Principles of Cell Energetics', 'Energy Sources for Cells' and 'Biosynthesis and Integration of Cell Metabolism'.

8.2

1) False. It is true that cells with few or no mitochondria are dependent on glucose, but it is catabolised anaerobically to lactate.

2) True. An example would be the high activities of the enzymes of gluconeogenesis in liver.

3) False. The very high power muscles have few mitochondria and operate largely anaerobically using glucose. Fatty acid oxidation requires oxygen.

4) False. It is true that the brain does mainly use glucose but it does use other fuels such as ketone bodies as well.

8.3

1) Liver. This is a very important point to realise, it illustrates the liver's central role.

2) High. In order to be able to help reduce the blood glucose concentration after a meal, and to store glycogen, the liver needs a high capacity for glucose uptake.

3) Urea, ketone body. By being the only organ to be able to make urea, the liver effectively controls the circulating concentrations of amino acids. Its ability to synthesise ketone bodies enables it to supply other tissues with a water soluble fuel to replace glucose.

4) High. The liver depends on the oxidation of fatty acids for most of its energy requirement.

5) High. A very important role of the liver is the synthesis of glucose when there is none available from the gut.

6) Glycogen. As part of the mechanism for regulating the blood glucose concentration, the liver stores glucose as glycogen which can be released when necessary.

8.4

1) True. VLDL is mainly synthesised in the liver and has a high content of triglyceride which can be taken up by adipose tissue and skeletal muscle.

2) True. As VLDL lose their triglyceride they become IDL and LDL, though the process also involves other components as well, including the gain of cholesterol.

3) True. LDL seems to be a major mechanism whereby several tissues gain cholesterol.

4) True. HDL does appear to function to remove cholesterol from many tissues and transport it to the liver.

5) False. Only the amino acids require active transport to cross cell membranes other than those of the gut epithelial cells. The others cross by facilitated diffusion, bound to a carrier molecule.

6) Partially true. Insulin stimulates the uptake of glucose by many cells, such as those in adipose tissue and skeletal muscle, but not in the liver, brain and red blood cells.

8.5

1) (↓) Liver glycolysis is decreased in order to conserve glucose for other tissues. In fact, gluconeogenesis, the opposite process, occurs.

2) (↑) Muscles able to use fatty acids do so, again to conserve glucose.

3) (↓) There is no spare glucose available to store as fat in adipose tissue.

4) (↓) The brain can use ketone bodies as a partial alternative to glucose.

5) (→) Red blood cells have a constant requirement for glucose that cannot be met by any alternative fuel.

6) (↓) Protein synthesis is decreased, whereas protein degradation is increased to provide precursors for gluconeogenesis.

7) (↑) Muscles are able to use ketone bodies as an alternative fuel when they become available.

8) (→) Fasting as such does not alter the muscle glycogen content since muscle glycogen cannot be broken down to release glucose into the blood. However, after exercise which might use up some of the glycogen, there would be a lower rate of replenishment of the store.

9) (↓) The lack of available glucose results in a reduction in fat synthesis.

10) (↑) The increased breakdown of muscle protein to amino acids, and their subsequent use for glucose synthesis, results in an increased supply of amino nitrogen to the liver, which it disposes of as urea.

8.6

1) Glucagon.

2) Insulin.

Both of these are pancreatic hormones, secreted in response to changes in blood glucose concentration.

3) Glucose.

4) Glycogen.

5) Gluconeogenesis.

The major function of the liver is the maintenance of blood glucose levels. It can provide glucose to the blood by breaking down glycogen or synthesising glucose.

6) Fatty acids. These are the major fuel of the liver except in the well fed state.

7) Adipose tissue. Although triglyceride is present in other tissues, adipose tissue is the major store able to release fatty acids into the blood.

8) Insulin. Insulin promotes esterification of fatty acids, ie storage, when blood glucose levels are high.

9) Glucagon.

10) Lipolysis

Glucagon, and also growth hormone, adrenaline and cortisol, stimulate lipolysis, the release of fatty acids from triglycerides.

11) Glycerol. Triglyceride is composed of glycerol and fatty acids.

12) Lactate. Lactate is the end product of the anaerobic catabolism of glucose.

13) Alanine and glutamine. These are the major amino acids which arise from the catabolism of skeletal muscle proteins.

14) Proteins.

15) Ketone bodies. These are the products of the partial oxidation of fatty acids, are only made in the liver, and can be oxidised by several tissues.

16) Brain. Since the brain is totally dependent on glucose, except in long term starvation and since its functioning is so important, it is vital to maintain its supply of glucose.

17) Adrenaline. The secretion of the hormone is greatly increased in anticipation of and during exercise.

18,19,20,21) Cortisol, adrenaline growth hormone, glucagon. All of these hormones show raised levels during stress, and all tend to increase the loss of muscle protein to provide precursors for glyconeogenesis.

22) Proteins.

8.7

1) c). If the adult person is in energy and nitrogen balance, any increase in energy intake, whether fat, carbohydrate or protein, will result in a positive energy balance, but this will be due to increased fat, not protein. a) will also be true.

2) h), j). For someone losing energy and nitrogen, any increase in energy intake will obviously result in decreased energy loss, but since some of the dietary protein will be being used for energy, part of this will now be spared and so there is decreased protein loss.

3) a). A really normal child should be in positive energy and nitrogen balance, but this one is just maintaining the energy and protein content of the body. An increase in the intake of fat will, therefore, lead to a positive energy balance. Whether this is accompanied by increased protein storage or just fat gain will depend on the contribution of the protein already in the diet to the energy. If the problem is inadequate energy, but with adequate protein, there should be an increase in protein storage.

4) a). In pregnancy there should be an energy and nitrogen gain, because of the growth of the foetus, placenta and mammary glands. An increase in protein intake must increase the energy gain. The distribution of this between fat and protein would depend on how adequate the protein intake was in relation to need.

5) c), f). The reduced intake of lysine, at a time when the intake of protein is barely adequate, must lead to a loss of nitrogen because lysine is an essential amino acid, so a less than adequate intake of this is equivalent to a less than adequate intake of nitrogen generally. There would then be an increased availability of the amino acids for metabolism. Since the total energy intake remains the same, amino acids would be available for fat synthesis.

If you got most of these correct, you must have a very good understanding of the interrelationships between the three major fuel nutrients, they are certainly not easy questions!

Responses to Chapter 9 SAQs

9.1

The list of events you produced should have contained the following:

- Action potential generated at the neuromuscular junction enters the transverse tubules and triggers the release of Ca^{2+} from the cisternae of the sarcoplasmic reticulum.

- Ca^{2+} binds to troponin on the actin myofilaments and tropomyosin moves to uncover the myosin cross-bridge binding sites on the actin.

- Myosin cross-bridges bind to the actin.

- Cross-bridge binding causes the release of energy stored in the myosin. There is a conformational change and each cross-bridge moves pulling the actin along with it during the power stroke.

- ATP binds to cross-bridge head causing detachment of the head from the active site.

- ATP is split by the powerful ATPase activity of the myosin head.

- Released energy is transferred to the myosin cross-bridge which is revoked ready to perform another power stroke.

- Cross-bridges repeat steps c) to g) producing movements of actin myofilaments past myosin myofilaments. Shortening of muscle is produced.

- If stimulation of muscle ceases, Ca^{2+} is actively transported into the sacroplasmic reticulum.

- Removal of Ca^{2+} from sarcoplasm and therefore from troponin causes tropomyosin to block the active sites on the actin.

- Cross-bridge cycle ceases and the muscle relaxes.

9.2 Your flow diagram should contain the following information:

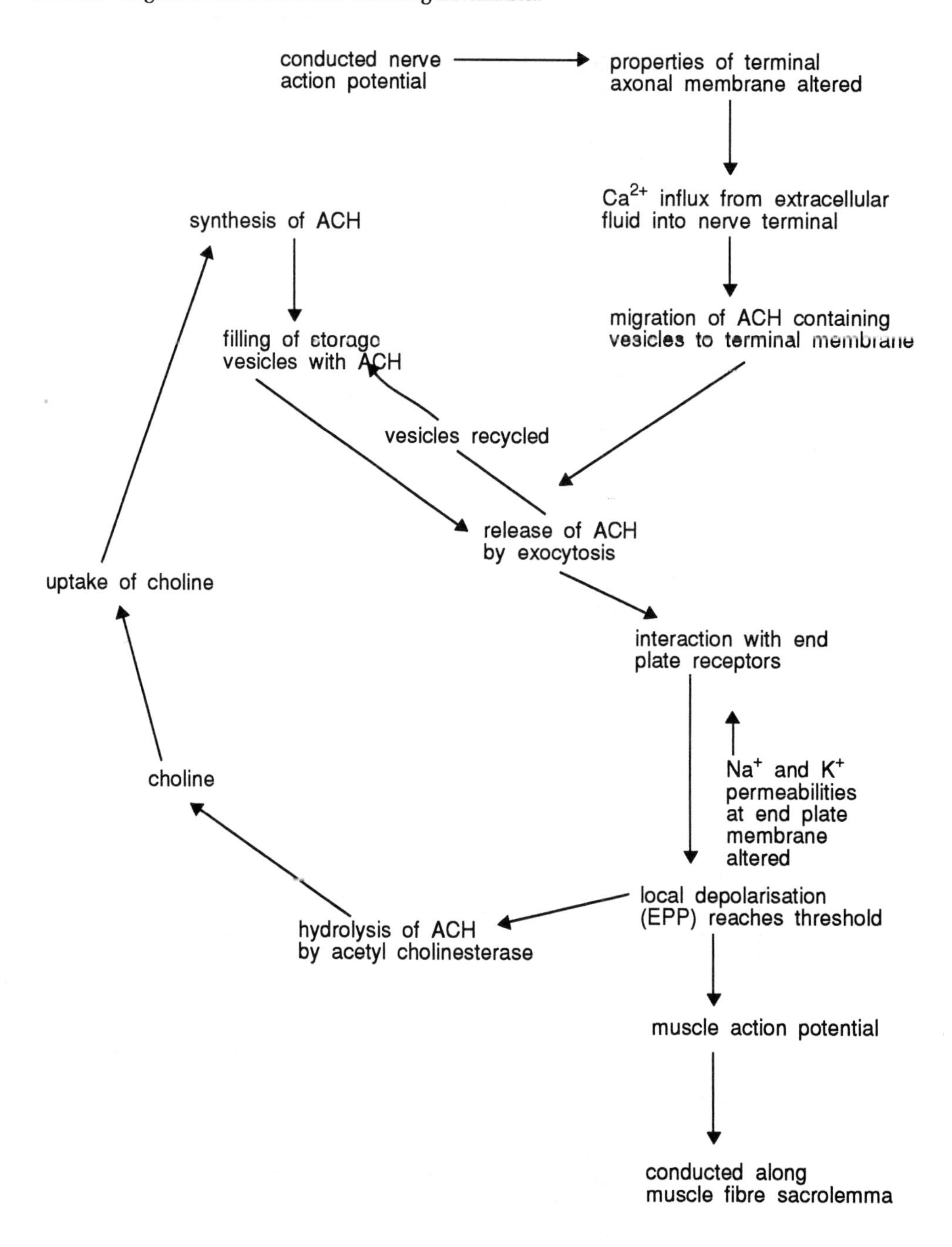

9.3 Your graph should have had the following shape.

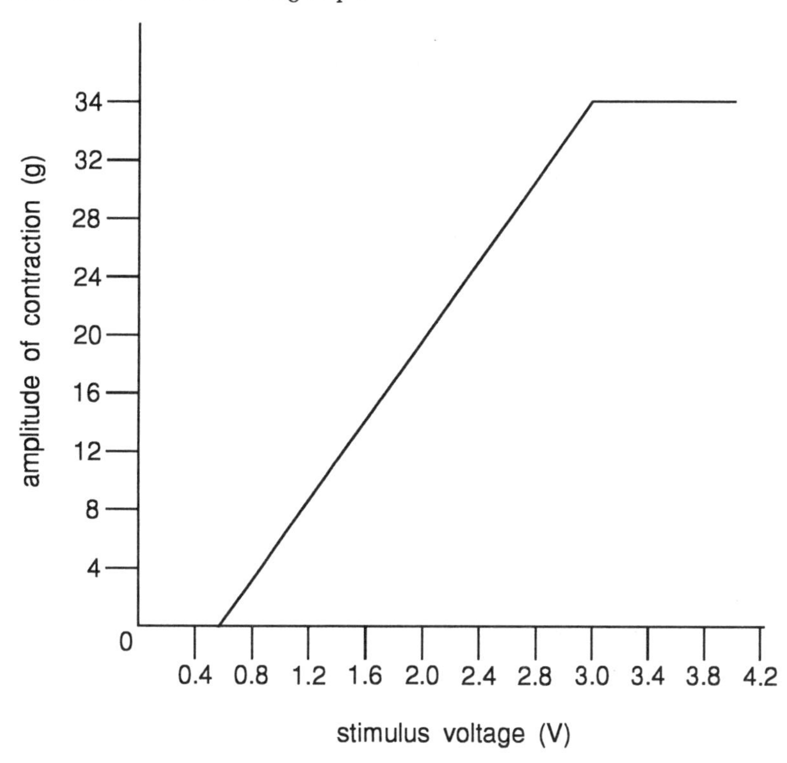

Explanation

There was no response to a stimulus of 0.6V, so this must be a subthreshold stimulus. From a stimulus of 0.8v the amplitude of the muscle contraction increased, linearly, until it reached a maximum at a stimulus voltage of 3V. Beyond this voltage, the muscle contraction remained at a constant value, indicating that 3V was the maximal stimulus.

Below the stimulus threshold the motor nerve was not excited so that no motor units in the muscle were stimulated to contract. At the threshold stimulus (0.8V), the nerve fibres with the lowest threshold to stimulation were excited and their accompanying muscle fibres contracted. As this involved only a few motor units, there was only a small response from the muscle. With the stimuli increasing from 1 to 3V an increasing number of motor units contracted as their thresholds were reached; hence the overall contraction from the muscle increased up to a maximum. At this point all of the constituent motor units had been stimulated to contract. No further increase occurred as the whole muscle was responding.

9.4 As blood brings oxygen to the muscle fibres, then those fibres which required oxygen to produce ATP would be the ones whose energy metabolism would be decreased. The class of fibre concerned is, therefore, the type I group. These fibres rely entirely in oxidative production of ATP. Type IIa fibres would be partly affected, but they would still be capable of producing ATP by anaerobic processes.

9.5 Type IIb fibres are used in activities where a short, sharp burst of high power is required, whereas the type IIa fibres provide a moderate power output that can be maintained for a considerable length of time.

Activity	Type of fibre used
squash	IIb
football	IIa and IIb
100m sprint	IIb
1000m race	IIa
high jump	IIb
jogging	IIa

9.6 The inclusion of two interneurons makes the resulting spinal reflex look as shown below.

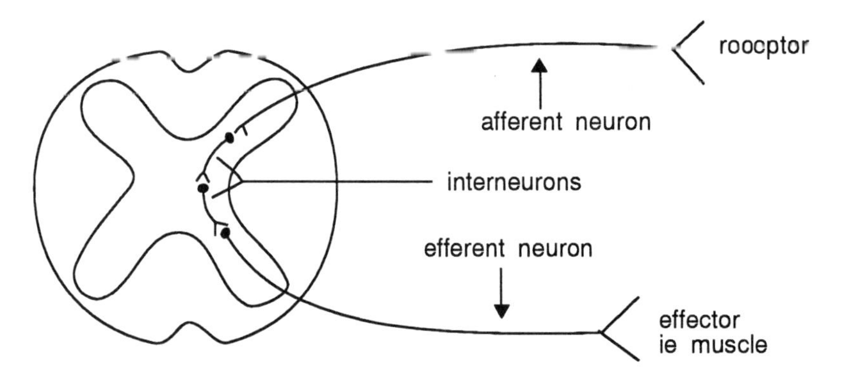

Responses to Chapter 10 SAQs

10.1 Your flow diagram of the hormonal control of male reproductive function should have the following relationships. Inhibitory effect is denoted by -.

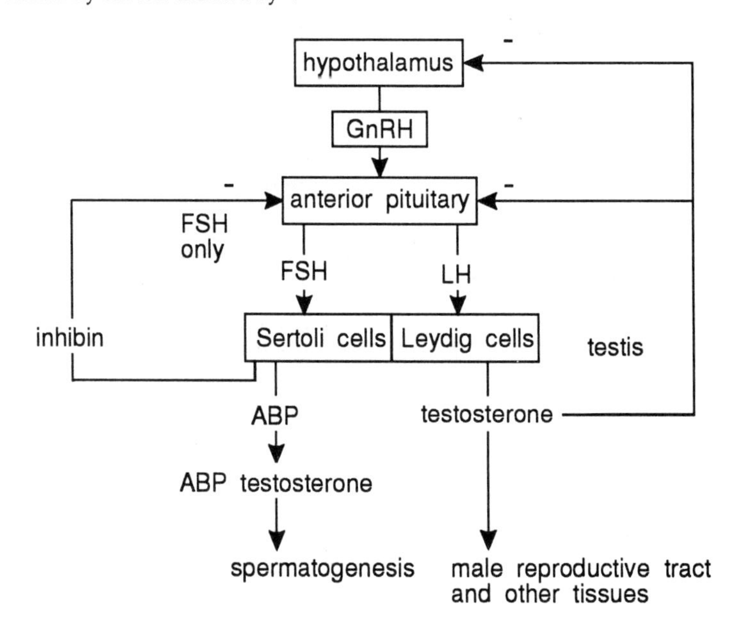

10.2 The effect would be answer (3) ie an increase in the output of testosterone, as LH stimulates Leydig cells to produce the hormone.

The other answers are wrong because:

1) Oestrogen output is not increased by LH in the male.

2) Spermatogenesis is stimulated by FSH and LH.

4) There would be an increase in testicular activity not a decrease.

5) GnRH output would be decreased, not increased, as a result of the negative feedback effect of the excess LH.

10.3 Flow diagram summarising ovarian events during the menstrual cycle.

Remember the cycle can be divided into two phases, the earlier follicular phase and the later luteal phase, so the events can be tabulated thus:

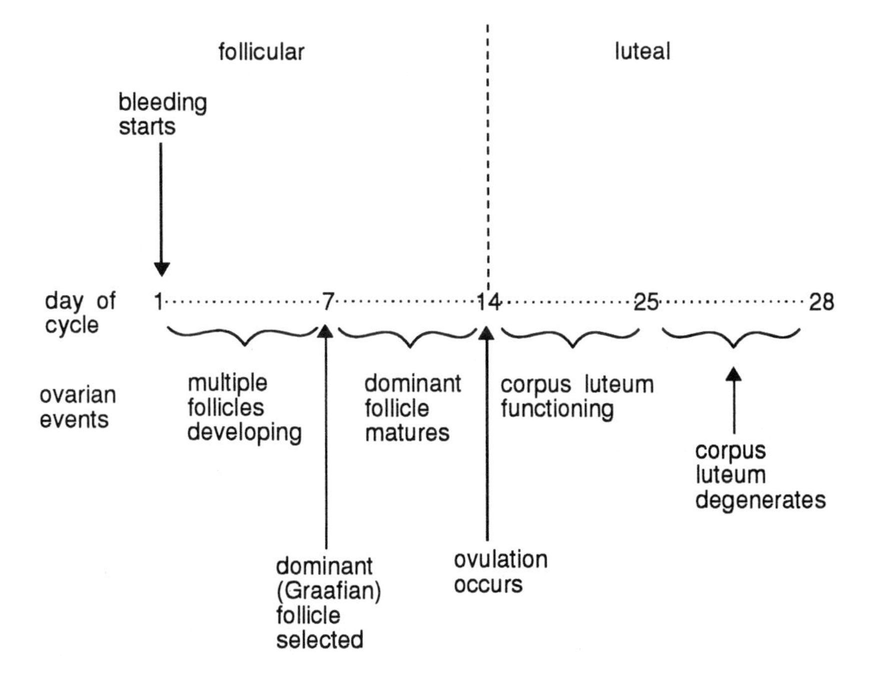

10.4 Your flow diagram may look quite different to ours, but it should contain the same information. Note we have represented a post stimulatory effect by + and an inhibitory effect by -.

Remember, as before, ovarian function is in two phases; the early follicular and the later luteal, each having a different hormone profile so the diagram ought to attempt to include both.

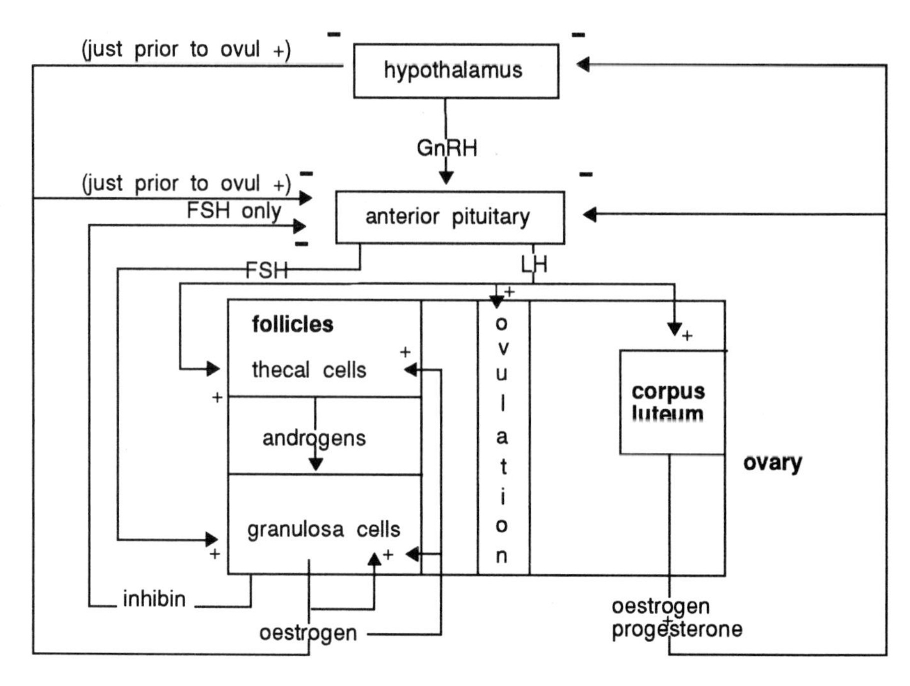

(based on Vander, Sherman and Luciano, (1990), Human Physiology, McGraw Hill Inc, USA, p622)

10.5 The true statements are 1) and 3).

2) Is incorrect because it is the blastocyst, not the morula, which is implanted into the endometrium.

4) Is incorrect as, normally, maternal and foetal blood is not mixed.

5) Is incorrect because gastrulation refers to the processes by which the blastocyst differentiates into the three primary germ layers prior to organ development.

10.6 Figure illustrating the factors influencing parturition.

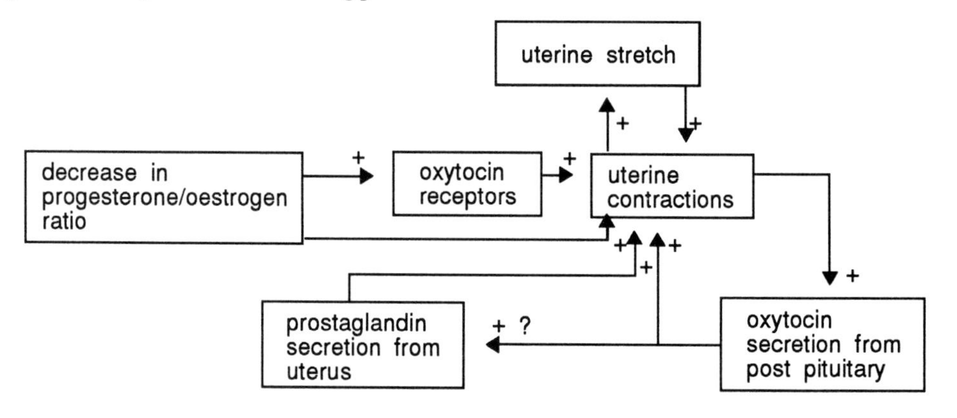

10.7 The following treatments would have a contraceptive action:

4) Because the effectiveness of GnRH would be decreased, thus the output of LH and FSH would be decreased thereby preventing ovulation.

5) This is the mechanism of the contraceptive pill.

6) High levels of progesterone would decrease the output of LH and FSH by negative feedback, therefore ovulation would be prevented.

The other treatments would not have a contraceptive effect for the following reasons:

1) LH would initiate ovulation.

2) FSH would stimulate follicular maturation.

3) Given orally, the peptide would be destroyed by digestive processes.

Suggestions for further reading

General texts

Carola, R., Hurley J.P. and Noback, C.R. (1992), Human Anatomy and Physiology, 2nd Edition, McGraw Hill, USA, ISBN 0-01711-2561-2

Scanlon, V.C. and Sanders, T. (1991), Essentials of Anatomy and Physiology, F.A. Davies, Philidelphia, ISBN 0-8036-7741-3

Tortora, G.J. and Anagnostakos, N.P. (1990), Principles of Anatomy and Physiology, Biological Sciences Textbook Inc, USA

Vander, A.J., Sherman, J.H. and Luciano, D.S. (1990), Human Physiology, McGraw Hill Inc., USA

Wheater, P.R., Burkitt, H.G. and Daniels, V.G. (1987), Functional Histology, Churchill Lingstone, London ISBN 0-443-02341-7

Text relating to specific physiological

Nerve and muscle function

Carpenter R.H.S. (1990) Neurophysiology, 2nd Edition, Edward Arnold, London, ISBN 0-340-50634-2

Fitzgerald, M.J.T. (1992), Neuroanatomy, Basic and Clinical, Bailliere-Tindall, London, ISBN 0-7020-1432-X

Fix, J.D. (1992), Neuroanatomy, Williams and Wilkins, Baltimore, London, ISBN 0-6830-03250-X

Wilkinson, J.L. (1992), Neuroanatomy for Medical Students, 2nd Edition, Butterworth Heinemann, Oxford, ISBN 0-7506-1487-1

Mathews, G.G. (1991), Cellular Physiology of Nerve and Muscle, Blackwell Scientific Publication, Oxford, ISBN 0-8654-2159-5

Endocrine system

Griffin, J.E. and Ojeda, S.R. (1992), Textbook of Endocrine Physiology, 2nd Edition, Oxford University Press, Oxford, ISBN 0-1950-7561-7

O' Riordan, J.L.H., Malan, P.G. and Gould, R.P. (1988), Endocrine Physiology, Blackwell Scientific Publications, Oxford, ISBN 0-6320-2112-8

Cardiovascular system

Berne, R.M. and Levy, M.N. (1992), Cardiovascular Physiology, 6th Edition, Mosby Yearbook, USA, ISBN 0-8016-6314-8

Renal functions

Lote, C.J. (1990), Principles of Renal Physiology, Chapman Hall, London, ISBN 0-4123-8200-8

Digestive system

Sanford, P.A. (1992), Digestive System Physiology, 2nd Edition, Edward Arnold, London, ISBN 0-3405-6020-7

Respiratory system

West, J.B. (1990), Respiratory Physiology - the essentials, 4th Edition, Williams & Wilkins, Baltimore and London, ISBN 0-683-08942-0

Widdicombe, J. and Davies, A. (1991), Respiratory Physiology, Edward Arnold, London, ISBN 0-3405-5253-0

Index

A

R

T

U